Design of Reinforced Concrete Foundations

Design of Reinforced Concrete Foundations

P.C. VARGHESE

Honorary Professor, Anna University, Chennai
Formerly, Professor and Head, Department of Civil Engineering
Indian Institute of Technology Madras, and
UNESCO Chief Technical Advisor, University of Moratuwa, Sri Lanka

PHI Learning Private Limited
Delhi-110092
2019

₹ 495.00

DESIGN OF REINFORCED CONCRETE FOUNDATIONS
P.C. Varghese

ISBN-978-81-203-3615-5

Eighth Printing ... ... **November, 2019**

Published by Asoke K. Ghosh, PHI Learning Private Limited, Rimjhim House, 111, Patparganj Industrial Estate, Delhi-110092 and Printed by Mudrak, D-61, Sector 63, Noida, U.P.-201301.

To the memory of

My father Puthenveetil Chandapilla

and

My mother Aliayamma

Contents

Preface

This book, a companion volume to my earlier book *Foundation Engineering*, exclusively deals with the structural design of reinforced concrete foundations. Both these books are meant as textbooks for undergraduates and postgraduates students of civil engineering specializing in Geotechnical Engineering and Structural Engineering. As the book explains the fundamental principles of the subject and the field practice, practising engineers will also find it very useful for a better understanding of the subject.

From my long years of experience in teaching, both Geotechnical Engineering and Structural Engineering at all levels, I feel that it is better for the students to learn the above two aspects of the same subject as two separate courses, even though they have to be combined together in actual practice. This has prompted me to write two separate textbooks on the twin aspects of the same subject.

My aim, as in my other textbooks, is to explain the principles involved so that it will be helpful for the young teachers to prepare their lectures, and the students to revise the lessons. Accordingly, the book is presented as lecture oriented, each chapter dealing with only one topic to be covered in one lecture in the classroom. In many cases, in order to reduce the volume of the book, examples have not been completely worked out as model designs. However, in all cases, the principles involved have been fully explained. Similarly, though many pieces of software are available for foundation designs, I have not included them in the book. After studying the basics, it will be easy for anyone to use the software and also check the results of the software output by conceptual methods of calculation. This aspect is very important in present-day education as there is a growing tendency among engineers to do all their design only through software without using conceptual methods to verify the results. This practice can lead to failures in civil engineering practice, with disastrous consequences, and should therefore be avoided.

I have also emphasized in this textbook the importance of being conservative both in design (dimensioning) of the foundations and strictly following the standard practices of detailing the reinforcements. This aspect is very important for many reasons, of which the following are very important:

- Foundations, compared to superstructure, are very difficult to modify or repair once they have been completed.
- In contrast to the exactness with which we can estimate the forces acting on the superstructure, the forces that we assume in our design to be acting on foundation structures can be very different from the actual forces acting on them due to the following two reasons. *Firstly*, we assume the soil deposit under the foundation is homogeneous. This is far from true. *Secondly*, in conventional design we assume the foundations are rigid, neglecting their relative flexibility with respect to the soil on which they rest. This is also not really true in the field in many cases.

Many topics such as pile foundations, well foundations stone columns are not included in this book as they have been fully explained in the companion volume, *Foundation Engineering*. However, topics such as beams on elastic foundation have been repeated in this book to make the students aware of the importance of the role of rigidity of foundation structures in their design.

I will be happy if students, teachers and professionals find this book interesting and useful. Any constructive suggestions to improve the contents will be highly appreciated.

P.C. Varghese

[P.S. For Soil Mechanics considerations for Design of Foundations and also Detailed Design of Ordinary Pile Foundations, Well Foundations etc. please refer author's book entitled *Foundation Engineering* published by PHI Learning]

Acknowledgements

I have taken help from a large number of books and other publications for writing this book. I gratefully acknowledge each of these publications.

I am thankful to Anna University, Chennai for giving me the opportunity to continue my academic pursuit after my retirement. I thankfully acknowledge the help and cooperation I have received from the staff of the Civil Engineering Department of Anna University, Chennai.

My special thanks go to Dr. H. Jane Helena of Structural Division of College of Engineering, Guindy, Anna University, Chennai, for checking the manuscript during its preparation. I thank Ms Rajeswari Sivaraman for word processing the manuscript, and Ms Uma of Anna University for drawing the figures.

I also wish to thank the editorial and production team of PHI Learning for providing excellent cooperation during the preparation and production of this book.

P.C. Varghese

Foundation Structures

1.1 INTRODUCTION

Structures constructed below the ground level are called *substructures*. They are generally divided into the following three groups:

- Foundation structures
- Retaining structures
- Other special systems such as tunnels, diaphragm walls, and buried pipes.

In this book, we will deal mostly with the first two groups.

Foundation structures transfer all the loads that come on the superstructure to the ground. These loads can be either vertical loads such as live and dead loads or horizontal loads such as wind loads and earthquake loads. Retaining structures are built to retain earth as in the case of basement of buildings. Foundation structures are themselves divided into the following subdivisions:

- Shallow foundations
- Deep foundations
- Special foundations

Shallow foundation can be of two types:

- Various types of footings
- Grids and rafts

Deep foundation can be of three types (see Chapter 15):

- Piles
- Piers
- Caissons

While piles are flexible members that bend under lateral loads, piers are rigid structures that undergo rotation under lateral loads. Caissons are usually those piers built under water using pneumatic pressure devices.

Special foundations are used for special structures such as transmission towers, cooling towers, and chimneys.

The geotechnical design of foundation (e.g. calculation of bearing capacity of soil, and also that of piles, and wells) is usually dealt with in books on Soil Mechanics, which is a part of Foundation Engineering [1]. In this book, we deal mainly with the *structural design* of commonly used foundations (e.g. footings, rafts, pile caps, and basement) for supporting superstructures.

1.2 RIGID AND FLEXIBLE FOUNDATIONS

Structures such as footings can be assumed to settle uniformly under a central concentrated load. Hence, if we assume that the ground reaction or contact pressure is proportional to deformation (the Winkler theory), the contact pressure will be uniform under the foundation. (However, we know that the elastic stress distribution under a rigid footing in clay or sand is not uniform even though at ultimate stage it tries to even out.) In certain calculations, we often make the assumption that the shallow foundations are rigid and the contact pressure is, therefore, uniform.

In reality, the soil pressure under a foundation depends on the rigidity of the foundation structure itself with respect to the foundation soil and, in many cases, the foundation structure we use may not be fully rigid. This soil structure interaction can be studied by modern theories [2], [3]. We will deal with its elementary principles by a study of "beams on elastic foundation" described in Chapter 22.

1.3 LOADS AND THEIR EFFECTS

The dead and live loads are gravity loads assumed to act vertically down. The effect of wind, earthquake, earth pressures, etc. is to produce horizontal loads. The effect of these vertical and horizontal loads can be divided into three groups:

1. Vertical loads that pass through the centre of gravity (CG) of the foundation. These can be assumed to produce uniform contact pressure under a rigid foundation.
2. Vertical loads that act with eccentricity. These also produce vertical loads and moments in the foundation and are again balanced by varying vertical reactions on the base of the foundation. Only vertical loads have to be balanced.
3. Vertical forces with horizontal forces such as wind load producing moments. These have to be balanced by vertical reactions. In addition, as different from the second

case, the horizontal forces have to be balanced or resisted by the corresponding horizontal reactions such as friction between foundation and ground or by passive earth pressure.

Thus, foundations of retaining walls and those of tall buildings resisting wind and earthquakes have to be designed against sliding and stability also. *This difference in action should be clearly understood in our analysis of foundations.*

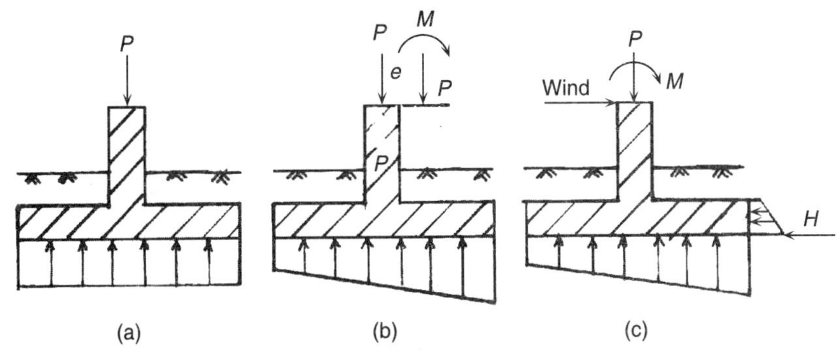

Figure 1.1 Loads and reaction from the ground (ground pressure) under rigid column footings. (a) Uniform vertical reaction due to a central load, (b) Linearly varying vertical reaction due to eccentric vertical load producing additional moment, (c) Linearly varying vertical reaction and also horizontal ground reactions due to vertical load and horizontal loads such as wind loads.

1.4 DESIGN REQUIREMENTS

The foundation is to be designed keeping the following conditions in mind:

1. All the applied vertical and horizontal loads on the structure should be resisted by foundation pressure, which does not exceed the safe bearing capacity of the soil.
2. The foundation should have adequate safety against sliding, overturning, or pull-out under the influence of the external loads.
3. The total settlement and the differential settlement of the structure should be within limits.

We should be aware that non-uniform pressure under the foundation in compressible (clay) soils can lead to the long-term effect of tilting of the foundation in clay soils. Hence it

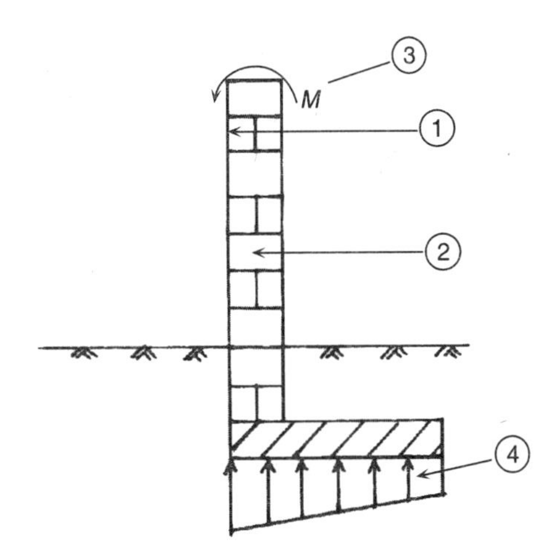

Figure 1.2 Tilting of compound walls built in clay soils due to non-uniform foundation pressure: (1) Property line, (2) Compound wall, (3) Moment produced by eccentricity of vertical load, and (4) Non-uniform loading of foundation.

is very common in compound walls built in clayey soils with eccentric load on foundation to tilt from the vertical after a length of time as shown in Figure 1.2. Thus, it is better to use short piles and grade beams for compound walls in clayey soils.

1.5 GEOTECHNICAL DESIGN

The problems of safe bearing capacity, total settlement, differential settlement, and so on are discussed in books on Soil Mechanics [1]. In this book, we restrict ourselves to structural design.

1.6 EMPIRICAL AND EXACT METHODS OF ANALYSIS OF FOUNDATIONS

Traditionally, because of the *high indeterminacy of foundation problems*, foundation analysis was mostly carried out by elementary or approximate methods of calculations and their design based on the theory of reinforced concrete. These calculations for analysis and design can be carried out by simple hand calculators. However, of late, more exact and theoretical methods of analysis based on soil-structure interaction have been evolved. All of them require not only *correct soil parameters but also use of modern computers*. However, a good estimation of the properties of the varying soils under foundation is rather impossible, and this considerably limits the application of these theories in real practice. However, such an analysis can give us a good indication of the nature of soil reaction, which can be used in our design.

The knowledge of the *existing empirical methods of analysis*, which we will be dealing in this book, instead of the theoretical soil-structure interaction methods, will be valuable to foundation engineers in three ways. *First*, in many simple cases, they can be directly used for construction. *Second*, they will serve as quick and good methods for initial estimation of the cost and comparison of the economy of the different types of foundations for a given situation. *Third*, they serve also as simple checks for the complex computer generated solutions. Such checks are always recommended whenever we use computer methods for structural solution.

This book principally deals with these approximate methods of analysis and design. More modern information dealing with soil structure interaction (which should be always used along with extensive soil investigation to get realistic soil parameters) is available in advanced books on the subject [2], [3]. Complex foundations can be designed by such methods. These methods become necessary only when we deal with highly complex situations. Also, as explained in Sec. 1.8, what we should aim for is a safe and conservative design.

1.7 DESIGN LOADS FOR FOUNDATIONS

Limit state design, according to IS 456-2000, is the method to be followed in India for the design of reinforced concrete members. According to this method, the structure has to be designed for different limit states which can be broadly classified into the following two groups:

- Limit state design of serviceability such as settlement and deflection
- Limit state of collapse or failure

The following load combinations are prescribed by Table 18 IS 456-2000 (Table 1.1 given below) for these two separate limit states. There are four combinations for limit state and three combinations for limit state of serviceability.

TABLE 1.1 Load Combination for Limit State Design (IS 456, Table 18)

Load combination	*Limit state of collapse*	*Limit state of serviceability*
DL + LL	1.5 (DL + LL)	DL + LL
DL + WL	1.5 DL + 1.5 WL 0.9 DL + 1.5 WL	DL + WL
DL + LL + WL	1.2(DL + LL) + 1.2 WL	DL + 0.8 LL + 0.8 WL

DL—Dead load, LL—Live load, WL—Wind load, EL—Earthquake load
(*Note:* When considering earthquake loads, we replace *WL* by *EL*)

1.8 RECOMMENDED APPROACH TO STRUCTURAL DESIGN OF FOUNDATIONS

We should always remember that our approach to design of foundation structures should be much more conservative than that of the design of a superstructure. In many cases of foundation design, we do not know the exact properties of the soil under the foundation. In most cases we assume that the structure is rigid and that the ground pressures are uniform. Soil deposits are never uniform with depth and may vary in the same site from place to place. Thus, many of the design assumptions that we make may not be true in the field.

Another important consideration is that any defect in a foundation is much more difficult to rectify than a defect found in the superstructure. Again, the considerations that we should give for the design of a foundation of a temple or church, which is to last many years, should be different from those we will give for a residential building. We must always be aware of the environmental conditions of the foundation and the likelihood of corrosion of steel by chlorides and deterioration of concrete by sulphates. Proper care should be taken not only for strength but also for durability. All these will need a conservative approach. In addition to the above general considerations, there are many other special considerations in detailing of steel in foundation.

As we are not sure of variation of soil conditions and the relative rigidity of our foundations, it is difficult to be sure about the reversal of moments. Thus, the conventional placement of steel as in beams and slabs may not provide for reversal of moments that may happen in foundation structure. Thus, we should always place some steel on both sides of members in all foundations.

Because of the above and other limitations, we should always check the sizes of members we get from our theoretical design with respect to the conventional size of members generally used in practice. Similarly, the amount of steel we get in our design should be checked with the conventional type of layout used in practice.

This section has been incorporated to draw the attention of designers to the necessity of giving extra consideration and being conservative in the design of foundation structures. We should always provide a strong foundation that will never fail. This is especially true nowadays when earthquake considerations require us to provide for very strong superstructures. Even though there are a number of case histories of failures of superstructures due to earthquakes, there have been very few cases of foundation failures due to earthquake, except those where the soil around failed due to liquefaction or other reasons. There are soil failures and not structural failures. Our aim should be to build foundations that are always stronger than the superstructure under all conditions.

1.9 SUMMARY

Many types of foundations are used in civil engineering structures. However, we must realise that foundations are very important part of a structure, and it is very difficult to rectify foundation defects once they occur in design or construction. We also know that it is difficult to estimate the exact soil properties of the foundation soil. Hence, we should always have a conservative approach in design, detailing and construction of foundation structures. The need for conservatism in the design of foundations should be always borne in mind.

REFERENCES

[1] Bowles, J.E., *Analytical and Computer Methods in Foundation Engineering*, McGraw Hill, New York, 1974.

[2] Bowles, J.E., *Foundation Analysis and Design*, McGraw Hill, New York, 1995.

[3] Varghese, P.C., *Foundation Engineering*, Prentice-Hall of India, New Delhi, 2005.

2

Review of Limit State Design of Reinforced Concrete

2.1 INTRODUCTION

In this chapter, a brief summary of the Limit State Design of Reinforced Concrete Members, as recommended by the latest Indian Code IS 456-2000 for the design of reinforced concrete members is given. This method is now being followed all over the world also. Only a brief account of the method applicable to the members of foundation structures is given in this chapter. More details of the subject can be obtained from textbooks on Reinforced Concrete [1].

Limit state design means designing the structure for many limit states such as durability, deflection, cracking, and stability, in addition to ultimate strength. Design with reference to the *limit state of ultimate collapse* is discussed in detail here while other limit states are dicussed in brief. The structure is generally designed by ultimate strength consideration; the other limit states are generally considered by *empirical methods* and *by detailing*. For example, for limit state of durability, the mix to be adopted (w/c ratio) and the cover to be provided depends on the environment. Deflection is usually covered by span/depth ratios. Cracking is controlled by detailing of reinforcement such as minimum ratio of steel and spacing of steel.

2.2 ULTIMATE STRENGTH DESIGN

For strength calculation, we use ultimate strength design of a section with suitable factors of safety. First of all, the loads used for design should be the *factored loads*. In IS 456, we use the same load factor for live loads (LL) and dead loads (DL). The following combination of live and dead loads is specified in IS 456, Table 18.

Factored load = 1.5DL + 1.5LL (IS Code)

(For other combination of loads, see Sec. 1.7 and Sec. 3.2).

In codes of other countries, the load factors used can be different for DL and LL. Thus, in the British Code, it is as follows:

Factored load = 1.4DL + 1.6LL (BS Code)

Usually, IS values give larger loads than BS values. A single factor is also more convenient to use in design calculations. Limit State Design of foundations involves checking with respect to the following six aspects.

1. **Safety against environment.** This is ensured by choosing a proper cover. Usually, a 75-mm clear cover is chosen for all foundation structures when cast directly on the ground. However, in India most foundations are cast on a blinding lean concrete 1:4:8 50 mm thick. In such cases we need to design for a cover of 50 mm only. The IS 456-2000, Cl. 26.4 requirements for cover are shown and described in Figure 2.1. The cement content, *w/c* ratio and strength to cater for the various environments have been specified in Table 5 IS 456-2000, as follows (Table 2.1).

TABLE 2.1 Requirements for Different Exposure Conditions with Normal Weight Aggregates of 20 mm (Table 5 IS 456-2000) for Superstructures

	Plain cement concrete			*Reinforced cement concrete*		
Exposure condition	*Min cement content* (kg/cm^3)	*Max w/c ratio*	*Min grade* (M)	*Min cement content* (kg/cm^3)	*Max w/c ratio*	*Min grade* (M)
Mild*	220	0.60	—	300	0.55	20
Moderate	240	0.60	15	300	0.50	25
Severe	250	0.50	20	320	0.45	30
Very severe	260	0.45	20	340	0.45	35
Extreme	280	0.40	25	360	0.40	40

*For description of the five levels of exposure conditions refer to IS 456-2000 Table 3.

Notes: (a) Cement content prescribed is irrespective of the grade of cement and is inclusive of additives like fly ash in the cement.

(b) Severe conditions for superstructure refer to surfaces which are exposed to alternate wetting and drying or conditions near coastal zones. As foundations are buried in soil "Normal foundation condition" should correspond to "Severe condition of superstructure".

(c) If sulphates or chlorides are found in the soil or in the ground water, then suitable type of cement as specified in IS 456 should be recommended for the site [1].

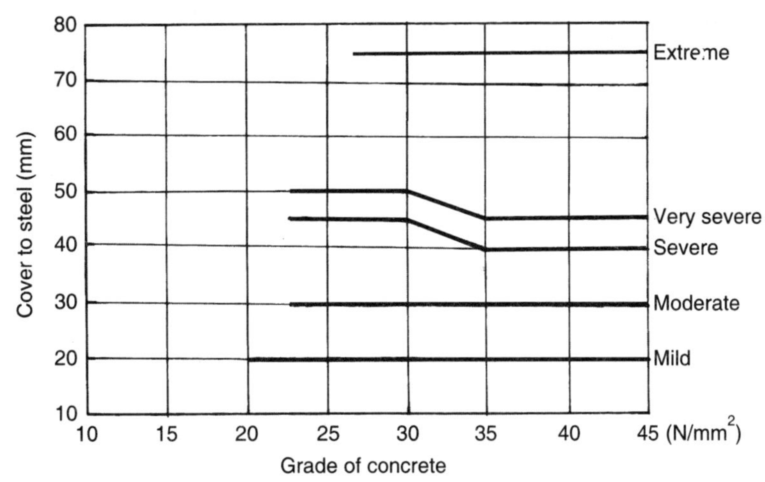

Figure 2.1 Nominal minimum cover to be provided to all steel reinforcement (including links) in R.C.C. members as specified in IS 456 (2000) Clause 26.4 [*Note:* In practice for foundation work we do not level the bottom of excavation accurately, but use a levelling course called "blinding concrete" or mud mat 50 to 80 mm thick with 1:5:10 lean concrete using 40 mm and above size coarse aggregate. IS 456 clause 26.4.2.2 specifies a minimum cover of 50 mm for footings with mud mat concrete (without mud mat concrete cover should be 75 mm). For rafts we always provide a mud mat to level the ground and provide 50 to 75 mm (at least 50 mm) cover (Table 16 IS 456-2000).

2. **Safety against bending failure.** The depth of the section and the area of steel provided should be designed for the bending moment at that section (see Sec. 2.3).
3. **Safety against bond failure.** The sizes of reinforcements should be so chosen that there is enough development length. Otherwise it should be bent up to provide the necessary development length (see Sec. 2.4).
4. **Safety against bending shear or one-way shear.** Slabs should be safe without shear reinforcements. Beams should always be provided with at least the minimum shear reinforcement (see Sec. 2.5).
5. **Safety in punching or two-way shear.** This is very important in foundations which carry concentrated column loads, walls, etc. It is always better to provide sufficient depths of slabs so that the punching shear is satisfied even though necessary punching shear can be resisted by special reinforcements such as shear head reinforcements (see Sec. 2.6). The depth should be sufficient without additional reinforcement.
6. **Safety against cracking.** Though we can calculate the probable crack width, etc., this aspect is usually automatically ensured by obeying the detailing rules such as maximum bar size, spacing of steel, minimum bar size, minimum and maximum steel ratios, etc. specified in the codes of practice for reinforced concrete structures (see Sec. 2.7). Hence standard detailing is very important in R.C. design.

These aspects are briefly dealt with in the following sections.

2.3 DESIGNING FOR MAXIMUM BENDING MOMENT

There are two aspects in maximum bending moment design. Generally, the depth provided for *foundation structures* should be sufficient enough to suit singly reinforced section. (If the depth is limited, we have to use doubly reinforced sections.) Secondly, the area of tension steel is to be determined. We use the following principles for design for bending moment.

2.3.1 Determination of the Bending Moment M_u

Generally, the positions of the points where maximum bending moment occurs are determined by using the simple principle of the theory of structures that they are the points where the shear force is zero or changes its sign. However, in the design of footings, etc., the positions where the bending moment is to be calculated are generally specified in the code.

2.3.2 Determination of Minimum Depth for M_u

Having determined the maximum bending moment, we determine the minimum depth of section needed in bending as given by the formula

$$M_u = Kf_{ck}bd^2$$

The value of K depends on the grade of steel being used. For the normal Fe 415 steel, $K = 0.138 \approx 0.14$. Hence for Fe 415 steel, effective depth d required is given by

$$d = \sqrt{\frac{M_u}{0.14 f_{ck} b}}$$

or

$$d = \sqrt{\frac{7.1 M_u}{f_{ck} b}} \qquad (2.1)$$

2.3.3 Determination of Steel Area Required

There are different methods to determine the steel area. We can find it from the fundamentals. However, it is much easier to use the tables published by Bureau of Indian Standards in SP 16. For singly reinforced beams of breadth b and depth d, we proceed as follows.

Method 1: The most convenient method is the use of SP 16 Design Aids for R.C. design published by BIS [2]. Find M_u/bd^2 (in N mm units). SP 16 gives table (see Table B.1) for different grades of steel and concrete the value for p, the percentage of steel required.

$$p = \frac{A_s}{bd} \times 100$$

or

$$A_s = \frac{p}{100}(b \times d) \qquad (2.2)$$

It is also good to remember (as can be found from SP 16) that for $M/bd^2 < 0.4$, we need to provide only 0.12% of steel with Fe 415 and M20 concrete. For 0.25% steel, the value of $M/bd^2 = 0.85$. Similarly, for the maximum (1%) of steel, the value of M/bd^2 is equal to 2.76.

Similar tables are available in SP 16 for the design of doubly reinforced sections also. (See Appendix B for Tables for design of singly reinforced beams for Fe 415 steel and M20 concrete)

Method 2: In this method we calculate the length of the lever arm. From the theory of limit state design, we can derive the following expression [1]:

Let z = depth of the lever arm

Let $z/d = L_a$ called lever arm factor

From the fundamentals, we can derive the following [1]:

$$\frac{1.16M_u}{f_{ck}bd^2} = (1 - L_a)L_a$$

Putting,

$$\frac{1.16M_u}{f_{ck}bd^2} = f_1$$

we obtain

$$L_a^2 - L_a + f_1 = 0$$

Solving this equation and taking the larger of the two values, we get

$$L_a = 0.5 + (0.25 - f_1)^{1/2} \tag{2.3}$$

The value of lever arm factor L_a can be calculated from Eq. (2.3). The area of steel is calculated by the equation

$$A_s = M/0.87f_y\,z \quad \text{where } z = L_a(d)$$

Method 3: An alternative method to find L_a is to read off from a graph constructed for $M_u/f_{ck}bd^2$ vs L_a, as shown in Figure 2.2. This graph has to be plotted for a range of values

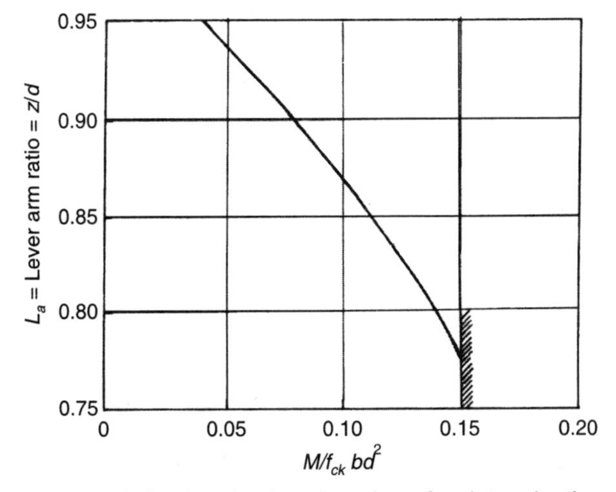

Figure 2.2 Lever arm/effective depth ratio values for determination of steel area in reinforced concrete design (alternate method).

starting from a minimum value of $M_u/(f_{ck}bd^2)$—corresponding to a maximum value of $L_a = 0.95$ and a minimum percentage of tension steel—to a maximum value of $M_u/(f_{ck}bd^2) = 0.15$. The area of steel is calculated by the relation

$$A_s = M/0.87 f_y z \tag{2.4}$$

where $z = L_a \times d$.

For slabs, we usually express steel as areas required per metre width and find steel spacing required. For beams and narrow slabs, we find the number of rods required.

2.3.4 Minimum Areas of Steel in R.C. Members

Table B.2 gives the IS 456 recommendations for minimum steel in R.C. members. In all R.C. members, this should be satisfied.

2.4 CHECKING FOR BOND

When we were using mild steel rods without ribs as reinforcement, we had to check both for *local bond* and *anchorage bond requirements*. However, it has been confirmed by tests that with the use of deformed bars, only average bond needs to be checked. This simply means that the length of the reinforcement rod buried in concrete from the point of maximum tension will have the development length L_d. The diameter of the rods chosen should satisfy this requirement. Table B.3 gives the L_d/ϕ values for various grades of steel and concrete. It is good to remember that for Fe 415 steel $L_d = 47\phi$ for tension steel. The required length can also be made less if we increase the area of steel required and thus reduce the stress level in steel. (In most cases we use smaller diameter bars to increase the perimeter area for bond.) This bond is very important in *foundation structures* as they are short members and the loads to be carried are heavy. The diameters of the reinforcement should be carefully chosen to satisfy this requirement. It is good to remember that for Fe 415 steel and M20 concrete, the development length L_d is [Refer Table B.3 in Appendix B]

$$L_d = 47\phi \tag{2.5}$$

2.5 DESIGN OF SLABS AND BEAMS FOR BENDING SHEAR (ONE-WAY SHEAR)

As we have already seen that slabs of reinforced concrete foundations should be designed to be safe in shear without shear reinforcements. In fact, usually it is shear, and not bending, that determines the depths of slabs in foundations except in very low allowable ground pressures of the order of only 50 kN/m^2. For this purpose, the shear at a distance equal to effective depth from edge of support only is to be considered. (For theory regarding this practice, refer any textbook on R.C design [1].)

2.5.1 Shear in Beams

All beams have to be provided at least with nominal shear (see Table B.7) Hence, in the case of beam design, unlike slabs where the depth is first designed for depth with no shear steel, we first find the depth required in bending and then we provide for shear. The allowable stress in shear depends on the percentage of tension steel as shown in Table B.4. This gives values for a depth of section of 300 mm or more. The value can be increased as shown in Table B.5 for depths of slabs smaller than 300 mm. The maximum values allowed are given in Table B.6.

It has also been observed in tests that in places where the face of support is in compression, we can use an enhanced shear strength near the supports. In order to simplify matters, the code specifies that the beam sections to be tested for shear need to be taken only at distances equal to the effective depth of the section. As the allowable shear stress τ_c is a function of the tension steel at the section, *it will depend on the amount of negative steel at support*. It is also mandatory that the specified tension steel percentage should be available for a distance equal to effective depth d on both sides of the section.

2.5.2 Design Procedure for Slabs in One-way Shear

The practices for design in bending shear in slabs are as follows:

1. *Where the face of support is in tension* as in the case of a bottom slab of an elevated water tank, the design shear is taken as shear V at the support.

$$\text{Shear stress } \tau = \frac{V}{bd} \quad \text{(IS 456, Figure 2)}$$

2. Where the face of support is in compression as in simply supported or continuous beams and slabs, we consider shear V at a distance equal to d, the effective depth, from the edge of the support and calculate the shear stress as $\tau = \dfrac{V}{bd}$.
3. For the design of footings and bases, we consider shear at a distance d from the edge of the column and find $\tau = \dfrac{V}{bd}$.

It should be noted that the maximum allowable values of shear τ_c with shear steel for slabs are the same as those given in Appendix B Table B.4 for beams. (It is interesting to note that in BS 8110 for footings the one-way shear is considered (BS 8110, Cl. 3.7.7) at a section 1.5d from the face of the load. Also, the allowable values in shear are slightly more than the IS values in Table B.4. Thus, the IS method of shear design is quite conservative.) It is good to remember that τ_c for $p \leq 0.15$ is 0.28 N/mm^2 for M20 concrete. However, for initial determination of the depth of slab, we may design slabs for $t_c = 0.35$ corresponding to about 0.25% of tension steel at that section as in practice because of the importance of foundations, a minimum steel of 0.25% tension steel is usually provided in all foundations (see Examples 2.3 and 2.4 for procedure for design).

Let us now derive an expression to find the depth of the slab required for shear = τ_c N/mm^2 in one-way shear.

For a section XX, as shown in Figure 2.3(a),

$$V = [L/2 - a/2 - d] \times q \times B$$

$$\tau_c = \frac{V}{Bd} = \frac{[L/2 - a/2 - d] \times q \times B}{Bd}$$

$$d = \frac{q(L/2 - a/2)}{\tau_c + q} = \frac{q(L - a)}{2\tau_c + 2q} = \frac{q(L/2 - a/2)}{\tau_c + q} \tag{2.6}$$

For τ_c = 350 kN/m^2, this reduces to the following expression in kN and m units.

$$d = \frac{q(L - a)}{700 + 2q} = \frac{q(L/2 - a/2)}{350 + q} \tag{2.7}$$

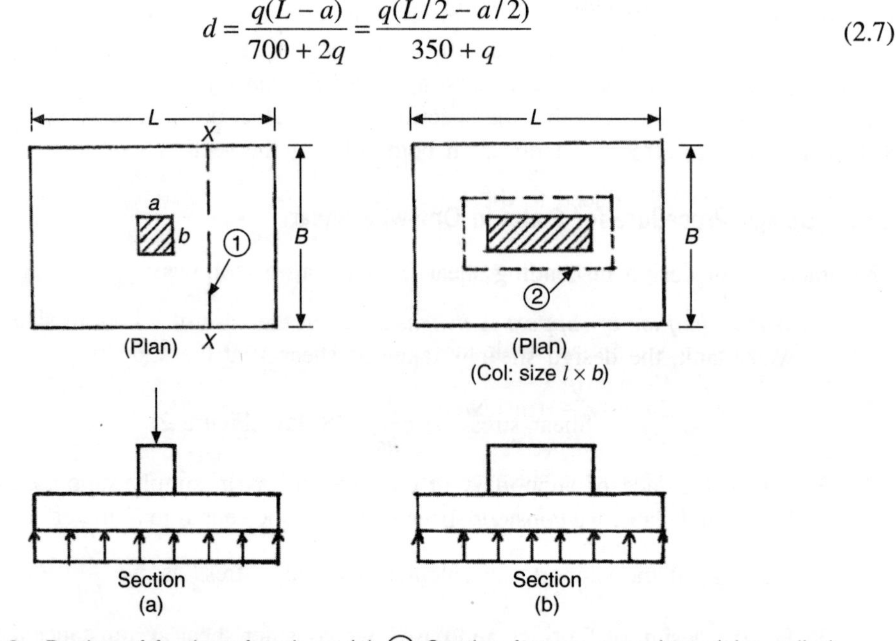

Figure 2.3 Design of footings for shear (a) ① Section for one-way shear (also called beam shear). Definition of *a* and *b* in Eq. (2.7). (b) ② Section for punching shear. Definition of *l* and *b* in Eq. (2.10).

2.5.3 Procedure for Design of Shear Steel in Beams

We have seen that when we design a slab, we choose its thickness to be safe in shear without any shear reinforcements. For slabs thinner than 300 mm thick, IS 456 allows an increased shear also as given in Table B.5. Thus, the thickness of the slab is so chosen that it requires no shear steel. In foundations, a minimum total thickness of 225 mm (150 + 75 mm) is usually provided.

However, beams carry heavy shear and it is mandatory that we should have at least the minimum (nominal) vertical stirrups in a beam. If the shear τ exceeds τ_c in beams, we can provide designed shear steel in the form of vertical stirrups or bent bars to take excess shear.

It is also specified that when we design for shear steel, at least 50% of the excess shear is to be taken by vertical stirrups. (For details, refer to books on reinforced concrete.) For shear design of a section ($b \times d$) in shear, proceed as follows.

Step 1: Calculate the shear stress at distance d from face of support. Let it be τ. If $\tau > \tau_c$, we have to design for shear. In any case, τ should not be greater than τ_{max} as given in Table B.4 of Appendix B. If $\tau < \tau_c$, we have to provide at least nominal shear steel.

Step 2: If $\tau > \tau_c$ but less than τ_{max} allowed, the excess shear over that allowed has to be carried by steel. To design for shear we proceed as follows. Let the area of the two legs of stirrups be A_{sv} and its spacing be s_v

$$V_s = (\tau - \tau_c)\, bd$$

Then

$$\frac{A_{sv}}{s_v} = \frac{V_s}{0.87 f_y d} \qquad \text{(IS 456, Cl. 40.4)}$$

This can be reduced to the following formula as is done in SP 16 (Design Aid).

$$\frac{V_s}{d} = \frac{A_{sv}(0.87 f_y)}{s_v} \tag{2.8}$$

We can make a readymade table, as shown in Table B.7, from where the spacing for a given value of $\frac{V_s}{d}$ (in kN/cm) can be read.

Step 3: In beams even if the shear is less than allowable, we have to provide at least nominal steel (IS 456, Cl. 26.5.1.6). The *nominal shear steel* can be obtained by one of the following methods.

Method 1: For Fe 415 steel, the spacing for nominal shear becomes

$$s_v = \frac{0.87 f_y}{0.4}\left(\frac{A_{sv}}{b}\right) = 902\left(\frac{A_{sv}}{b}\right) \tag{2.9}$$

Method 2: For minimum shear reinforcement in beams, we can read the value from Table B.7 (Appendix B) assuming that the nominal shear to be taken by the section is 0.4 N/mm^2.

Step 4: IS 456 also specifies that the maximum spacings of shear steel should not be more than $0.75d$ or 300 mm. Hence, we choose a spacing to satisfy this rule (see Table B.7).

2.6 PUNCHING SHEAR (TWO-WAY SHEAR) IN SLABS

Punching shear is the shear produced by a concentrated load punching through the slab around it. It is also called *two-way shear.* This type of shear occurs around columns in footings and flat slabs. In IS 456, this stress is calculated around the *column load at a distance d/2 from the edge of the column.* The allowable value is

$\tau_p = k(0.25\sqrt{f_{ck}})$ (Thus for M20 concrete $\tau_p = k \times 1.118$ or $k \times 1.12$ N/mm^2)

where $k = (0.5 + \beta) \not> 1$ and β = ratio of the short to long side of the column.
(For a ratio β, less than half the shear value gets reduced.)

In foundation slabs, the thickness of concrete around columns should be so chosen that they can withstand the above punching shear without shear reinforcement. If the shear value is high, there are methods, such as shear head reinforcement, for reinforcing for punching shear as described in books on R.C. They are mostly used in flat slab construction and rarely used in foundation slabs.

Earlier (and in some modern codes other than IS), the full thickness of concrete including cover was taken as depth resisting punching shear. However, IS 456-2000 takes only effective depth d as resisting this shear as shown in Figure 2.3(b). Taking P as column load, ground pressure as p, column size as $l \times b$, and effective depth of slab as d,

$$\tau = \frac{P - q(b+d)(l+d)}{(2l + 2b + 4d)\, d} \not> (k \times 1.12) \quad \text{[For M20 concrete]} \tag{2.10}$$

As the above equation reduces to a quadratic in d, it is easier to find the depth first in one-way shear and then check whether that depth is sufficient for two-way shear. See Example 2.5 for the procedure for design.

Alternatively, with the value of known d obtained from one-way shear design, we can find the punching shear capacity V. If it is equal to or greater than the value of load P, it is safe in punching shear.

As $V = P - q(l + d)(b + d)$, the punching shear resistance for M20 concrete is

$$V_P = \tau_p[2l + 2b + 4d] = 1.12[2l + 2b + 4d]$$

If $V_P > P$, it is safe against punching shear.

Hence, for safety we use the following formula:

$$\frac{P - q(l+d)(b+d)}{(2l + 2b + 4d)\, d} \not> \tau_p \tag{2.11}$$

(When we work in metre units we usually take τ_p = 1120 kN/m^2 for M20 concrete.)

2.7 DETAILING OF STEEL

Crack control is deemed to be satisfied if we follow all the simple rules of detailing of steel such as cover minimum and maximum percentage of steel, minimum and maximum spacing of reinforcements, anchorage lengths, and lap lengths.

2.8 WIDTH OF FLANGE OF T-BEAMS

IS 456, Clause 23.1.2 defines effective flange width for compression for isolated beams as follows:

$$b_f = \frac{L_0}{L_0/b + 4} + b_w \text{ or actual width} \tag{2.12}$$

where L_0 = 0.7 × (effective span for continuous beams and distance between supports for simply supported beams)

b_w = Breadth of the web

[However, when planning T-beams the projection beyond web can be three times as much as the depth of the flange.]

2.9 SUMMARY

We have listed in this chapter the formulae to be used in the design of R.C. members. The following are some of the important ones:

1. To find the depth for a bending moment (BM) = M

$$d = \left(\frac{M_u}{0.14 f_{ck} b}\right)^{1/2} = \left(\frac{7.1 M_u}{f_{ck} b}\right)^{1/2} \quad \text{[Eq (2.1)]}$$

2. To find the required steel, we proceed as follows:
   Find M/bd^2 and use Table B.1.
3. The required bond length for Fe 415 steel and M20 concrete is 47ϕ [Eq. (2.5)].
4. For design of shear in beams, we use Table B.7. For nominal steel we assume that the nominal shear to be taken by section is 0.4 N/mm^2.
5. For depth of footing slab in one-way shear, we use the following formula (see Example 2.3):

$$d = \frac{q(L-b)}{2\tau_c + 2q} = \frac{q(L/2 - b/2)}{\tau_c + q} \quad \text{[Eq. (2.7)]}$$

6. For punching shear, we use the depth from bending shear and check the following formulae (see Example 2.5):

$$\frac{P - q(a+d)(b+d)}{(2a + 2b + 4d)\,d} \ngtr \tau_p \quad \text{[Eq. (2.11)]}$$

EXAMPLE 2.1 (Design for moments)

Find the percentage of steel required for the following:

1. A beam of b = 250 mm and d = 400 for carrying a factored moment = 80 kNm.
2. A slab of depth 650 mm and width 3000 mm with factored moment 380 kNm.

Reference	*Step*	*Calculation*
		(a) Design of beam
	1	*Find M_u/bd^2*
		$\frac{M_u}{bd^2} = \frac{80 \times 10^6}{250 \times 400 \times 400} = 2.0$
Table B.1		(As this value is < 2.76, the depth is enough to act as a singly reinforced beam)

Reference	Step	Calculation
	2	*Find percentage steel required*
Table B.1		For Fe 415 steel and M20 concrete
		p for $M_u/bd^2 = 2.0$ is 0.64%
Eq. (2.2)		$A_s = \dfrac{0.64 \times 250 \times 400}{100} = 640 \text{ mm}^2$
		Alternate method using Figure 2.1
Eq. (2.3)		$\dfrac{M}{f_{ck}bd^2} = \dfrac{2.0}{20} = 0.1$
		Value of lever arm = z = 0.86 × 400 = 344 mm
		$A_s = 80 \times 10^6/(0.87 \times 415 \times 344) = 644 \text{ mm}^2$
		(b) Design of slab
	1	*Find M_u/bd^2*
		$\dfrac{M_u}{bd^2} = \dfrac{380 \times 10^6}{3000 \times 650 \times 650} = 0.30$
Table B.1		This value is less than 0.42 for minimum steel of 0.12%.
		Hence provide the minimum steel
		Note: A minimum percentage steel of about 0.25% and a maximum of 0.75 are generally recommended for *slabs in foundations*.

EXAMPLE 2.2 (Design for one-way or bending shear in slabs)

What allowable shear value can be adopted for preliminary design of foundation slabs?

Reference	Step	Calculation
	1	*IS Recommendation*
		For footings in one-way shear, IS 456 recommends that the shear V be considered at a section equal to the effective depth of the slab from the column face. It also recommends that the thickness of slabs should be so designed that the shear is less than that is allowed in concrete for the given percentage of tension steel so that no extra shear steel is provided for slabs.
		$\tau = \dfrac{V}{bd} < \tau_c$ for the tension steel in slab
		For $p \le 0.15\%$; $\tau_c = 0.28$ for M20 concrete
Table B.4		For $p = 0.2\%$; $\tau_c = 0.33$ N/mm^2 for M20 concrete
		For $p = 0.25\%$; $\tau_c = 0.36$ (say, 0.35) N/mm^2 for M20 concrete

Reference	*Step*	*Calculation*
		As we usually provide not less than 0.25% steel in foundation structures, we can safely assume $\tau_c = 0.35$ N/mm^2 in practical designs. It can also be noted that IS values in shear are very conservative compared to BS values as shown below.
	2	*BS Recommendations*
B.S code		BS is more liberal in shear consideration in footing slabs. The shear to be considered is at 1.5d from the face of the column and the values recommended for M20 concrete are
Table 2.9		For $p \le 0.15\%$; $\tau_c = 0.28$
B.S code		$p = 0.20\%$; $\tau_c = 0.33$
	3	*Increased value of* τ_c *for slabs less than* 300 mm
		IS also recommends increased value of τ_c for slabs less than 300 mm in depth as given in Table B.5
	4	*Recommended value for design*
		Thus, a value $\tau_c = 0.35$ N/mm^2 can be recommended for practical design assuming 0.25% steel in the foundation.

EXAMPLE 2.3 (Preliminary design of thickness of foundation slabs in one-way shear in centrally loaded footings with uniform soil pressure)

1. Derive an equation for the determination of depth of foundation slab in bending shear for a single footing.
2. Find the depth required for a footing 3 × 3 with a factored load of 3450 kN on a 300 × 300 mm column.

Reference	*Step*	*Calculation*
		1. Derive a general equation
		Let q = foundation pressure in kN/m^2
		Consider a section distant d from the column face
		$V = (L/2 - a/2 - d) \times q \times B$
		$\tau = \dfrac{V}{Bd} = \dfrac{(L/2 - a/2 - d)q}{d} = 350$ in metre units
		Taking $\tau_c = 0.35$ N/mm^2 = 350 kN/mm^2 for 0.25% steel
		$350d + qd = q(L/2 - a/2)$
		$d = \dfrac{q(L/2 - a/2)}{350 + q} = \dfrac{q(L - a)}{700 + 2q}$ (a)
		In general, we may write the equation in metre units as

Reference	Step	Calculation
		$$d = \frac{q(L-a)}{2\tau_c + 2q} \quad \text{(b) [Eq. (2.6)]}$$
		where τ_c = allowable shear in kN/mm².
		We may also multiply the area of footing A. Or $\tau_c = 0.35$, as $qA = P$, the equation can be simplified as
		$$d = \frac{P(L-a)}{700A + 2P} \quad \text{(c) [Eq. (2.7)]}$$
		2. Calculation of slab depth for one-way shear for given data
	1	*Find q, the base pressure*
		$$q = \frac{3450}{3 \times 3} = 383 \text{ kN/m}^2$$
		$L = 3$ m; $b = 0.3$ m
		Assuming 0.25% steel and $\tau_c = 0.35$
	2	*Find depth for one-way shear*
Eq. (2.7)		$$d = \frac{q(L-a)}{700 + 2q} = \frac{383(2.7)}{700 + 766} = 0.7 \text{ m} = 700 \text{ mm}$$

EXAMPLE 2.4 (Approximate determination of the slab thickness for bending shear for one-way shear for a footing under varying ground pressure)

A footing for a column 500 × 300 mm is 2.45 × 2.0 m in size. The column is at the centre of the footing. Under a load and moment, the ground pressure varies from 260 kN/m to 140 kN/m along the longer base due to a moment along the long direction. Find the approximate depth of the slab required to resist one-way shear.

Reference	Step	Calculation
	1	*Conservative design using maximum pressure*
		Take $q = q_{max} = 260$ kN/m²
Eq. (2.7)		$$d = \frac{q(L/2 - a/2)}{350 + q} = \frac{260(1.23 - 0.25)}{350 + 260} = 0.42 \text{ m}$$
	2	*Approximate design can be made using average pressure*
		Take $q = q_{\text{average}} = \frac{260 + 140}{2} = 200$ kN/mm²
Eq. (2.7)		$$d = \frac{q(L-a)}{700 + 2q} = \frac{200(2.45 - 0.5)}{700 + 400} = \frac{200 \times 1.95}{700 + 400} = 0.35 \text{ m} = 350 \text{ mm}$$
		Any of the above values is a fair estimate.
	3	*More exact calculation*
		From the above depths, a reasonable value can be assumed and we check for the shear to be resisted

EXAMPLE 2.5 (Checking for punching shear)

A column 500 × 300 mm is centrally placed on a 2.45 × 2.0 m footing. It is subjected to a factored vertical load and moments which produce an average ground pressure on a base of 200 kN/m². The depth required for one-way shear is 400 mm. Check whether this is ample for two-way shear (punching shear) also.

Reference	*Step*	*Calculation*
	1	*Derive general formula (column size a × b)* Shear to be resisted = Col. load – $(a + d)(b + d)q$ Shear resistance = $(2a + 2b + 4d)\tau_p$ The equation to be satisfied is $$\tau_p' = \frac{P - q(a+d)(b+d)}{(2a+2b+4d)d} \ngtr \tau_p \qquad \text{[Eq. 2.11]}$$
	2	*Find punching shear force* Column load = 2.45 × 2 × 200 = 980 kN $V = (980) - (0.5 + 0.4)(0.3 + 0.4)200 = 854$ kN
Eq. (2.11)	3	*Method 1: Assume the depth of the slab obtained from bending shear and use Eq. (2.11) to find shear strength required is not more than the allowable value* $$\tau_p' = \frac{P - q(a+d)(b+d)}{(2a+2b+4d)d} \text{ must be } \ngtr \tau_p$$ i.e. $$\frac{980 - 200(0.9)(0.7)}{3.2 \times 0.4} = \frac{854}{1.28} = 667 \ngtr 1120 \text{ kN/m}^2$$ Hence, the depth is suitable.
Eq. (2.11)	4	*Method 2: Find depth from the following equation* $$\frac{P - q(l+d)(b+d)}{2(l+b+2d)d} = 1120 \text{ kN/m}^2$$ To find depth d, we can use the above quadratic equation d in for the allowable punching shear.
	5	*Method 3: (For large safety factor) Calculate resistance o perimeter area at d/2 and check whether it is greater than the load I from column.* *Perimeter area resisting punching shear* $V_P = (2a + 2b + 4d) \times d = (1 + 0.6 + 1.6) \times 0.4 = 1.28$ m² Let $\tau_p = 1120$ kN/m² *Calculate punching shear resistance* $V_R = 1.28 \times 1120 = 1434$ kN > 980 kN. Hence safe. The first and third methods are easier than the second method.

EXAMPLE 2.6 (Design of nominal shear steel and designed shear steel for beams)

1. A beam of section b = 350 and d = 550 is subjected to a shear V_s = 125 kN. If f_{ck} = 25 N/mm², f_y = 415 N/mm², and percentage of steel 1.60%, find the shear steel required.
2. If in the above section V_s = 400 kN, design the shear reinforcement if the percentage of steel provided is 1.67%.

Reference	*Step*	*Calculation*
		1. Design for nominal shear
	1	*Find shear stress and check its value*
		$$\tau = \frac{V}{bd} = \frac{125 \times 10^3}{350 \times 550} = 0.65 \text{ N/mm}^2$$
		τ_c = 0.76 N/mm² for M25 and p = 1.6%
		Nominal shear steel is required for beams.
	2	*Provide nominal steel*
Sec. 2.5 Eq. (2.9)		*Method 1:* By formula for Fe 415 steel
		$$s_v = \frac{902(A_{sv})}{b}$$
		Choose 10 mm links; A_{sv} = 157 mm²
		$$s_v = \frac{902 \times 157}{350} = 404 \text{ mm}$$
		However, max spacing allowed is 0.75d or 300 mm
		0.75d = 0.75 × 550 = 412.5
		Hence, adopt 12 mm links @ 300 mm spacing
		Method 2: Design for nominal shear = 0.4 N/mm²
		$$\frac{V_s}{d}\left(\frac{\text{kN}}{\text{cm}}\right) = \frac{0.4 \times 350 \times 550}{10^3 \times 55} = 1.40$$
Table B.7		From Table B.7, spacing required = 400 mm
		Spacing allowed = 12 mm links @ 300 mm
		2. Design for shear steel for V_s = 400 kN
Table B.6		$$\tau = \frac{400 \times 10^3}{350 \times 550} = 2.08 > \tau_c = 0.76 < \tau_{max} = 3.1 \text{ N/mm}^2$$
Table B.7		We can design by formula (a) $A_{sv}/s_v = b(\tau - \tau_c)/0.87f_y$ or
		Using Table B.7
		Balance shear/depth in N/cm
		$$\frac{V_s}{d} = \frac{400 - (0.76 \times 350 \times 500) \times 10^{-3}}{55} = 4.85$$
		Choose 10 mm links, s_v = 12 cm (giving 4.726)
Table B.7		Provide 10 mm links at 12 cm spacing.

EXAMPLE 2.7 (Expression for maximum bending moment at face of the column for varying pressures at the base of a footing)

1. Find the bending moment of the pressures at the face of the column with base pressure as shown in Figure E2.7.

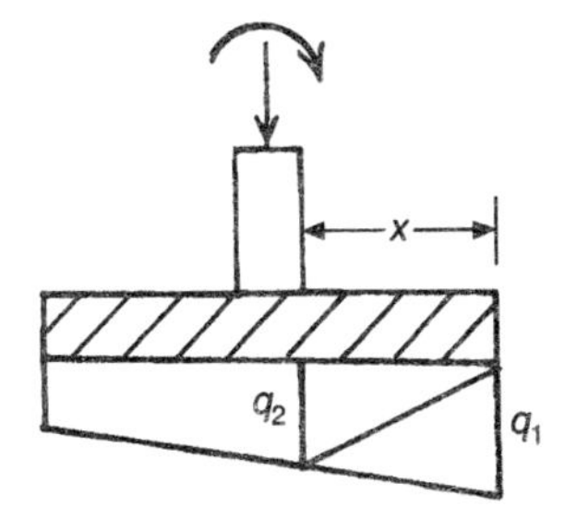

Figure E2.7

Reference	*Step*	*Calculation*
	1	*Calculation of CG* Moment = (Area of CG) of base pressure Let x be the length of projection from the face of the column q_1 is max. pressure at the end of x q_2 is the pressure at section BM per *unit breadth* *Method 1* $$\text{BM} = \left[\left(\frac{q_1 x}{2} \times \frac{2}{3}x\right) + \left(\frac{q_2 x}{2}\right)\left(\frac{1}{3}x\right)\right]$$ $$= \frac{x^2}{6}(2q_1 + q_2) \qquad \text{(a)}$$ *Method 2.* Find CG of trapezium $$= \frac{x}{3}\left(\frac{2q_1 + q_2}{q_1 + q_2}\right) \text{ from } q_2 \qquad \text{(b)}$$ $$\text{BM} = x\left(\frac{q_1 + q_2}{2}\right)\left(\frac{x}{3}\right)\left(\frac{2q_1 + q_2}{q_1 + q_2}\right)$$ $$= \frac{x^2}{6}(2q_1 + q_2) \qquad \text{(c)}$$ These formulae are quite useful in the design of footings.

REFERENCES

[1] Varghese, P.C., *Limit State Design of Reinforced Concrete*, Prentice-Hall of India, New Delhi, 2005.

[2] SP 16, *Design Aids for Reinforced Concrete*, Bureau of Indian Standards, New Delhi.

IS 456 Provisions for Design of Footings and Pedestals

3.1 INTRODUCTION*

In this chapter, we examine the IS 456 provision for design of R.C. footings and pedestals. Footings are structural elements carrying the loads from walls, columns, etc. and their own self-weight to the ground. They can be made of simple lime concrete, plain cement concrete or reinforced concrete, depending on the loading and site conditions.

Pedestals are short compression members whose height is usually less than three times their lateral dimensions. They are usually placed at the base of steel columns to transfer their load to the foundation [see Sec. 4.8].

In this chapter, we deal with the general principles of design of *simple footings* and *pedestals* given in IS 456[1]. More detailed discussion about the design of the various types of footings and examples of design will be given in subsequent chapters of this book.

3.2 DESIGN LOADS FOR FOUNDATION DESIGN

1. **Loads for determination of size of foundation.** The condition to be satisfied by the subsoil in the design of foundation is that its safe bearing capacity (which is based on both strength and settlement) should not be exceeded by the loads from the structure. As the

*Design charts for empirical design of isolated columns and footings are given in Appendix D.

safe bearing capacity is obtained (from the principles of soil mechanics) by dividing the ultimate capacity of the soil by a suitable factor of safety, its value represents the serviceability condition, and not the limit state condition. Accordingly, the loads to be used to determine *the size of the foundation should be the service loads, and not the factored loads*. The three combinations of loads to be used are as follows:

1. Dead + imposed loads	1.0 DL + 1.0 LL	
2. Dead + wind or earthquake loads	1.0 DL + 1.0 WL	(3.1)
3. Dead + imposed + wind (or earthquake)	1.0 DL + 0.8 LL + 0.8 WL (EL)	

2. **Loads for limit state design of foundations.** For reinforced concrete *structural design by limit state design*, we use factored load. IS 456, Table 18 gives the four combinations of loads to be used as follows:

   (a) 1.5 (DL + LL)
   (b) 1.2 (DL + LL + WL)
   (c) 1.5 (DL + WL) (3.2)
   (d) 0.9 DL + 1.5 WL

In multi-storeyed buildings, one should also consider the allowable *reduction in live load* for residential and office buildings as given in Table 3.1.

TABLE 3.1 Reduction of LL on Columns and Foundations (IS 875 Part 2)

No. of floors/roofs supported by column	*Per cent reduction of load on all floors*
1	0
2	10
3	20
4	30
5–10	40
Above 10	50

Note: No reduction for GF to basement and *no deduction for dead loads* are allowed.

3.3 BASIS OF STRUCTURAL DESIGN OF R.C. FOOTINGS

Footings under walls are called *one-way footings* and those under columns, *two-way footings*. The first step in the design of footings is to calculate the necessary area from the formula using service loads. Hence,

$$\text{Area of footings} = \frac{\text{Service load on the column or wall above [loads in Eq. (3.1)]}}{\text{Safe bearing capacity of the soil below}}$$

In the next step, its structural design is carried out by using factored loads and principles of limit state design as already discussed in Chapter 2. The main items to be designed are the *thickness of slab footings* and *its reinforcement*. The thickness should be sufficient to

1. resist the shear force without shear steel and the steel should satisfy the bending moment without compression steel;
2. give the structure the required structural rigidity so that settlement is uniform and the foundation reaction below can be assumed to be uniform (see Chapter 22);
3. withstand the environmental conditions that can be caused from the ground (see Figure 2.1).

It is also important to remember that the percentage of longitudinal steel provided should not be less than 0.15 for Fe 250 steels and 0.12 for Fe 415 steel as specified for slabs in IS 456, Cl. 26.5.2.1. This is only minimum shrinkage steel. However, many authorities recommend 0.2–0.25 per cent steel as main steel for *foundations* as foundation is a very important part of any structure. The 0.12 per cent value can be used for distribution steel only.

3.4 SOIL PRESSURE ON FOUNDATIONS

In most designs of foundations, especially in individual footing design, the soil is considered elastic and the R.C. foundation structure infinitely rigid. Hence, if foundation pressures on these rigid structures are to be assumed as uniformly distributed on the base, it is necessary that the CG of the external load system should always coincide with the CG of the loaded area. Otherwise, there will be a variation of pressure on the base of the foundation which, for rigid foundations, may be assumed as linearly varying. In all layouts of foundations, these basic principles should always be borne in mind.

3.5 CONVENTIONAL ANALYSIS OF FOOTINGS SUBJECTED TO VERTICAL LOAD AND MOMENTS

As already discussed in Chapter 1 and shown in Figure 1.1 with vertical load P and moment M acting on a column, the base pressure under the rigid footing will be non-uniform but will be linearly varying from a maximum value q_1 to minimum value of q_2 as shown in Figure 3.1. As already mentioned, two different cases can arise.

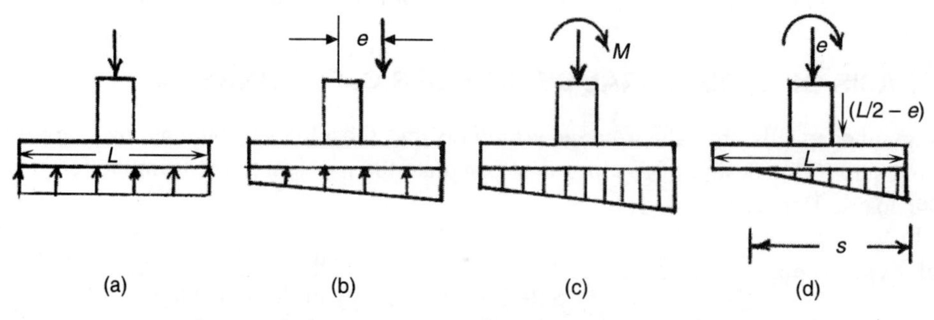

Figure 3.1 (a) to (c) Ground pressures under rigid column footing, (d) With large moments the pressure extends only to $s = (1.5L - 3e)$, where $e = M/P$.

Case 1: The centre of gravity of the foundation and the line of application of P the vertical loads coincide and M due to external forces such as wind forces. The values of q_1 and q_2 with moment M are given by Figure 3.1(a) as

$$q_1 = \frac{P}{BL} + \frac{6M}{BL^2}, \qquad q_2 = \frac{P}{BL} - \frac{6M}{BL^2} \tag{3.3}$$

where

L = Dimensions along the plane of action of the moment

B = Breadth

q_1 and q_2 = Foundation pressures

As explained in Chapter 1 Section 1.3 the horizontal loads have to be resisted by the ground reaction and friction.

Case 2: If the moment produced is by the resultant *vertical load P acting with an eccentricity e*, the resultant base reaction will counteract the applied moment. The moment M produced is

$$M = eP$$

3.5.1 General Comments

In soils such as clays, large non-uniform base pressures can lead to differential settlements and consequent tilt in the column. Hence the ratio of (q_1/q_2) should not exceed 2. They should be small under dead load. In the case of small eccentricities due to P and M, the centre of gravity of the foundation itself may be offset from the line of action of the load P to produce uniform resultant distribution on the base as described in Chapter 6.

The value of $e = M/P$ gives the indication of the probable variation of q_1 and q_2. If, $e = L/6$, according to the middle third rule, $q_2 = 0$ and q_1 will be twice the average pressures. If e is greater than $L/6$, part of the base will not be in contact. We always try to avoid this condition. When the base is not fully in contact with the ground, the maximum soil pressures q on the foundation can be calculated from the following equation (see Figure 3.1(c)).

$$P = \frac{1}{2} qsB \tag{3.4}$$

where B is the breadth of the footing and s is the length of the foundation in contact with the ground and q the ground pressure. The value of the eccentricity e is given by the equation

$$\frac{M}{P} = e = \left(\frac{L}{2} - \frac{s}{3}\right) \tag{3.5}$$

Hence, $s = 1.5L - 3e$.

In footings subjected to biaxial bending, the effect of variation of base pressures in both X and Y directions should be considered and combined to get the final values.

3.6 GENERAL PLANNING AND DESIGN OF INDEPENDENT FOOTINGS

IS 456, Cl. 34 deals with the design of independent footings. The main recommendations are the following:

1. There can be three types of reinforced concrete individual footings: rectangular, sloped or stepped as shown in Figure 3.2. They may rest on soil, rock or on piles. *The minimum thickness of the edge of the footing on soil and rock should be* 150 *mm and that on top of piles should not be less than* 300 *mm* [IS Cl. 34.1.2].

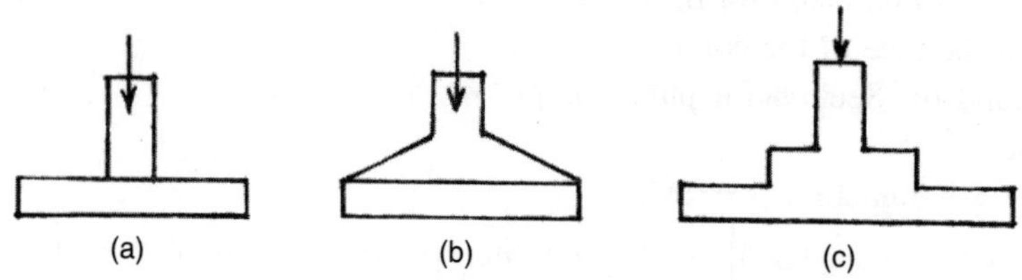

Figure 3.2 Types of isolated footings: (a) simple slab footing, (b) sloped footing, (c) stepped footing.

2. The column transfers the load on top of the footing by bearing. In limit state design, the value of the pressure allowed under direct compression on an unreinforced loaded area of same size is to be limited to $0.45f_{ck}$ as given by IS 456, Cl. 34.4. However, when the supporting area is larger than the loading area on all sides, it may be increased by the factor A_1/A_2 (where in the case of a footing A_1 = area of the footing and A_2 = area of the column base), but the factor should not be greater than 2.
3. If the above permissible stresses are exceeded, the transfer of forces should be with steel reinforcement, by extending the reinforcement into the footing or by providing dowel (starter) bars. According to IS 456, if dowel bars are provided, they should *extend into the column* a distance equal to the full development length of the column bars (IS 456, Cl. 34.4.4). However, this requirement for the development length has been relaxed in BS 8110 (1985): Cl. 3.12.8.8. According to BS code, compression bond stress that develops on starter bars within bases need not be checked, provided the starter bars extend down to the level of the bottom reinforcement. The application of this clause can reduce the depth required in footings and save on steel requirement. According to IS, as explained in Chapter 2, the depth required for anchorage in compression for grade 20 concrete and Fe 415 steel works out to (47ϕ for tension). With 30% reduction, it will be equal to 37.6ϕ (say, 40ϕ) in compression.

   When the depth of the footing required (to satisfy the above development length of *the starter bars* or because of any other causes) is very large, it is more economical to adopt a stepped or sloped footing so as to reduce the amount of concrete that should go into the footing.
4. According to IS 456, Cl. 34.4.3, the extended longitudinal column bars or dowels (started bars) should be at least 0.5% of the cross-sectional area of the supported columns or pedestal, and a minimum of four bars should be provided. The diameter of the dowels should not exceed the diameter of the column bars by 3 mm. [The minimum area of compression steel to be provided in columns according to IS 456 is 0.8%. The minimum size of bars is taken as 12 mm].

As seen from the above discussions, the depth of the footings needs to be generally taken as that obtained from the point of shear and bending moment only. In all cases, the depth should be such that extra shear reinforcement or compression steel for bending can be avoided.

5. The sections to be taken for design are the following as shown in Figure 3.3: (a) Section for BM at the face of the column, (b) Section for one-way shear at d from the face of the column, and (c) Section for punching shear around the column at $d/2$.

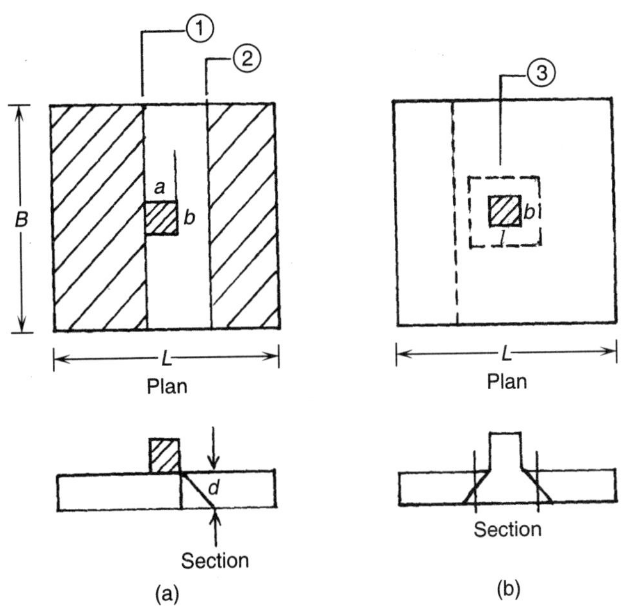

Figure 3.3 Design of independent footing sections for: ① bending moment, ② one-way shear, ③ punching shear.

6. Rectangular footings are generally used with rectangular columns. The most economical proportions of the footing are given if the rectangular base projects the same distance beyond all the column faces so that the footing requires the minimum amount of materials.

3.6.1 Calculation of Shear for Design of Slab Footings

In many cases the thickness of the footing will be determined by the requirements for shear. Both one-way shear (wide-beam shear) and two-way shear (punching shear) requirements are prevalent as discussed now (see Figure 3.3).

1. **One-way shear (wide-beam shear).** One-way shear is similar to shear in bending in slabs. Considering the footing as a wide beam, the shear is taken along a vertical plane extending the full width of the base located at a distance equal to the effective depth of

the footing (i.e. a dispersion of 45°) in the case of footings on soil and a distance equal to half the effective depth of footings for footings on piles.

The allowable shear stress is the same as in beams. The tension reinforcement to be considered for estimating the allowable shear should continue for a distance equal to the full effective depth beyond the section. For routine design, the lowest value of allowable shear in Table 19 of IS 456, i.e. 0.28 N/mm^2 (or BS value of 0.35 N/mm^2) is recommended. In one-way shear, the shear force to be resisted is the sum of the upward forces in the foundation area from the critical section to the edge of the footing. The consequent shear per unit area is given by

$$\tau_v = \frac{V}{bd} \tag{3.6}$$

This τ_v should not be more than the τ_c specified as given in Table B.4.

Design depth for one-way shear. We generally find the depth required from shear consideration before we find the depth required from bending moment consideration in pad footings. (In sloped footing we may first find a safe depth for bending as in Example 4.6.)

In both IS 456 and BS 110, the value of allowable shear in bending τ_c is a function of the percentage of steel on the tension side (Table B.4). As we usually place tension steel of about 0.2–0.25% steel, we can assume an allowable τ_c = 0.35 N/mm^2 for M20 and M25 concrete.

From Sec. 2.5.2,

$$d = \frac{q(L-a)}{2\tau + 2q} \text{ in kN and m units} \tag{3.7}$$

By assuming τ_c = 0.35 N/mm^2 or 350 kN/m^2 for 0.25% minimum steel, we have

$$d = \frac{q[L-a]}{700 + 2q} \text{ (in kN and m units)} \tag{3.8}$$

We will use this approach in our design procedure. Note that the shear force at both ends of L is nil and that in Eqs. (3.7) and (3.8), the size of the column in the direction of L is a.

2. **Checking for two-way or punching shear.** Punching shear indicates the tendency of the column to punch through the footing slab, as illustrated in Figure 3.3(b). As it has two-way section, it is also called *two-way shear.* This type of shear is similar in flat slabs around the supporting columns. As already stated, it is easier to check than design a section for two-way shear.

   The method of checking for punching shear has been explained in Sec. 2.6. The formula to be used is Eq. (2.11). Taking P as load from the column, the shear to be resisted is obtained as

$$V = P - q(l + d)(b + d)$$

$$V_p = \tau_p(2l + 2b + 4d) \tag{3.9}$$

is the available shear resistance.

(Values of τ_p = 1.12 N/mm^2 for M20 concrete); if $V_p > V$, the footing is safe in punching shear.

Hence $V_p > P$, it is very safe in punching (two-way) shear. (3.10)

It should be noted that the British practice in punching shear is to use the same design values for punching shear as those for one-way shear. These values are much lower than those recommended for punching shear in IS and ACI codes. In order to adjust to the low value τ_c of the BS code, the critical perimeter is taken at a distance equal to the effective depth d (instead of $d/2$) from the column face. This will result in larger perimeter length and reduced requirement for shear strength. Both these approaches are found to give safe values for design, especially in footings.

3.6.2 Bending Moment for Design

We must find the steel for bending. IS 456, Cl. 33.2.3 states that the bending moment to be taken for design of footings is the moment of the reaction forces due to the applied load *(excluding the weight of the foundation itself)* at the following sections:

1. At the face of the column for footings supporting a reinforced concrete column as shown in Figure 3.2(a).
2. Half-way between the centre line and the edge of the wall for footings under masonry walls.
3. Half-way between the face of the column and the edge of the gusseted base for footings under gusseted bases.

It should be specially noted that moments should be considered both in X and Y directions and the necessary areas of *steel provided in both directions.*

The steel for the above bending moment is placed as detailed under placement of reinforcement (IS 456, Sec. 22.7). *The footing is to be considered as a slab and the rules for minimum reinforcement for solid slabs should apply to these slabs also* (IS 456, Cl. 34.5). This recommendation is very important as in many cases of design of footings, the reinforcement calculated from bending moment consideration can be less than the minimum required as a slab of 0.12% for Fe 415 steel as specified in IS 456, Cl. 26.5.2.1.

3.7 MINIMUM DEPTH OF FOOTING AND DETAILING OF STEEL

As already pointed out, the depths of footings should be such as to make them rigid. These are governed by the following considerations:

1. The depth should be safe in one-way shear without shear reinforcements.
2. The depth should be safe in two-way shear without any shear reinforcement.
3. The depth should be safe for the bending moment without compression reinforcement.
4. The depth, according to IS 456, should develop the necessary transfer bond length by the main column bars or dowel bars if it is necessary (see Sec. 3.7.1).

According to Clause 34.5 of IS 456, the minimum steel and maximum spacing of bars in footings should be as in R.C. slabs: 0.15 per cent for Fe 250 and 0.12 per cent for Fe 415. If the column bars are bent at the bottom and properly extended into the footing, such bars can also be considered to act with the footing. In a sloped footing, the check for minimum steel area then needs to be made only at the middle of the footing. The maximum spacing is $3d$ or 300 mm. The steel required in the *X*- and *Y*-directions should be distributed across the cross-section of the footing as follows.

1. According to IS 456, Cl. 26.5.2.2, the maximum diameter of the bar should not exceed one-eighth the total thickness of the slab.
2. In two-way *square* reinforced column footings, the steel is distributed uniformly across the full width of the footing.
3. In two-way *rectangular* column footings, the steel in the long direction is distributed uniformly over the width of the footing. But as regards the *steel in the short direction*, more of this steel is placed on the portion near the column, commonly known as *column band (CB)*, than in the outer portion. *For this purpose, a column band equal to the width of the footing is marked beside the column, along the length of the footing, as shown in* Figure 3.4. The portion of the reinforcement to be placed at equal spacing in this band is determined by the equation (IS, Cl. 34.3.1, Figure 3.3)

$$\text{Reinforcement in column band} = \frac{2}{(y/x+1)} \times A_t = \frac{2}{\beta+1} A_t \qquad (3.11)$$

where,

A_t = Total area of reinforcement in the short direction

y/x = Ratio of the length to breadth of the footing = β

The remaining steel is placed at uniform spacing outside the column band.

(*Note:* In a long footing as in combined footings, we plan the transverse steel by assuming a column strip (instead in column band) as explained in Sec. 7.8.1. Details of placing steel in a footing is shown in Figure 3.5.)

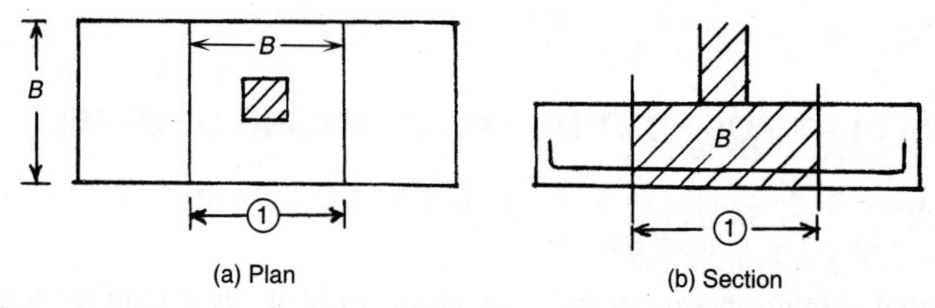

Figure 3.4 Detailing of transverse steel in column band in rectangular isolated footing ① column band equal to the width of the footing [IS 456 (2000), Cl. 34.3].

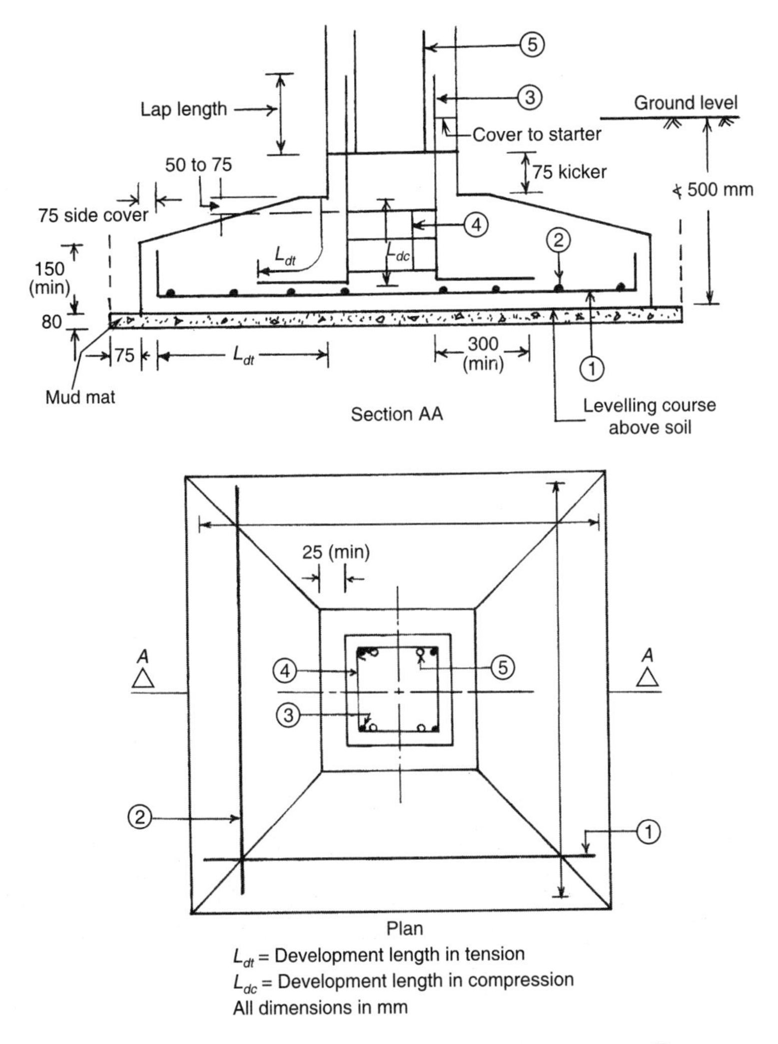

Figure 3.5 Reinforcement drawing of a footing. (1–5: Bar marks. (1) Main reinforcement bent upwards if required; (2) Main reinforcement in other direction; (3) Starter bars; (4) Stirrup, unless specified use T8 @ 300, minimum 3; (5) column bars.)

3.7.1 Transfer of Load at Base of Column

The standard field practice in India is to extend the column bars or its equivalent dowel bars and to bend them at the bottom of the footing. However, we should be aware, as already pointed out in Sec. 3.6, that it is not necessary to have the depth of the slab equal to the development length in all cases. IS 456, Cl. 34.4 specifies only the following conditions for design of junction between the column and the footing (refer Sec. 3.6).

1. The compression stress in concrete at the base of the column should not exceed $0.45 f_{ck}\sqrt{A_1 / A_2}$, where

A_1 = Supporting area for a rectangular base and an area at 45° to a column for a sloping footing

A_2 = Base of the column

2. If the permissible stress is exceeded, *the balance* should be taken by either extending the longitudinal bar into supporting member or by dowel bars.
3. *For these reinforcements only*, the development length must be deemed to be satisfied by extending that steel to the full depth of the footing.
4. Starter bars or dowels equal to at least 0.5 per cent of the cross-sectional area of the column should be provided. The diameter of dowels should not exceed that of the columns. (The minimum percentage of steel for columns is 0.8%.)
5. Bending of bars by 90° increases the anchorage length of tension bars only by 4ϕ for every 45°, but not for compression bars. As bond in compression is 25 per cent higher, we need less anchorage length for compression bars. For M20 concrete, tension anchorage is 47ϕ and compression anchorage is 37ϕ.
6. For column bars of 36 mm, dowels of the necessary area as above can be used, but they should extend into the column and the footing the full development length.

3.8 CHECKING FOR DEVELOPMENT LENGTHS OF MAIN BARS IN FOOTINGS

It is more important to check the development length of *the main tension bars* of the footing than that of the dowel bars. Individual footings are small sized members with large bending moments. The diameter or the sizes of the bars selected should have the required surface area to develop the full development length in the available dimension of the footing. For Fe 415 and M20, $L_d = 47\phi$. Thus, as shown in Figure 3.5, the anchorage length L_d required for the sizes of the bars selected in the long and short direction should be less than one-half of the lengths for the footing in the respective directions. One way of increasing the development of tension bars in footings is to bend them up at 90° as shown in Figure 3.3. Another method is to increase the number of bars (steel area) and thus reduce the stress in the steel. But this is not an economic solution. (Anchorage length is different from development length for mild steel plain bars we usually bend the bar at the end for anchorage).

3.9 DESIGN OF PEDESTALS

The area of steel to be provided in a pedestal will depend on the base pressure of the column on concrete. Their design is discussed in Sec. 4.8.

3.10 DESIGN CHARTS FOR PRELIMINARY DESIGN OF COLUMN AND WALL FOOTINGS

The principles explained in this chapter will be applied in the examples worked out in the various chapters of this book. For preliminary design of columns, column footings and wall footings for given loads, we may use Charts D.1 to D.8 of Appendix D. Example 4.2 illustrates its use for footing design. The chart is for Fe 415 steel and M20 concrete.

3.11 DESIGN CHARTS FOR DESIGN OF COLUMNS AND FOOTINGS

In many cases we can use tables and charts for preliminary design of columns and footings. Charts D.1 to D.8 (Appendix) can be used for this purpose [2].

3.12 SUMMARY

In this chapter we have examined the provisions of IS 456 for the design of foundations in one-way shear, two-way shear and bending moment. We also examined the detailing to be followed in joining columns to the footings. Examples of design are given in Chapter 4.

EXAMPLE 3.1 (Use of charts given in Appendix D for preliminary design of R.C. column footing)

1. A column is to carry a load of 45 tons. If the bearing capacity of the soil is 7.5 t/m^2, find the dimensions and reinforcements of a suitable footing.
2. A column is to carry a load of 50 tons. If the bearing capacity of the soil is 7.5 t/m^2, find the dimensions and reinforcements of a suitable footing.

Use Fe 415 steel and M20 concrete.

Reference	*Step*	*Calculation*
		Case 1: For column to carry 45 tons
Table in	1	*Find size of column*
Chart D.1		Use 230 × 230 mm column
	2	*Find the footing size for 45 tons load*
Chart D.1		Footing 2550 × 2400 mm size
	3	*Find top and bottom layers of steel*
		Top layer 16 Y 10 (16 Nos. Fe 415 steel)
		Bottom layer 14 Y 10 (14 Nos. Fe 415 steel)
		Case 2: For column to carry 50 tons
Table in	1	*Find size of column*
Chart D.1		Use 300 × 300 mm column
	2	*Find footing size for 50 tons load*
		Footing 2700 × 2400 mm size
	3	*Find top and bottom layers of steel*
		Top 18 Y 10 (18 Nos. of Fe 415 steel)
		Bottom 16 Y 10 (16 Nos. of Fe 415 steel)

REFERENCES

[1] IS 456, 2000, *Code of Practice for Plain and Reinforced Concrete*, Bureau of Indian Standards, New Delhi, 2000.

[2] Zacheria George, *Common Building Frame Design*, Indian Concrete Institute, Chennai, Special Publication.

Design of Centrally Loaded Isolated Footings and Column Pedestals

4.1 INTRODUCTION

The principles of design of centrally loaded footings have been already explained in Chapter 3. In this chapter, we will deal with a few examples to illustrate these principles.

4.2 GENERAL PROCEDURE FOR DESIGN

The major steps in the design of a column footing square or rectangular can be summarized as follows. For circular or square columns, we adopt a square footing and for a rectangular column, a rectangular footing is adopted.

Step 1: Determine the plan area from the allowable bearing capacity and *service loads* from the column, assuming a reasonable (10% of loading) as weight for the footing.

Step 2: Taking the factored dead and live loads, determine the ultimate soil reaction for factored design load (see Sec. 3.2).

Step 3: Determine the depth for one-way shear, assuming a design shear strength value τ_c. Theoretically, this value depends on the percentage of steel in the slab. However, for preliminary design, a value of $\tau_c = 0.35$ N/mm^2 corresponding to 0.25% steel may be assumed.

Step 4: Determine the depth from bending considerations.

Step 5: Check the depth adopted for safety against punching shear. If it is not sufficient, increase the depth so that the footing is safe in punching shear.

Step 6: Choose the largest depth required considering steps 3, 4 and 5 and provide the necessary cover.

Step 7: Calculate the reinforcement required in the X- and Y-directions from bending moment considerations. The steel provided at the section for maximum moment should not be less than the minimum specified for slabs.

Step 8: Check the development length required and choose the proper diameter of bars.

Step 9: Detail the steel as specified in the code (see Sec. 3.7).

Step 10: Provide the necessary cover to reinforcement and find the total depth of footing required.

Step 11: Verify design by charts given in Appendix D.

4.3 DESIGN OF SQUARE FOOTING OF UNIFORM DEPTH (PAD FOOTING)

Square footings are often met with in practice, and it is worthwhile to derive the expression for their exclusive design. Let the data for design be as follows (Figure 3.3):

Size of footing = $L \times L$

Column size = $a \times a$

Factored load = P

Service load + wt. of footing = P_1

Bearing capacity = q_a

One-way shear value = τ_c

Two-way shear value = τ_p

Step 1: Find plan size of footing.

$$\text{Area, } A = \frac{\text{Service load}}{\text{Allowable bearing capacity}}$$

Step 2: Soil reaction for limit state design.

$$q = \frac{\text{Factored load}}{\text{Area}} = \frac{P}{L^2}$$

Step 3: Find the depth for one-way shear. By considering one-way shear, the depth d is obtained from the shear at section at X_1X_1 at d from the face of the column for a square footing ($L \times L$). As seen in Sec. 3.6.1, it can be evaluated from the following formula.

Assuming τ_c corresponding to 0.25% steel and M20 concrete in Table B.4, $\tau_c = 0.35$ N/mm^2 (350 kN/m^2) and using metre and kN units ($P/L^2 = q$)

$$d = \frac{q(L-a)}{700+2q} \tag{4.1}$$

[Note: The actual value of τ_c will depend on the percentage of main steel present at the section continued for a distance d on both sides of the section. Some recommend adopting the lowest value of τ_c (namely, the one corresponding to a percentage of steel equal to 0.15%) for routine calculations, others recommend that the minimum main steel in all foundations should be 0.2 to 0.25% as it is difficult to repair once foundations are built. Distribution steel may be only 0.12% for Fe 415 steel.]

Step 4: Find the depth for resistance in the bending. The depth from bending moment consideration is obtained by taking moments at the face of the column XX.

$$M = \frac{P}{L^2}\left[\frac{L(L-a)^2}{8}\right] = \frac{P}{8L}(L-a)^2 \tag{4.2}$$

$M = Kf_{ck}Ld^2$ (as single reinforced beam). [Note the whole length L is assumed to resist bending.]
Hence,

$$d = \left(\frac{M}{Kf_{ck}L}\right)^{1/2} \qquad \text{[For Fe 415, } K = 0.138 \text{ or} \approx 0.14] \tag{4.3}$$

Step 5: The depth should satisfy two-way shear or punching shear at section $d/2$ from the column face.
Critical parameter = $4(a + d)$. Considering equilibrium of forces, we get

$$(P/L^2)\lfloor L^2 - (a+d)^2 \rfloor = 4(a+d)\, d\tau_p$$

Hence

$$q(L^2 - (a+d)^2) < 4(a+d)\, d\tau_p$$

where $\tau_p = 0.25\sqrt{f_{ck}}$. It is easier to check for the value of τ_p in this expression for the value of d obtained from Eqs. (4.1) and (4.3) than to solve the equation for d. Alternately, if the R.H.S is equal to or greater than P, then the design is safe against punching shear.

Step 6: Take the larger of the depths as obtained from steps 3 to 5. Provide the cover.

Step 7: Find the area of steel required from the value of M/Ld^2 and percentage of steel p using Table B.1 or by $M = A_s f_s jd$ using Figure 2.2.

(Use M/Ld^2 SP 16 for easy determination of the area of steel.)

Also, check for minimum steel percentage and steel spacing as given in Sec. 22.7.

Step 8: Check the development length. Select the size of the bar whose development length is less than $l/2(L - a)$; otherwise, provide the development length by 90° bend at the end.

Step 9: Detail steel as discussed in Sec. 3.7.

(One of the methods to decrease steel in footings is to adopt a liberal depth for the footing.)

4.4 DESIGN OF SLOPED RECTANGULAR FOOTINGS

In India, to save steel we commonly use thick footings, designed and constructed as stepped footing or sloped footing, sloping from the column face to the edge. Sloped footings generally require more depth, and hence less steel, than block footings. Figure 4.1 shows such a sloped footing. According to IS 456, Cl. 34.1.4, the edge thickness should not be less than 150 mm. The slope should not exceed one vertical to three horizontal if top forms are to be avoided. In most cases, the concrete to be placed on the slope has to be relatively dry, so that it does not slide down along the slope. It is very important to remember that as the strength of the footing depends on the compressive strength of this concrete along the slope, *special care should be taken in the placement, compaction and curing to get it free from voids.*

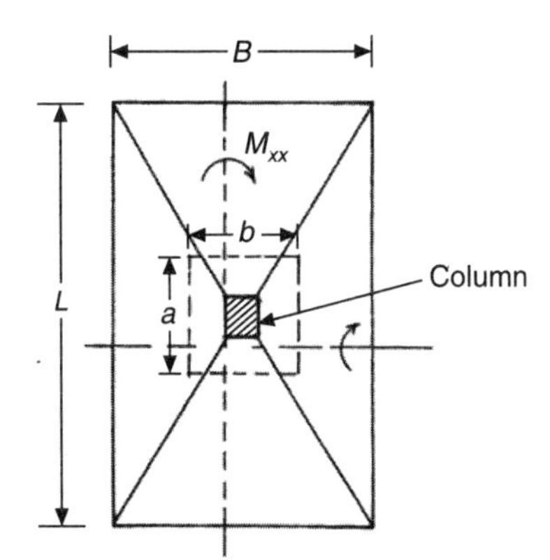

Figure 4.1 Plan of sloped rectangular column footing. Calculation of bending moment and shear for design (Example 4.6). Moment M_{xx} denotes the bending moment for which steel is placed in the X direction.

The procedure for the design of sloped footings varies among designers. This is due to the fact that there have not been many large-scale tests on sloped footings, and an accepted yield line pattern at the failure of centrally loaded and eccentrically loaded sloped footings is not yet available. As also stated in Sec. 3.7, the minimum steel needs to be checked at the mid-depth of sloped footings after it is designed as an ordinary footing.

4.4.1 Design Procedure

A sloped footing can be designed by using the following steps: Let,

L = Length of the footing in the Y-direction
a = Dimension of the column in the Y-direction
B = Length of the footing in the X-direction
b = Dimension of the column in X-direction

[*Note:* Symbol b is also used in $M = Kf_{ck}bd^2$, the formula used in R.C. design and use of SP 16.]

Step 1: *Determination of the required depth in bending:* The aim here is to find a reasonable value which will be larger than that required for a block footing. This depth is to be checked for shear and is used for the calculation of the steel area. The bending moment is taken at the face of the column and any of the following three procedures can be used for this purpose.

Method (a): The first method is to assume that the bending moment is the same as that due to a pad footing in the XX and YY planes. In order to get a liberal value for depth, we assume it is resisted by the column face line only. The expressions are simple and can be derived as follows:

Footing size = $(L \times B)$ with $L > B$

Column size = $(a \times b)$ with $a > b$ [base dimensions of the column]

Taking M_{XX} as moment for steel in the X-direction on section normal to shorter span and M_{YY} as moment on section normal to larger span, we get

$$M_{XX} = \left(\frac{P}{LB}\right) L \left(\frac{B-b}{2}\right)\left(\frac{B-b}{4}\right)$$

$$= \frac{P}{8B}(B-b)^2 \text{ (for steel in the } X\text{-direction) resisted by length } a \qquad (4.4)$$

Similarly,

$$M_{YY} = \frac{P}{8L}(L-a)^2 \text{ (for steel in the } Y\text{-direction) resisted by length } b \qquad (4.4a)$$

Method (b): A less conservative method, which will give a lesser depth of footing is to assume that the moment to be resisted by column faces is that due to the loads in the trapezoidal area only, as shown in Figure 4.1. Accordingly,

$$M_{XX} = \text{(Pressure)} \times \text{(CG of the area)}$$

$$M_{XX} = \frac{P}{LB}\left[\left(\frac{L+a}{2}\right)\left(\frac{B-b}{2}\right)\right] \times \left[\left(\frac{2L+a}{L+a}\right)\frac{1}{3}\left(\frac{B-b}{2}\right)\right]$$

$$M_{XX} = \frac{P}{24LB}(2L+a)(B-b)^2 \text{ (for steel in the } X\text{-direction)} \qquad (4.5)$$

Similarly,

$$M_{YY} = \frac{P}{24LB}(2B+b)(L-a)^2 \text{ (for steel in the } Y\text{-direction)} \qquad (4.5a)$$

In both cases, the moment is to be resisted by the dimensions a and b respectively at the *base of the column.*

$$M_{XX} \text{ for steel in the } X\text{-direction along } B,\ M_{XX} = Kf_{ck}ad^2$$

$$M_{YY} \text{ for steel in the } Y\text{-direction along } L,\ M_{YY} = Kf_{ck}bd^2 \tag{4.6}$$

[*Note:* We assume that the column dimensions a and b only resist the bending. For Fe 415 steel, $k = 0.138$.]

Method (c): As the first method gives very conservative results, some designers use method (a) for the calculation of moments but assume that the moment is resisted by an effective breadth larger than that of the dimension at the column base. One such approximation made is to use the formula

$$b_{\text{eff}} = b + \frac{1}{8}(B - b), \qquad a_{\text{eff}} = a + \frac{1}{8}(L - a) \tag{4.7}$$

Step 2: *Checking for one-way shear:* As shear failure is assumed at 45° from the base of the column. One-way shear is checked by taking a section at a distance equal to the effective depth (as obtained from step 1) from the face of the column. We find b_1 and d_1 corresponding to the quadrant to that section.

The shear force V_1 to be resisted is taken as that acting in the corresponding quadrant into which the slope footing is divided as shown in Figure 4.1. *The area of concrete resisting the shear is taken as the breadth of the quadrant b_1 at the section multiplied by the depth of the concrete at that section d_1.* Accordingly, for equilibrium at failure,

$$V_1 = b_1 d_1 \tau_c \tag{4.8}$$

Step 3: *Checking for two-way shear:* The depth of the column should also be enough to resist two-way (punching) shear. The section for two-way shear is $d/2$ from the face of the column (where d is the effective depth at the face of the column).

$$a_2 = a + d \qquad \text{and} \qquad b_2 = b + d$$

The lengths a_2, b_2 as well as the depth d_2 at a distance $d/2$ from the face of the column resisting the perimeter shear are calculated. The shear force to be resisted, i.e. V_2, is given by

$$V_2 = \frac{P}{LB}(LB - a_2 b_2) \tag{4.9}$$

should be less than mobilizable resistance. The condition to be satisfied is

$$V_2 \le 2(a_2 + b_2)\tau_p d_2 \tag{4.9a}$$

Step 4: *Calculation of the area of steel required:* The area of the steel required is to be calculated from moment considerations by any of the following methods.

(a) M calculated as in usual pad footing but resisted only by the column face only.

(b) M calculated from pressures from the trapezium and resisted by the column face only.

(c) M calculated as in (a) above and resisted by the whole section with mean depth of section. This value will be less than the above two.

The area of the steel is obtained from the calculated moment by the equation

$$M_u = 0.87 f_y A_{st} d \left(1 - \frac{A_{st} f_y}{bdf_{ck}}\right)$$

from fundamentals, or much more easily by calculating M/bd^2 and finding percentage of steel p by using SP 16 or any other method.

The step-by-step procedure for design is illustrated in Example 4.6.

4.5 DETAILING OF STEEL

All types of column footings are detailed in the same way. Typical detailing of steel in footings is shown in Figure 3.5. The IS specifications have already been discussed in Sec. 3.7.

4.6 DESIGN OF RECTANGULAR PAD FOOTINGS

Let the length L be in the Y-direction and B in the X-direction (Figure 4.1). The considerations are the same as already derived before for sloped footings. The corresponding moments are as in Eq. (4.4)

$$M_{XX} = \frac{P}{8B}(B - b)^2, \qquad M_{YY} = \frac{P}{8L}(L - a)^2 \tag{4.10}$$

where M_{XX} is the bending moment for steel in the X-direction and M_{YY} the bending moment for steel in the Y-direction. We assume this moment is taken by the full length of the section on which the moment acts (see Figure 10.3 also).

4.7 DESIGN OF PLAIN CONCRETE FOOTINGS

Plain concrete footings are used under brick walls, brick pillars, etc. where the pressures transmitted at the foundation level are small. The principles of design are explained in Chapter 5.

4.8 DESIGN OF PEDESTALS

As already explained in Chapter 3, the pedestal is a short compression member (the height being usually less than three times its least lateral dimension) placed at the base of columns to transfer the load of the column to a footing pile cap, mat, etc. as shown in Figure 4.2. Pedestals become essential in the layout of steel columns for many reasons. It is also provided for steel columns in factory buildings as a precaution against possible corrosion of steel, from foundations and wet floors. Thus, in industrial buildings where the floors are washed regularly, the column bases should be 50 to 100 mm above the floor level. Otherwise, special precautions should be taken to encase the bottom part of the column in concrete.

Another factor that makes the compulsory introduction of pedestals under steel columns is the large tolerances that have to be provided in civil engineering constructions. Steel columns are fabricated in workshops to exact sizes. The use of a pedestal makes it convenient to make adjustments for the variation of the foundation levels in construction.

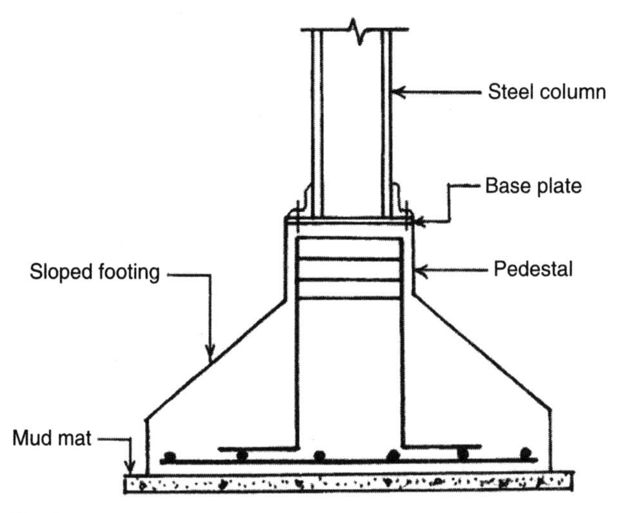

Figure 4.2 Pedestal as a compression member of reinforced concrete. For pedestals the effective length should be less than three times its lateral dimension (Cl. 26.5.3).

In reinforced column construction also the enlargement of its base as a pedestal is practised by many engineers to make-up levels, supply larger bearing areas to foundations, and provide enough development length for reinforcements. In such cases, pedestals and the base footing together act as a stepped footing as shown in Figure 4.2.

If these pedestals are reinforced as in columns with longitudinal and lateral steels, they should be treated as enlargement of the column. If they are provided with longitudinal steel, then they should be treated as part of the footings.

4.8.1 Design Calculation for Pedestals

Based on the theory of transfer of loads at column base as explained in Sec. 3.7.1, one can arrive at a pedestal which may be designed as a plain pedestal or a reinforced pedestal.

As explained in Sec. 3.7.1, when the stress on the top of the pedestals is lesser than $(0.45f_{ck}\ (A_2/A_1))$, theoretically no steel is required in the pedestal. However, in practice, for pedestals used for R.C. and steel columns, it is advisable to have at least 0.4 per cent of the area of the pedestal as nominal longitudinal steel with 12 mm laterals binding them together as in columns.

When the stress on the base is greater than $(0.45f_{ck}\ (A_2/A_1))$, the reinforcement shall be provided for developing the excess force either by extending the longitudinal bars from the column into the base or by means of dowels from the base to columns (IS 456, Cl. 34.4.1).

An interpretation of this rule obviously means that it is sufficient that steel in the base be less than in the column, i.e. the total area of the dowels need not be the same as the area of longitudinal steel in the columns. Accordingly, dowels (starter bars) can be of the same number but of less area. However, the Indian construction practice in R.C. columns is to continue all the longitudinal bars from the column to the base of the foundation structure and provide laterals for these bars as in the top column. This is a very safe and conservative practice.

4.9 SUMMARY

This chapter explains the procedure to be adopted in the design of reinforced concrete isolated footings and also the design of pedestals which are commonly used for transfer of heavy column loads to isolated footings.

EXAMPLE 4.1 (Design of square footing)

A solid footing has to transfer a dead load of 1000 kN and an imposed load 400 kN from a square column 400 × 400 mm (with 16 mm bars). Assuming $f_y = 415$, $f_{ck} = 20$ N/mm^2, and safe bearing capacity to be 200 kN/m^2 (20 t/m^2) design the footing.

Reference	*Step*	*Calculation*
	1	*Required plan area*
IS 456		Load = 1.0 DL + 1.0 LL + wt. of footing
36.4		= 1000 + 400 + 100 (say) = 1500 kN
		$\text{Area} = \frac{1500}{BC} = \frac{1500}{200} = 7.5\ \text{m}^2$
		Adopt 2.8 m square base of constant depth.
	2	*Ultimate soil reaction (only* DL + LL *to be taken)*
		Design load = 1.5 DL + 1.5 LL
		= 1.5 (1400) = 2100 kN
		Reaction = 2100/(2.8)^2 = 268 = 270 kN/m^2
	3	*Depth for one-way shear at d from column*
Table B.4		(Assuming min. shear = 0.35 N/mm^2 corresponding to 0.2% steel)
		$d = \frac{P(L-a)}{2P + 700L^2}$ (in metres)
		$= \frac{2100(2.8-0.4)}{2 \times 2100 + 700(2.8)^2} = 0.520\ \text{m} = 520\ \text{mm}$
		D = 520 + 10 + 75 = 605 mm (Assume ϕ = 20 mm)
	4	*Depth for two-way shear*
		IS critical section at $d/2$ from face as in flat slabs.

Reference	*Step*	*Calculation*
		Perimeter = $4(b + d) = 4(0.4 + 0.52) = 3.68$ m
		Shear = $270(2.8^2 - 0.92^2) = 1890$ kN
		Shear value = $0.25\sqrt{f_{ck}} = 0.25\sqrt{20} = 1.12$ N/mm^2
		$d = \dfrac{1890 \times 10^3}{1.12 \times 3680} = 459$ mm
	5	*Depth from bending (for square of L)*
Eq. (4.2)		$M = \dfrac{P(L-a)^2}{8L} = \dfrac{2100(2.8-0.4)^2}{8 \times 2.8} = 540$ kNm
		$Kf_{ck} = 0.138 \times 20 = 2.76$
		$d = \sqrt{\dfrac{M}{2.76L}} = \sqrt{\dfrac{540 \times 10^6}{2.76 \times 2800}} = 264$ mm
	6	*Depth adopted*
		Adopt greater of the values $d = 520$ and $D = 605$ mm.
		(*Note*: According to BS, the critical perimeter is at $1.5d$ from the column face, and the allowable shear is, as in slabs, a function of percentage of steel present.)
	7	*Reinforcement required*
		$\dfrac{M}{Ld^2} = \dfrac{540 \times 10^6}{2800 \times (520)^2} = 0.71$ (top layer)
SP 16 Table 2		Steel percentage = 0.206 (> min. 0.15)
		$A_s = \dfrac{0.206 \times 2800 \times 520}{100} = 3000$ mm^2 [10Y20 = 3140 mm]
Text Table B.1		(10Y20 means 10 nos high yield bars of 20 mm dia)
	8	*Check development length (for bond) of main steel*
Text Table B.3		Length from the face of the column = $\dfrac{1}{2}(2800 - 400) = 1200$ mm
		Development length for 20 mm $\phi = 47 \times 20 = 940$ mm (less than 1125 mm), i.e. 1/2 (length of the footing – col.) – 75 mm
	9	*Distribution of steel* (Cover = 75 mm)
IS 456 Cl. 26.3.3		Square footing: steel uniformly distributed
		Spacing = $\dfrac{2800 - 150(\text{cover}) - 20}{9} = 290$ mm
		Spacing < $3d$ and 450 mm
	10	*Anchorage of compression bars from columns (According to IS)*

Reference	Step	Calculation
Table B.3		Anchorage for 16 mm bars = 35ϕ = 560 mm
		Available d = 520 (satisfactory)
		(This condition need not be satisfied in IS or in BS 8110, see Sec. 3.7.1.)
	11	*Total depth* (concrete cast on ground)
		Assume cover to steel = (27 + 50) = 75 mm
IS 456 Cl. 34.4.2		With two layers of steel, D = 520 + 20 + 75 = 615 mm
		[*Note*: According to IS 456, with large diameter column bars, the depth for compression anchorage will be large. However, in BS 8110 the compression anchorage of steel in R.C. columns does not require checking, if the started bars extend to the bottom of the footing.]

EXAMPLE 4.2 (Design the column and footing by use of charts)

Design a column and footing for the loads given in Example 4.1.

Reference	Step	Calculation
Refer Chart D.6	1	*Find dimensions of column and footing*
		Column load = 1400 kN, footing load = 1500 kN
(Appendix D)		Adopt col. size 300 × 600 or 230 × 750 (mm)
		(Column adopted in Example 4.1 = 400 × 400 mm)
		From the Chart D.6, size of footing for 1500 kN = 300 × 255 mm
	2	*Depth of footing required*
		From the Chart D.6 = 900 mm (Example 4.1 = 620 mm)
	3	*Steel to be provided*
		From the Chart D.6 $= \dfrac{\text{26Y10 or 19Y12 at top}}{\text{24Y10 or 18Y12 at bottom}} = \dfrac{2041\ \text{mm}^2}{1884\ \text{mm}^2}$
		In Example 4.1 we got the steel as, 10Y20 = 3140 at the top
		(*Note:* The smaller area from the chart is because we have adopted a larger depth in the chart design. Similarly, one or two rods less are necessary at the bottom as the depth of lower steel is larger than the top steel.)

EXAMPLE 4.3 (Design of dowels in column)

A square footing 3 × 3 m and 60 cm thickness supports a 450 × 450 mm column; f_y = 415 and f_{ck} = 25 N/mm^2. If D.L. = 780 kN and L.L. = 1000 kN design the column tooling junction.

Reference	Step	Calculation
	1	*Max. load on base of column*
		Design load = 1.5(DL + LL)
		= 1.5(780 + 1000) = 2670 kN
IS 456	2	*Max. allowable stress on top of the footing*
Cl. 34.4		Permissible bearing = $0.45f_{ck}$ = 0.45 × 25 = 11.25 N/mm^2
		Increase the value due to increased area
		$\sqrt{\frac{A_1}{A_2}} = \left(\frac{3000}{450}\right) > 2.0$
		Limiting value = 2 × 11.25 = 22.5 N/mm^2
	3	*Load taken by concrete*
		P_c = 22.5 × 450 × 450 = 4556 kN
	4	*Area of dowel steel*
IS 456 Cl. 34.4.3		The entire column load can be transferred by the concrete above. However, codes and field practice require minimum dowel area of 0.5% of the area of the column.
		$A_s = 0.005(450)^2 = 1012$ mm^2
		Provide 4 Nos. of 20 mm diameter.
		.*Notes*: (a) The general practice is to provide dowel bars of the same diameter as the column bars.
		(b) IS and ACI require that the development length of dowels into the footing be checked. BS 8110 does not require this checking if the other requirements for the thickness of footing (BM and shear) are satisfied. However, the development length of bars inside the column should be satisfied in all cases.

EXAMPLE 4.4 (Design of rectangular footing)

Design a footing for a 500 × 350 mm column using 20 mm bars as dowels to transmit characteristic loads of 600 kN as dead load and 400 kN as live load to a foundation with safe bearing capacity of 120 kN/m^2. Assume grade 20 concrete and Fe 415 steel.

Reference	Step	Calculation
	1	*Plan area of footing*
		A = characteristic load/safe BC
		$= \frac{600 + 400 + 100(\text{wt. of footing})}{120} = 9.17 \text{ m}^2$
		$\frac{a}{b}$ ratio of the column $\frac{500}{350} = 1.42$

Reference	Step	Calculation
		$1.42B^2 = 9.17$. Hence adopt $B = 2.55$ m, $L = 3.60$ m
		Area provided = 9.18 m^2
	2	*Ultimate soil reaction*
		$$q = \frac{1.5(\text{DL} + \text{LL})}{\text{Area}} = \frac{1.5(600 + 400)}{9.18} = 163 \text{ kN/m}^2$$
	3	*Depth from one-way shear*
		For max V, take section along breadth in the YY-direction at a distance d from the column.
		Assume $\tau = 0.36$ (corresponding to 0.25% steel)
Eq. (3.7)		$$d = \frac{q(L-a)}{2(q+\tau)} = \frac{163(3.6-0.5)}{2(163+360)} = 0.45 \text{ m} = 450 \text{ mm}$$
		(Check section in XX-direction also)
IS 456 Cl. 31.6.3.1	4	*Check depth for two-way shear*
		Shear strength $= 0.25\sqrt{f_{ck}} = 0.25\sqrt{20} = 1.12 \text{ N/mm}^2$
		As $b > 1/2a$, no correction is needed for β.
		Taking section at $d/2$ around the column, we get
		$V = 163[9.18 - (a + d)(b + d)] = 163[9.18 - 1.1 \times 0.95] = 1326$ kN
		$$\tau_p = \frac{1326}{2(a+d+b+d)d} = \frac{1326 \times 10^3}{2(1750) \times 450} = 0.84 < 1.11 \text{ N/mm}^2$$
Text	5	*Depth from bending*
Figure 22.6(b)		*Section YY*
		$$M_{\text{long}} = \frac{P}{LB} B\left(\frac{L-a}{2}\right)\left(\frac{L-a}{4}\right) = \frac{P(L-a)^2}{8L}$$
		$$= \frac{1500(3.6-0.5)^2}{8 \times 3.6} = 500 \text{ kNm}$$
Text Eq. (22.9)		*Section XX*
		$$M_{\text{short}} = \frac{P(B-b)^2}{8B} = \frac{1500(2.55-0.35)^2}{8 \times 2.55} = 355 \text{ kNm}$$
	6	*Reinforcement required*
		Longitudinal direction
		For A_{st}, $\dfrac{M}{bd^2} = \dfrac{500 \times 10^6}{2550 \times (450)^2} = 0.96, \quad p = 0.282,$

Reference	Step	Calculation
Appendix B		$$A_s = \frac{0.282 \times 2550 \times 450}{100} = 3236\ \text{mm}^2$$
Table B.1		*Short direction*
		For A_{sb}, $\frac{M}{bd^2} = \frac{3500 \times 10^6}{3600(450)^2} = 0.49, \quad p = 0.143,$
		$$A_s = \frac{0.143 \times 3600 \times 450}{100} = 2317\ \text{mm}^2$$
	7	*Development length in short direction*
		16 mm rods = 47ϕ = 752 mm
		$$\frac{1}{2}(B-b) = \frac{1}{2}(2550-350) = 1100\ \text{mm}$$
		1100 – 2(cover) = 1020 mm > L_d
	8	*Placing of steel*
		(a) Reinforcement in long direction along the width is placed at uniform spacing.
		$$\text{Spacing} = \frac{2550 - 2(40) - 16}{15} = 163\ \text{mm}$$
Chapter 3 Sec. 3.7		(b) For reinforcement in short direction,
		$$\beta = \frac{3.6}{2.55} = 1.41, \quad \frac{2}{\beta+1} = \frac{2}{2.41} = 0.83$$
		As this percentage is high, place the steel uniformly on the shorter side also.
	9	*Transfer of load to base of column*
Chapter 3 Sec. 3.7.1		Capacity = $0.45 f_{ck}\sqrt{A_1/A_2} = (0.45 f_{ck})2(\text{area})$
		= 0.45 × 20 × 2 × 500 × 350
		= 3150 kN > 1500 kN
IS 456 Cl. 34.4.3		Thus dowels are not theoretically needed, but at least four rods (equal to 0.5 per cent area of the column) are extended to the footing. *Note*: If 20 mm bars are provided as dowels to transfer the load, then, according to IS, the depth of the footing should be equal to its development length.
		L_d of 20 mm bars = 752 mm
		This depth is too large for a footing of constant depth.
		In such cases use a stepped footing or a sloped footing to reduce the amount of concrete. Otherwise use a pedestal 350 mm high around the column with an offset of 200 mm around the column. Pedestal will be (350 + 400) × (550 + 400) in plan. The rest of the footing will be stepped and will be of constant depth of 750 – 350 = 400 mm. However, the provision that the depth of the footing should satisfy the development length of dowels is not specified in BS 8110.

EXAMPLE 4.5 (Design of pedestals)

Design a concrete pedestal for supporting a steel column carrying a total factored load of 1700 kN. The size of the base plate is 300 mm square. Assume grade 25 concrete and Fe 415 steel (refer Figure 4.2).

Reference	*Step*	*Calculation*
		Case 1: Design for unreinforced pedestal
	1	*Size of pedestal*
IS 456		Bearing strength $f_{cb} = 0.45 f_{ck} = 0.45 \times 25 = 11.25$ N/mm^2
Cl. 34.4		Max. allowed strength = 2 × 11.25 = 22.5 N/mm^2
		Pressure on base plate $= \dfrac{1700 \times 10^3}{300 \times 300} = 19 \text{ N/mm}^2$
		Min. size ($L \times L$) to carry this pressure is given by the condition
		$11.25\left(\dfrac{L}{300}\right) = 19, \quad L = 507 \text{ mm}$
		Choose pedestal size 510 × 510 mm
	2	*Provision of steel*
		Theoretically, the above pedestal need not be reinforced.
		To avoid brittle failure, 0.4 per cent is usually provided.
		$A_s = \dfrac{0.4 \times 510 \times 510}{100} = 1040 \text{ mm}^2$
		Provide 4 Nos. of 20 mm rods and the usual laterals.
		Case 2: Design as a reinforced pedestal
		Adopt minimum size for pedestal (10 mm clearance) = 310 × 310 mm
IS 456		Safe pressure = 11.25(3.1/3.0) = 11.63
Cl. 34.4		Load carried by pedestal is equal to
		$11.63 \times 310 \times 310 \times 10^{-3} = 1117$ kN
		Balance load = 1700 – 1117 = 583 kN
		$A_s \text{ required} = \dfrac{583 \times 10^3}{0.87 \times 415} = 1614 \text{ mm}^2$
		$\text{Percentage of steel} = \dfrac{1614 \times 100}{310 \times 310} = 1.68\%$
		This is greater than the minimum, i.e. 0.4%.

EXAMPLE 4.6 (Design of sloped footing)

Design a sloped square footing for a circular column 500 mm in diameter and intended to carry a characteristic load of 1000 kN. The safe bearing capacity of the soil is 200 kN/m². Assume that grade 15 concrete and Fe 415 steel are used for the construction.

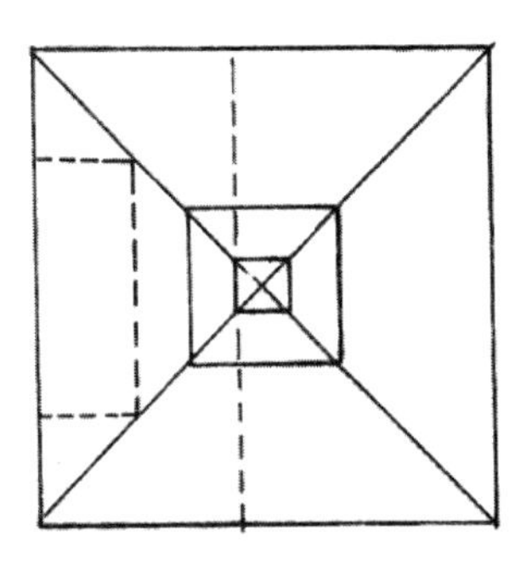
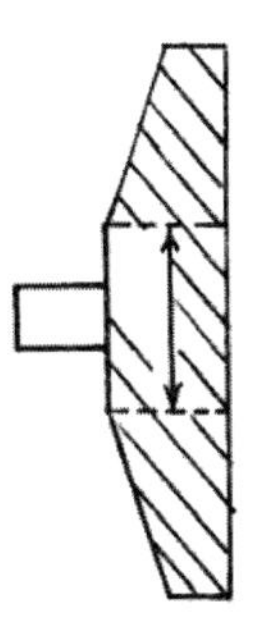
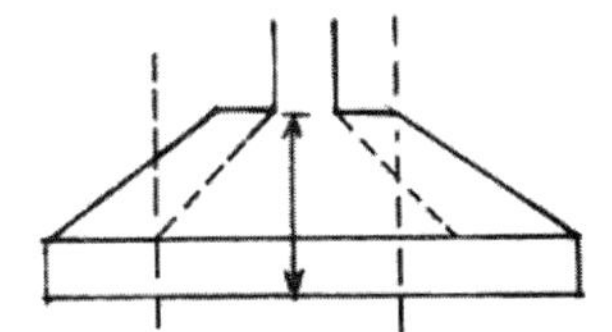

Figure E4.6

Reference	Step	Calculation
	1	*Calculate size of footing required* (self wt. 10%)
		$$\sqrt{A} = \left(\frac{1000+100}{200}\right)^{1/2} = 2.34 \text{ m}$$
		Adopt 2.4 × 2.4 m footing.
	2	*Size of equivalent square column for circular column*
		$$\frac{500}{\sqrt{2}} = 354 \text{ mm}$$
IS 456 Cl. 34.2.2		Provide a square ledge 600 × 600 mm around the column. Calculations are made on column size 354 × 354 mm with a ledge 600 × 600 mm
	3	*Ultimate loads from column*
		Factored load = 1.5 × 1000 = 1500 kN
		$$\text{Upward pressure} = \frac{1500 \times 10^3}{2400 \times 2400} = 0.26 \text{ N/mm}^2$$
	4	*Depth from moment consideration as in plain footing and resisted by ledge face*

Reference	Step	Calculation
Text Sec. 22.11		Taking moment about the face of the column as in plain footings. For a *conservative estimate of* (d),
Eq. (4.4)		$M_{xx} = \frac{P}{8L}(B-b)^2 = \frac{1500 \times 10^3}{8 \times 2400}(2400 - 354)^2 = 327 \times 10^6$ N/mm^2
SP 16		We assume breadth resisting moment = 600 mm as ledge face.
Table C IS 456 Cl. 34.1.2		$M_u = 0.138 f_{ck} b d^2$ for Fe 415
		$d = \sqrt{\frac{327 \times 10^6}{0.138 \times 15 \times 354}} = 668$ mm
		Provide effective depth of 600 mm at the column to ledge face and 250 mm at the end. Provide cover to steel of 75 mm. This is a conservative estimate of d.
	5	*Check depth from one-way shear*
IS 456		Take section at effective depth from the face of column (d = 600)
Cl. 34.2.4.1(a) Figure E4.6(b)		The distance of section from the edge of the footing is equal to
		$\left(\frac{2400 - 354}{2}\right) - 600 = 1023 - 600 = 423$ mm
		The breadth of the column face at this section with 45° diagonal is
		b_1 = 2400 – 2 × 423 = 1554 mm
		$V_1 = \left(\frac{2400 + 1554}{2}\right) = (423)(0.26)\,\text{N} = 217$ kN
		The effective depth of section, d_1 at the end of 45° line is
		$d_1 = 250 + \frac{(600 - 250) \times 423}{(2400 - 600) \times 0.5} = 414$ mm
		$\tau_c\ (\text{reqd.}) = \frac{217 \times 10^3}{1554 \times 414} = 0.33$ N/mm^2
IS 456 Table 13		Equal to τ_c for grade 20 concrete with 0.2% steel.
	6	*Check for two-way shear*
Figure E4.6(b)		Section at d/2 from the face of the column
		$d/2 = \left(\frac{600}{2}\right) = 300$ mm from the column face
IS 456 Cl. 34.2.4.1(b)		Distance of section from the edge of the footing is equal to
		$\left(\frac{2400 - 354}{2}\right) - 300 = 723$ mm

Reference	Step	Calculation
		b_2 = 2400 – 2 × 723 = 954 mm
		d_2 = at 723 mm from col. = $250 + \frac{(600-250)\times 723}{900} = 531$
		$V_2 = 0.26\lfloor (2400)^2 - (954)^2 \rfloor = 126\text{ kN}$
IS 456 Cl. 31.6.3.2		$\tau_p = \frac{1261\times 10^3}{4\times 954\times 531} = 0.62\text{ N/mm}^2$
		Allowable punching shear = $0.25\sqrt{f_{ck}} = 0.25\sqrt{15} = 0.97\text{ N/mm}^2$
	7	*Find area of steel required*
		The effective depth d in step 1 was calculated on b = 600. The steel area is calculated by any of the following methods:
Sec. 4.4.1		*Method (a):* According to the conventional method, A_s is calculated from M, obtained from step 4 above (b = ledge face).
		$\frac{M}{bd^2} = \frac{327\times 10^6}{600\times (600)^2} = 1.51, \quad p = 0.484\% > 0.25\%$
		$A_s = \frac{0.484}{100}(600\times 600) = 1742$
		Method (b): A more economical solution is to take moments of the pressures inside the diagonals only and take b = 600 mm.
		$M_x = \frac{P}{24LB}(2L+a)(B-b)^2$
		$= \frac{0.26}{24}(4800+354)(2400-354)^2 = 234\text{ kNm}$
SP 16 Table 1		$\frac{M}{bd^2} = \frac{234\times 10^6}{600(600)^2} = 1.08$ and $p = 0.329\% > 0.25\%$
		$A_s = \frac{0.329}{100}\times 600\times 600 = 1185\text{ mm}^2$
		Alternative method of calculation of A_s
		$\frac{M}{f_{ck}bd^2} = \frac{234\times 10^6}{15\times 600(600)^2} = 0.07$
Text Chapter 2 Figure 2.2		LA factor = 0.92 (approx.)
		$A_s = \frac{234\times 10^6}{0.87\times 415\times 0.92\times 600} = 1174\text{ mm}^2$ (same as above)

Reference	Step	Calculation
		(*Note:* We may use method (c) of using a larger width to resist the moment. It will require lesser steel.)
		Method (c): By calculating A_s on full breadth and mean depth across section along $X'X'$ of Figure E4.6. Thus, the mean depth considering the whole width along the ledge is equal to,
Figure E4.6(c)		$\left[600(600) + (2400 - 600)\dfrac{(600 + 250)}{2}\right] \Big/ 2400 = 469 \text{ mm}$
		$\dfrac{M}{bd^2} = \dfrac{327 \times 10^6}{2400(469)^2} = 0.61, \quad p = 0.175\%$
SP 16 Table I		Put minimum 0.25%
		$A_s = \dfrac{0.25 \times 2400 \times 469}{100} = 2814 \text{ mm}^2$
		(*Note*: This method will give less steel than the other two methods.)
	8	*Distribution of steel—Check for spacing*
		Distribute total steel uniformly as per IS. Hence,
IS 456 Cl. 34.5		Spacing $= \dfrac{[2400 - (75 \times 2)]}{10} = 225 \text{ mm}$
		The spacing is less than 300 and $3d$.
	9	*Check for development length*
SP 16 Table 65		L_d for 12 mm bars in M15 concrete = 677 mm
		Length available $= \dfrac{2400 - 600}{2} = 900 \text{ mm}$
		This is just sufficient and, if necessary, the bars can be given a standard bend upwards at the ends.
	10	*Overall dimensions of footing*
		Provide cover to effective depth of 75 mm; $L = B = 240$ mm The footing depth varies from (250 + 75) at the edge to (600 + 75) at the column face.

EXAMPLE 4.7 (Design a pedestal (column base) and plain concrete footing)

A 400 × 400 mm column carrying a service load of 1280 kN is to be supported on a plain concrete footing resting on soil where safe bearing capacity is 275 kN/m^2. Design a suitable foundation assuming a dispersion load of 60° with the horizontal on plain footing.

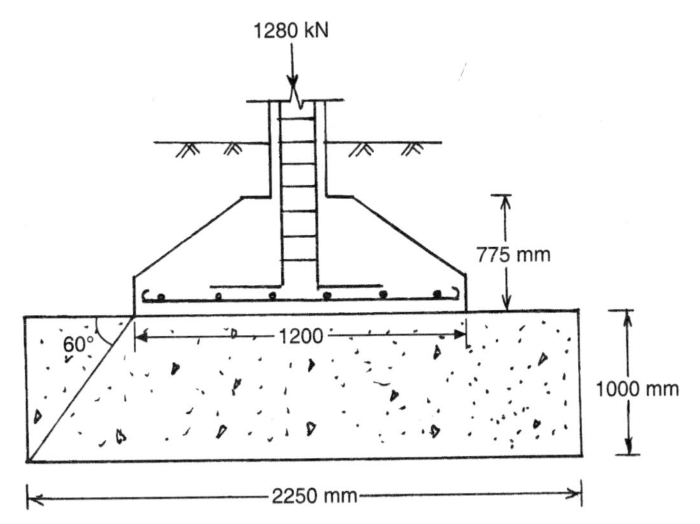

Figure E4.7

Reference	*Step*	*Calculation*
See Sec. 3.7.1	1	*Check stress at bottom of column*
		$0.45 f_{ck} = 0.45 \times 20 = 9$ N/mm^2
		Stress = $\dfrac{1.5 \times 1280 \times 10^3}{400 \times 400} = 12 \text{ N/mm}^2 > 9 \text{ N/mm}^2$
		Hence, we will put a column base above the footing.
	2	*Find required area of footing for bearing capacity* = 275 kN/m^2
		$A = \dfrac{1280 + 128}{275} = 5.12 \text{ m}^2$ (Assuming 10% D.L.)
		Adopt a 2.25 × 2.25 square footing above plain concrete.
	3	*Planning footing*
		The column size is 400 × 400 mm.
		Base is 2250 × 2250 mm
		We introduce a sub-base (pedestal) over the mass concrete as shown in Figure E4.7. Assume allowable load on mass concrete is 1 N/mm^2 (1000 kN/m^2) only.
		Required area of sub-base on mass concrete $= \dfrac{1280}{1000} = 1.28 \text{ m}^2$
		Adopt square base 1.2 × 1.2 m.
	4	*Find depth of sub-base at 60° dispersion with the vertical*
See Sec. 5.2.1		We carry load from 400 × 400 mm to 1200 × 1200 mm base
		Overhang = 400 mm on each side.

Reference	*Step*	*Calculation*
		Depth for 60° dispersion = 400 × tan 60 = 492 mm
		(We will provide a depth D = 692 + 75 = 767 mm or 775 mm depth.)
	5	*Find depth of base under sub-base*
		Area of foundation = 2250 × 2250
		Area of sub-base = 1200 × 1200
		Overhang of foundation = 525 mm
		Depth required = 525 tan 60
		= 909 mm (provide depth of 1000 mm).
	6	*Provision of steel*
		Provide nominal steel at the bottom of sub-base and continue column steel into the sub-base as shown in Figure E4.7. The plain base concrete will be detailed like plain wall footings described in Sec. 5.2.1.

5

Wall Footings

5.1 INTRODUCTION

Wall footings are continuous strip footings under masonry walls. Strip footings under two columns are called *combined footings*. Combined footings are described in Chapter 7. Strip footing under a series of columns is dealt with in Chapter 9. In ordinary low rise buildings, we use masonry walls to carry the superstructure loads to the ground. The types of foundations generally used in them are the following:

- Simple plain (or stepped) concrete continuous strip wall footing
- Reinforced concrete continuous strip wall footing
- Reinforced concrete continuous T or U beam wall foundations (when the beam is upstanding, it is called a T-beam and when it is downstanding, it is called a U-beam) as shown in Figure 5.1

We will deal with the design of these footings in this chapter.

5.2 SIMPLE PLAIN CONCRETE WALL FOOTINGS

In all wall footings if the foundation soil is not sandy, it is preferable to have a 300–750 mm thick layer of sand or hardcore laid over the soil. Greater thickness of this base course is to be provided in clay soils. Over this hardcore, a block of 1:2:6 lime brick jelly concrete (preferred for clay soils) or plain concrete 1:4:8 to 1:2:4 with large size (50 mm) aggregate is laid as plain concrete footing. The thickness of this plain concrete footing should be 150–450 mm, depending on the site. The construction of brick work over this plain concrete is to be undertaken only after this concrete layer has properly set and hardened. (It is very important that excavations near the foundations should not be made below the sand layer, after the construction of the foundation.)

Brick walls are built below the ground with offsets equal to quarter size of the brick. (Thus, for the 9-inch brick wall commonly used, we give 2¼ in offset on both sides with a total of half brick or 4½ inches offsets.) Each course consists of 2 to 3 or more brick in height. (Thus a 9-inch wall is carried through 13½ inches, 18 inches, 22½ inches in steps, each step being 2- or 3-course brick work, depending on the depth of the foundation.) Generally, more than three offsets are provided in the foundation, starting from the plinth.

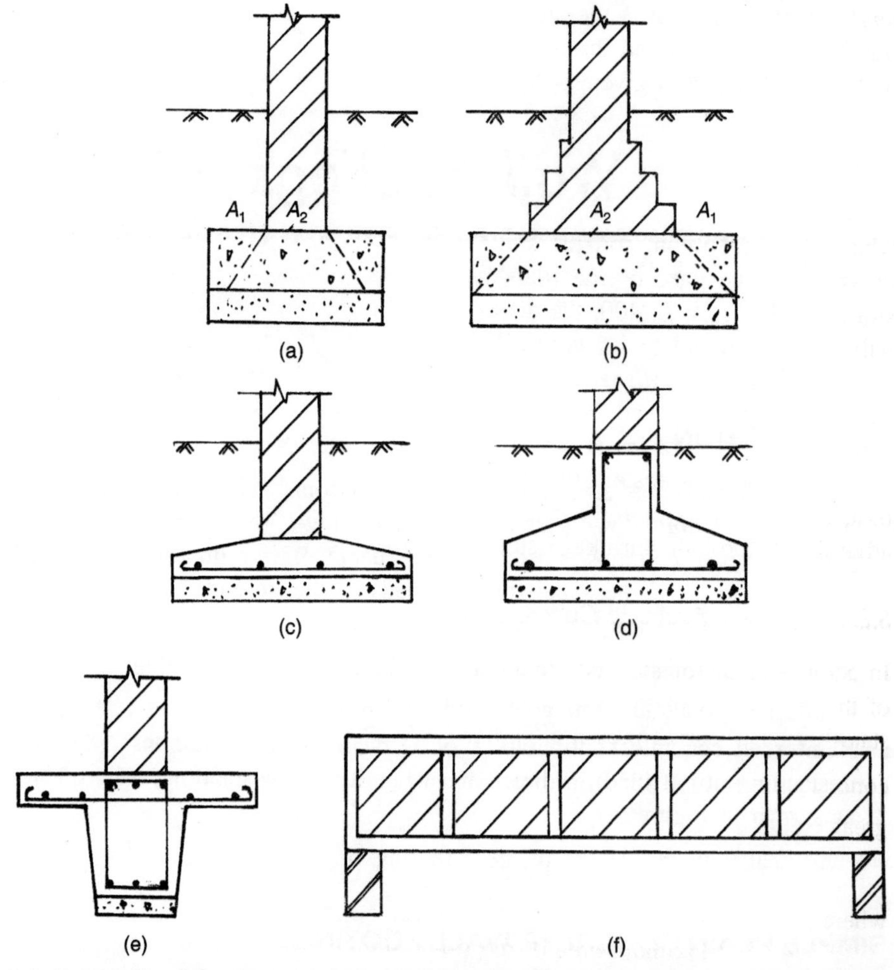

Figure 5.1 Types of wall footings: (a) Plain concrete footing under a straight wall; (b) Plain concrete footing under a stepped wall; (c) Reinforced concrete slab footing; (d) R.C. T-beam footing; (e) R.C. U-beam footing; and (f) Vierendeel frame footing which acts as a girder when supported at the end on rigid supports like a pile or pier.

In places where the ground water is high or the subsoil water is saline, ordinary brick work does not last long as it tends to deteriorate with time. It is better to use stone (rubble) walls or reinforced concrete Virendeel beams with brick infill as described in Chapter 20, in such places [Figure 5.1(b)]. The minimum thickness to which random rubble work can be built is 375 mm (15 inches). The offsets provided for rubble work are usually 150 mm (6 inches) as against 4½ inches in brick construction.

Note: Tables for design of simple wall footings are given in Sec. 5.3.1.

5.2.1 Dispersion of Load in Plain Concrete

For designing plain concrete footings, we should follow the recommendation of IS 456, Cl. 34.1.3. It specifies the dispersion of the load from base concrete to hardcore expressed as *angle of dispersion.* It is the angle at which the load from the footing can be assumed to be transferred to the ground through (a hard stratum like) plain concrete. The value of the *angle with the horizontal* is given as follows

$$\tan \alpha \geq 0.9\left[\frac{100q_a}{f_{ck}} + 1\right]^{1/2} \tag{5.1}$$

where q_a = maximum soil pressure, with q_a/f_{ck} as the ratio of the calculated maximum soil pressure on the base of the foundation to the characteristic strength of the concrete (hard stratum). [For $q_a = f_{ck}/400$, the dispersion will be 1 horizontal to 1 vertical; for $q_a = f_{ck}/50$, it will be 1 horizontal to 1.5 vertical].

In practice, the dispersion can be taken as 45 degrees 1 horizontal to 1 vertical in cement concrete (say M20) and 1 horizontal to 1.5 vertical (56.3 degrees to the vertical) in lime concrete, which is of lesser strength.

When full dispersion rules are satisfied, it is assumed that there will be no tension at the base and no steel reinforcements are needed in plain footings. However, in practice, it is advisable to provide shrinkage steel in mass concrete footings also.

5.2.2 Transfer Stress to Concrete

In addition to the dispersion from concrete to soil, we are required to check the transfer stress of the brick work of the wall to concrete also. IS 456 (2000)[1], Cl. 34.4 and Figure 4.2 of IS states that the bearing pressure on concrete should not exceed the permissible bearing stress in concrete (taken as $0.45f_{ck}$) in direct compression multiplied by $\sqrt{A_1 / A_2}$, which should not be greater than 2.

$$\text{Bearing stress} \not> 0.45 f_{ck} \sqrt{A_1 / A_2} \tag{5.2}$$

where

A_1 = Maximum area of the supporting area for bearing of the footing which is taken as shown in Figure 5.1(a).

A_2 = Loaded area of the column base

These design principles of plain concrete footings are illustrated in Example 5.1.

5.3 REINFORCED CONCRETE CONTINUOUS STRIP WALL FOOTINGS

In most present-day constructions which progress rapidly, we provide reinforced concrete strip footings. As already stated, there can be mainly two conventional types of reinforced concrete continuous wall footings, namely,

- Continuous R.C. slab *strip wall footings* shown in Figures 5.1(a) to (e)
- Continuous R.C. *T or U beam wall footings* shown in Figure 5.1(d) and (e) respectively

The first is a *slab, cantilevering* on both sides of the wall, and the second is a beam and slab construction with a slab on both sides of the wall and a beam being built under and along the wall. To these, we may add one more, not so conventional type, namely,

- Virendeel frame or girder wall footing shown in Figure 5.1(f).

5.3.1 Design of Continuous Strip Wall Footings

The method of design of these footings is given in IS 456, Cl. 34. The thickness of the edge of the footing should not be less than 150 mm for footings in soil (and 300 mm above the top of the piles for footings on regular piles).

For the wall constructed directly above the footing the following conditions should be satisfied.

1. The width should be such that the bearing capacity should not be exceeded. As a rule it is made at least three times the width of the wall.
2. The bending moment for design of transverse steel is taken about a section half way between the *edge of the wall* and the centre line of the footing. (For concrete columns, it is taken the edge of the column and for gusseted bases half way between the face of the column or pedestal and the edge of the gusseted bases.)
3. The critical section for shear is to be taken at a distance equal to the effective depth of the strip from the wall. (If the section is taken within this depth. i.e. at a lesser distance, enhanced shear can be assumed in the action of the footing.)
4. The development length for the diameter of steel selected from the section for B.M. should be satisfied. If necessary, the steel can also be bent up to get the development length.

5.3.2 Design for Longitudinal Steel

For good soil condition, as in sandy soils, only the minimum longitudinal steel along the length of the wall is to be provided to tie these transverse steel in place. However, for design in clayey or shrinkable soils where long-time settlements occur, special care should be taken in the design of longitudinal steel. This is dealt with in Sec. 5.5. Example 4.2 gives the steps to be followed in the design. Table 5.1 gives data which may be used for a quick preliminary design of these footings.

TABLE 5.1 Design of Strip Wall Footings After Finding Size of Footing [Figure 5.2]

Overhang on each side = A (mm)	*Safe bearing capacity of foundation* 10 T/m²		15 T/m²		20 T/m²		A_{st2} (on each side)
	A_{st1}	*D/d*	A_{st1}	*D/d*	A_{st1}	*D/d*	
375	8 mm @ 200	300/150	8 mm @ 200	300/150	8 mm @ 200	300/150	3 of 8 mm
500	8 mm @ 200	300/150	8 mm @ 200	300/150	8 mm @ 150	375/150	4 of 8 mm
600	8 mm @ 200	375/150	8 mm @ 150	375/150	8 mm @ 125	450/150	5 of 8 mm
750	10 mm @ 175	450/150	10 mm @ 125	450/150	10 mm @ 150	600/150	4 of 10 mm
900	10 mm @ 125	450/150	12 mm @ 125	450/200	12 mm @ 150	600/200	6 of 10 mm

A_{st1} gives the transverse steel to be provided along the width.
A_{st2} gives the longitudinal steel to be provided *on each side* in overhang.

5.4 R.C. T BEAM OR U WALL FOOTINGS IN SHRINKABLE SOILS

With clay soils, especially the shrinkable (or expansive clays), ordinary wall footings do not work well as there will be differential settlements along the length of the beam. The walls will tend to crack. The foundation has to be more rigid lengthwise. Some of the solutions usually adopted are:

1. In moderately shrinkable soils, a rigid T beam or U beam, as shown in Figure 5.3, gives good results.
2. In more severe cases, we may adopt R.C. slab footings with stub column at intervals with a connecting plinth beam or grade beam placed at the ground level with infill between the stub columns with brick work (or rubble work in submerged conditions) to form Virendeel frame or truss like construction, which is found to be more rigid than the conventional wall footings. (This type of foundation is dealt with in Chapter 20.) In this section, we deal with T and U beams as wall foundations.

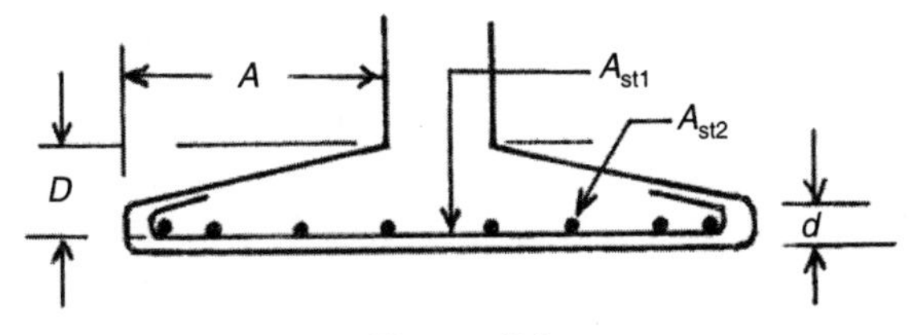

Figure 5.2

5.4.1 Design of R.C. T or U Continuous Beam Footings

The beams, as shown in Figures 5.1 and 5.3, built on shrinkable soils should be distinguished from strip foundations used for columns spaced at intervals described in Chapter 9 (where also we may use T and U beams). The concepts of design of these continuous wall foundations on shrinkable soils are based on differential settlement or loss of support.

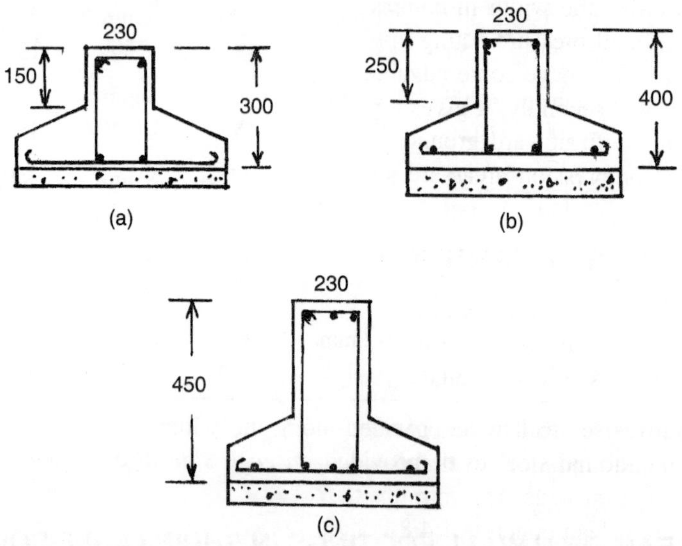

Figure 5.3 T beam and slab foundations for expansive soils.

IS 2911 on pile foundations recommends that grade beams supported on soil between piles can be designed for a B.M. = $wl^2/30$, where w is the distributed load per unit strength along the beam. *This is a very safe value*. Investigation conducted at Department of Civil Engineering, College of Engineering, Guindy, Anna University has recommended a loss of support for beams on shrinkable soils as given in Table 5.2. It depends on the activity index of soil. The suggested percent loss of support, moment coefficient, suggested depth of T beams and increase in ground pressure due to loss of support are given in Table 5.2. It can be seen that the value of B.M. = $wl^2/30$ can be safely used for the design of these beams also.

TABLE 5.2 Characteristics of Shrinkable Soils

Activity index of foundation soil	*Degree of expansion*	*Loss of support*	*Suggested moment coefficient*	*Minimum recommended depth of T beam in mm*	*Increase in pressure*
0.5–0.75	Low	10	$wl^2/166$	350	1.25
0.75–1.00	Moderate	15	$wl^2/88$	400	1.60
1.00–1.2	High	20	$wl^2/53$	450	1.65
>1.2	Very high	30	$wl^2/32$	As required	2.00

Note: Activity = $\dfrac{\text{Plasticity index}}{\text{Clay fraction}}$; A, B, C are shown in Figure 5.3.

5.5 DESIGN OF U BEAM WALL FOOTINGS

As shown in Figure 5.1, in a T beam the rib is upstanding above the slab, whereas in a U beam the rib is downstanding below the top slab. (Such beams are commonly used for strip foundations of a series of columns also, as described in Chapter 9.) The design principles for design of T and U beams are the same, except that in the U beam under positive moment (with soil pressure upwards), the width in compression is that of the beam only. The slab of the U beam due to the ground pressure acting upwards is under tension in the longitudinal direction.

However, U beams have some advantages over T beams. First, in clay soils it is easy to excavate in slopes and secondly, it gives a plane surface on top and is easy to build on. As the maximum moments with upward ground pressure occur at the supports, in beams continuous over supports the compression areas are also in the right positions.

5.6 FOUNDATIONS OF PARTITION WALLS IN GROUND FLOORS

Foundations of partition walls and walls for shelves, platforms, etc. of buildings need not be as strong as the foundations of load bearing walls. Tamil Nadu Public Works Department (PWD) recommends different types of foundations for the following situations [2].

1. **Foundations for half brick partition walls that are to be taken up to the ceiling level.** The foundation level should be about 0.9 m below the ground level. It should have a 300 mm wide, 150 mm thick sand fill; over it, a 300 mm wide, 150 mm thick p.c.c. slab of lean concrete 1:4:8 is laid. This forms the base. A *one brick wall is built up to the level of the bottom of the floor* over which the partition wall 100 mm can be built. (For the half brick wall construction, it is good to provide steel in the masonry in the form of steel bars or wire mesh enclosed in mortar at intervals of every sixth course.)
2. **Foundations for half brick walls which are only up to 117 cm (3 ft 10 in) meant to support shelves in store rooms.** We provide a foundation similar to (1) above with the half brick partition wall of 100 mm. In good soils we can build the half brick wall from the p.c.c. concrete without the one brick wall base.
3. **Foundations for half brick partition walls in framed buildings with grade beams between columns.** In framed buildings, the partition walls can be built on connecting beams laid just below the floor concrete level, with their ends resting on the walls built on the grade beams between columns. Every sixth course of all half brick walls should be reinforced as stated above.
4. **Foundations for light loads.** Partition walls of wardrobes and shelves in bed rooms, etc. can be built directly on the floor concrete, with a minimum of 150 mm sand filling below the floor.

5.7 SUMMARY

The layout and design of continuous wall footings should depend on the type of foundation soil. Plain concrete footings and ordinary reinforced concrete slab footings perform well in good soils. However, in clay soils and expansive soils, T or U beam and Virendeel frame type footings (see Chapter 20) are found to be more suitable than the others.

EXAMPLE 5.1 (Design of plain concrete wall footing on sandy soil)

Design a plain concrete footing for a wall stepped to 450 mm width to carry a load of 300 kN per metre length of the wall. Assume grade 20 concrete and the bearing capacity of the soil to be 200 kN/m^2. (This loading is rather high as indicated in Step 1, Notes of Example 5.2.)

Reference	*Step*	*Calculation*
	1	*Plan area required*
		P = 300 + self wt. of foundation at 10 per cent = 330 kN per m
		Base area required = $\dfrac{330}{200} = 1.65\ \text{m}^2$; required width = 1650 mm
	2	*Checking width for transfer stress from the wall to concrete*
		Transfer stress $= \dfrac{300 \times 10^3}{1000 \times 450} = 0.67\ \text{N/mm}^2$
Eq. (5.2)		Allowable bearing stress $= 0.45 f_{ck} \sqrt{A_1 / A_2}$
		$0.45 f_{ck} = 0.45 \times 20 = 9\ \text{N/mm}^2 > 0.7\ \text{N/mm}^2$
		Hence safe in bearing stress.
	3	*Find angle of dispersion*
Sec. 5.2		We have $\dfrac{f_{ck}}{q} = \dfrac{20}{0.67} = 29$ (intensity of load is high)
Eq. (5.1)		$\tan \alpha = 0.9 \sqrt{\dfrac{100q}{f_{ck}} + 1} = 1.9$ (1.9 vertical to 1 horizontal)
	4	*Find depth of footing*
		Length of projection $= \dfrac{1}{2}(1650 - 450) = 600$ mm (each side)
		$\dfrac{\text{Vertical}}{\text{Horizontal}} = \dfrac{D}{600} = 1.9$. Hence D = 1140 mm
		[*Notes:* (1) This depth is too large. We can decrease this depth of the footing by increasing the width of the foundation by providing more stepping and thus reduce the offset.
		(2) If we adopt large mass concrete, provide some steel for shrinkage.
Sec. 5.2.1		(3) According to IS 456, the angle of dispersion is taken as a function of the ratio f_{ck}/q. Some designers assume the angle as a constant value of 45°. Theoretically, this is true only for $f_{ck}/q \approx 400$.]

EXAMPLE 5.2 (Design of a R.C. wall footing on sandy soil)

A brick wall of 250 mm thick of a two-storeyed building is to rest directly on a R.C. strip footing. Design the footing assuming the soil is sandy and its safe bearing capacity to be 100 kN/mm^2. Check the design by Table 5.1.

Reference	*Step*	*Calculation*
	1	*Calculate characteristic load on the footing one storey height*
		(*Note:* Generally, in residential buildings we may assume a load of 30–40 kN per metre length (0.9–1.2 t per foot) of wall per storey height)
		Assume W = 80 kN/m of wall length for two storeys
	2	*Find required width of foundation*
		$b = \frac{80}{100} = 0.8$ m (Adopt one metre width)
		Pressure $= \frac{80}{1} = 80$ kN/m^2
		Hence safe for BC.
		Factored ground pressure = 1.5 × 80 = 120 kN/m^2
		As the soil is good, let us adopt an R.C. slab wall footing cantilevering on either side of the wall.
	3	*Find cantilever moment*
		B.M. section taken half way between the edge of the wall and the centre of the wall – (Overhang = 375 mm and width of wall 250 mm)
		$l = \frac{1000 - 250}{2} + \frac{125}{2} = 375 + 62.5 = 437.5$ mm
IS 456 34.2.3		Cantilever moment $= \frac{120 \times (0.4375)^2}{2} = 11.46$ kN-m
	4	*Find depth of footing required*
		$d = \left(\frac{11.46 \times 10^6}{0.14 \times 20 \times 1000}\right)^{1/2}$ as $M = 0.14 f_{ck} bd^2$ = 63.9 mm
IS 456		Adopt minimum allowed depth = 150 mm at the end with the slab sloping to thickness of 300 mm at the wall. (A slab of 150 mm can also be assumed.)
	5	*Area of steel required assuming D = 300 mm*
		Cover to be provided = 75 mm (cast directly on the ground)
		d at the section of the wall = 300 – 75 = 225 mm

Reference	Step	Calculation
Table B.1 (Chapter 2)		$\frac{M}{bd^2} = \frac{11.46 \times 10^6}{1000 \times 225 \times 225} = 0.226 < 0.42$ for p = 0.12% steel
		Provide nominal steel of 0.12% longitudinally and transversely.
		$A_s = \frac{0.12 \times 1000 \times 225}{100} = 270 \text{ mm}^2$
		Provide 8 mm rods @ 180 mm gives 279 mm^2.
		Distribution steel provides 3 Nos. of 8 mm rods on each side of the wall as distribution steel.
	6	*Check development length*
		$L_d = 47\phi = 47 \times 8 = 376$ mm
		Length from the edge of the wall to the end of the slab
		$\frac{1000 - 250}{2} - 75 = 300$ mm only
		We have to hook the bar at the ends to give the development length.
	7	*Check for bending (one-way) shear*
		Section d from wall (let us assume d as the average depth)
		Average depth $= \left(\frac{150 + 300}{2} - 75\right) = 150$ mm
Step 2 Table B.2		V @ 150 mm from the wall $= \left(\frac{1}{1000}\right)\left(\frac{1000 - 250}{2} - 150\right) \times 120$
		$= 0.226 \times 120 = 27.12$ kN
(Chapter 2)		$\tau = \frac{27.12 \times 10^3}{1000 \times 150} = 0.18 \text{ N/mm}^2 < 0.28$ (Hence safe)
	8	*Check design by Table 5.1*
		Overhang on each side $= \frac{1000 - 250}{2} = 375$ mm
		Safe BC = 10 t/m^2
Table 5.1		We get, $A_{st_1} = 8$ @ 200 mm centre
		Along length $A_{st_2} = 3$ of 8 mm on each side as distribution steel

(*Note:* The footing is cast on a 75 mm thick 1:4:8 mat concrete.)

EXAMPLE 5.3 (R.C. wall footing in clayey soils)

Design an inverted T strip foundation for a 230 mm wall for a single storey building. The foundation soil can be classified as medium shrinkable soil and its safe bearing capacity can be assumed as 10 t/m^2 (100 kN/m^2). Assume the length of the wall is 10 m and load at foundation is 40 kN/m.

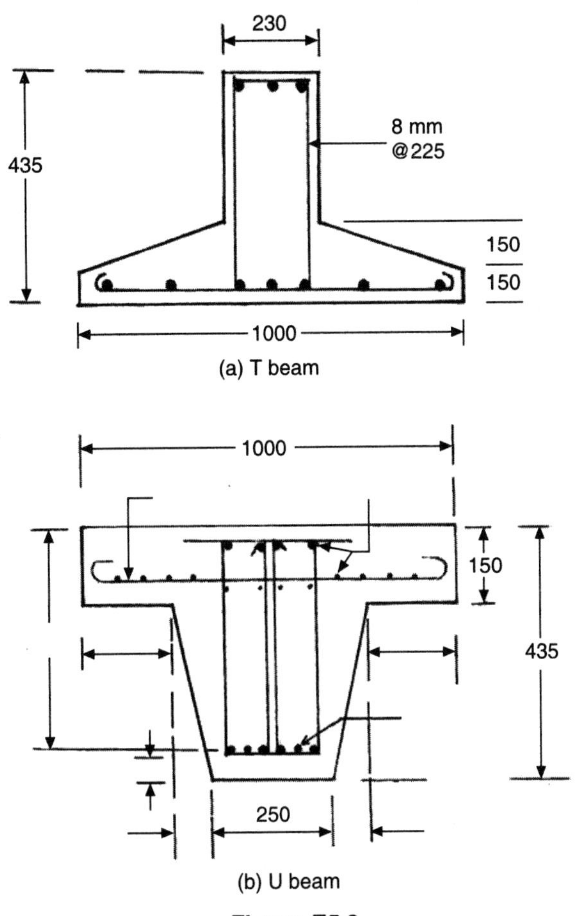

(a) T beam

(b) U beam

Figure E5.3

Reference	Step	Calculation
Table 5.2	1	*Calculate width of foundation* As the pressure on the base can increase by 1.6 times by loss of contact, we need more base width than otherwise. Load = 40 kN/m (4 t/m) Width required $= \dfrac{40 \times 1.6}{100} = 0.64$ m. Adopt—1000 m Adopt a T beam with a slab of depth 150 mm at the end to 300 mm at the edge of the beam
	2	*Design cantilever slab* Length of slab = (1000 – 230)/2 = 385 mm on each side

Reference	Step	Calculation
	3	*Design overhanging slab*
		Factored load = 1.5 × 40 = 60 kN per metre length
Limit state design		Upward reaction = 60/1 = 60 kN/m²
		Increased value for loss of support 1.6 × 60 = 96 kN/m²
		Design as in Example 2.3 or by Table 5.1
Table 5.1		A_{st1} = 8 mm @ 200 mm
		A_{st2} = Distribution steel (0.12%)4 of 8 mm
	4	*Design T beam—longitudinal steel*
Table 5.2		For moderately expansive coefficient $\frac{wl^2}{88}$
Approx. T beam design		$M_U = 60 \times (10)^2/88 = 68$ kNm
		$M_U = 0.14 f_{ck} bd^2 = 68$ kNm; $d = \left(\frac{68 \times 10^6}{0.14 \times 20 \times 230}\right)^{1/2}$
		d = 325 mm (Assume 375 mm). Let 16 mm main steel
		D = 375 + 50 + 8 = 435 mm (approx.)
		Design of T beam
		We can design the beam as a simple rectangular section or proceed with an approximate design assuming the lever arm depth—(½ slab depth)
	5	*Approximate design of T beam using approximate length of L.A.*
		$z = d - \frac{1}{2}(\text{slab depth}) = 375 - 75 = 300$ mm
	6	*Approx. capacity on concrete strength – assuming flange is in compression*
		$M_U = 0.36 f_{ck} b_f D_f$ (LA), where b_f, D_f refer to the flange
		= 0.36 × 20 × 1000 × 150 × 300 = 324 kNm
Step 4		= 324 kNm > 68 kNm required
	7	*Find approximate steel area required*
		$A_s = \frac{68 \times 10^6}{0.87 \times 415 \times 300} = 628 \text{ mm}^2$
		Provide 3 Nos. 16 mm at top giving 603 mm².
		(As we will provide the same steel at the bottom and also as the design is very conservative, we can provide slightly lower than the theoretical value.)

Reference	Step	Calculation
	8	*Bottom steel* In foundation design where we do not know where the settlement may occur, the usual practice is to provide the same steel at top and bottom without use of bent bars. Therefore, provide 3–16 mm bars at the bottom also.
	9	*Design of nominal stirrups to be provided* As the beam is continuously supported we design for nominal steel. Assume 2 legged 6 mm stirrups at max spacing Spacing = $0.75d = 0.75 \times 375 = 280$ mm (say, 275 mm) (For detailed design, proceed as explained in Chapter 2.)

REFERENCES

[1] IS 456, 2000, *Plain and Reinforced Concrete, Code of Practice*, Bureau of Indian Standards, New Delhi, 2000.

[2] Public Works Department, *Design of Foundations and Detailing*, Association of Engineers and Assistant Engineering, Tamil Nadu, 1991.

Design of Isolated Footings with Vertical Loads and Moments

6.1 INTRODUCTION

There are many instances where foundations are subjected to vertical loads W and moments M. We have to distinguish between the following three cases.

Case 1: Foundation is always acted on by W and M, where the moment M is due to the eccentricity of vertical loads and these loads always act on the structure and the foundation is always under W and M. (In this case, we design for uniform pressure by providing eccentricity in the foundation.)

Case 2: The moment M is large and acts only temporarily. This happens, for example, in the case of large wind loads acting always in one direction. This is also the case with a cantilever retaining wall foundation subjected to a large value of M due to earth pressure.

Case 3: Where the moment can reverse itself as happens when the wind can blow from any direction.

(*Note:* As has already been pointed out in Chapter 1, any external horizontal load acting on the structure will have to be balanced by horizontal ground reactions also. Moments due to vertical loads can be balanced by vertical reactions only.)

6.2 PLANNING LAYOUT OF ISOLATED COLUMN FOOTING WITH CONSTANT *W* AND *M* TO PRODUCE UNIFORM BASE PRESSURE (Case 1)

The variation of pressure on the base of a footing in direction L along which moment M acts is given by the following expression:

$$q = \frac{W}{A} \pm \frac{M}{I} y, \quad I = \frac{BL^3}{12} \quad \text{and} \quad Z = \frac{BL^2}{6} \tag{6.1}$$

where M acts in the direction of length L.

If we express $M = We_1$, then in the rectangular footing ($B \times L$), the pressure produced will be as follows:

$$q = \frac{W}{BL}\left[1 \pm \frac{6e_1}{L}\right] \tag{6.2}$$

If e is more than $L/6$, q will be negative and, as the foundation cannot take negative pressure, the pressure will readjust itself so that it is not greater than $L/6$. If eccentricity is $L/6$ towards one end, then the base pressure will be zero at the other end.

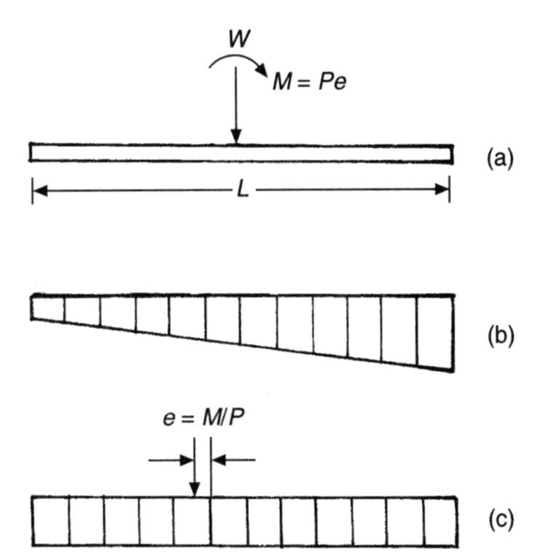

Figure 6.1 Case 1: Planning of footing with vertical load and constant moment *M*. (a) Footing with load and moment, (b) Effect of *W* and *M*, (c) Placing column with eccentricity *e* opposing effect of *M*, thus producing uniform ground pressure.

It can also be seen from the expression that if we place the footing eccentrically so as to produce a moment opposite in direction to the acting moment, the pressures under the footing will be uniform (as shown in Figure 6.1).

Thus, for this case, we place the column on the footing with an eccentricity = e_1 equal and opposite to $e_1 = M/W$.

6.3 PLANNING LAYOUT OF ISOLATED COLUMN FOOTING WITH CONSTANT *W* AND VARYING *M* IN ONE DIRECTION ONLY (Case 2)

In planning layout of isolated footing, if we place the footing with the full eccentricity M/W, then, without the moment acting, the distribution of the base pressure will not be favourable. The eccentricity to be given for the case of varying moment can be found as follows. Taking $B \times L$ as the footing, $I/y = z$ value; for moment along L, the value of $z = BL^2/6$.

As shown in Figure 6.2, the moment produces equivalent eccentricity $e_1 = M/W$ *and the column is placed at an eccentricity* e_2. The conditions to be satisfied by the two ends of the footing are as follows.

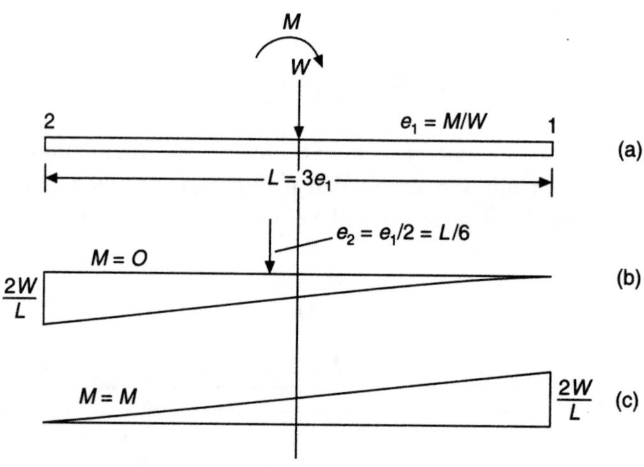

Figure 6.2 Case 2: Planning isolated footing for a column subjected to vertical load and varying moment (varying from O to M) acting in one direction only to produce allowable ground pressures. (a) *W* and *M* acting on foundation. Make $L = 3e_1$ and place column with $e_2 = e_1/2$ on opposite side of e_1, (b) Ground pressure with $M = O$, (c) Resultant ground pressure with moment equal to maximum moment *M*. (Provide area of foundation for 2*W*.)

Let $e_1 = M/W$, e_2 = eccentricity we adopt on the opposite side.

At end 1,

$$M = 0; \quad \frac{W}{BL} - \frac{We_2}{z} > 0 \tag{6.3}$$

$$M = M; \quad \frac{W}{BL} - \frac{We_2}{z} + \frac{We_1}{z} < q \text{ safe}$$

At end 2,

$$M = 0; \quad \frac{W}{BL} + \frac{We_2}{z} < q$$

$$M = M; \quad \frac{W}{BL} + \frac{We_2}{z} - \frac{We_1}{z} > 0 \tag{6.4}$$

From Eqs. (6.3) and (6.4),

$$\frac{We_2}{z} - \frac{We_1}{z} = -\frac{We_2}{z} \tag{6.5}$$

Therefore, $e_1 = 2e_2$, or we adopt $e_2 = e_1/2$.

As e_2 should not be $\not> L/6$ for no tension, we adopt a length of footing

$$L = 6e_2 \text{ or } 3e_1 \tag{6.6}$$

Thus, we adopt a length of footing 3 times the value M/W of eccentricity and place the column with an eccentricity L/6. There will be no tension in the base with M and without M. (*Note:* In the next case when the wind can blow in the opposite direction as well, we will set $L = 6e_1$.)

The necessary width B is found as follows: From Eqs. (6.3) and (6.5) and putting $e_2 = L/6$, we obtain

$$\frac{W}{BL} + W\left(\frac{L}{6}\right)\left(\frac{6}{BL^2}\right) = \frac{2W}{BL} = q \tag{6.7}$$

We plan the footing in such a way that $B = 2W/Lq$.

6.3.1 Procedure for Planning Layout of Footings *W* with and Varying *M*

Step 1: Calculate the area of the footing for twice the load.

$$A = \frac{2W}{q} = LB$$

Calculate $e_1 = M/W$ and adopt L_x, the length of the footing as 3 times this eccentricity. Find also the breadth $B = A/L$. Place the column load with an eccentricity

$$e_2 = \frac{L}{6} \tag{6.8}$$

Step 2: Find factored loads for design.

Step 3: Check for the maximum ground pressure.

Step 4: Find the depth of the footing for one-way shear for conditions of only W acting.

Step 5: Repeat check for the depth for one-way shear for W and M acting.

Step 6: Check for punching shear.

Step 7: Find the maximum bending moment for longitudinal steel.

Step 8: Check the depth for adequacy in bending.

Step 9: Determine longitudinal steel required.

Step 10: Check the size of the rod selected for anchorage.

Steps 11 and 12: Show the procedure of determining transverse steel. There are two methods in this procedure.

Method 1: Proceed as in rectangular footing and place steel as specified in IS 456 for rectangular footings.

Method 2: Alternately, if the footing is very long, we may assume that the column load is transmitted only in a column band of width equal to column size + 0.75 times the depth on either side (a total of column width + 1.5 times the depth). Design this strip as a beam of this width in the transverse direction. In the remaining transverse region, put either nominal steel or the same steel (see Sec. 7.8.1).

This concept is further explained in Chapter 7 under combined footings.

6.4 ISOLATED COLUMN FOOTINGS WITH CONSTANT *W* AND MOMENTS IN ANY DIRECTION (Case 3)

In a structure such as a water tank built on top of a single column the wind can blow in any direction. Thus, we design the foundation for winds in the *X*- and *Y*-axis. If we adopt a square footing, the wind acting along its diagonal direction will be critical (see Example 6.3). Therefore, we proceed as follows:

As shown in Figure 6.4, assume the effect of a diagonal wind on the ground pressure to be due to the wind acting both in *X*- and *Y*-directions simultaneously. Thus,

$$q_{max} = \sqrt{2} \text{ (wind acting on the } X\text{- or } Y\text{-direction)}$$

We combine this load with dead and live loads to check that the pressures are not exceeded. We proceed as follows.

Step 1: Find the three cases of characteristic load combinations of service loads and four cases for limit state.

Step 2: Find the base area required. A square foundation of liberal size is preferred. The side of the square should exceed 6 times the eccentricity e_1 due to M, and the column is placed at the centre so that the ground pressure is always positive in whichever direction the wind blows. (In case 2, it was 3 times e_1 and, in the present case, as the wind can blow in opposite side also, it should be 6 times e_1.)

Step 3: Check for safety for maximum ground pressure. The effect of wind load can be considered as $\sqrt{2} \times$ wind in one direction.

Step 4: Take the *four cases of factored loads* [Sec. (3.2)] and find the maximum pressures for limit state design for each case.

Step 5: Select the maximum pressure for design.

Step 6: Find the depth for one-way shear.

Step 7: Check the above depth for punching shear.

Step 8: Find the steel for bending in the *X*- and *Y*-directions.

Step 9: Check the diameter of steel chosen to satisfy bond.

Step 10: Check against diagonal wind.

The procedure is illustrated in Example 6.3.

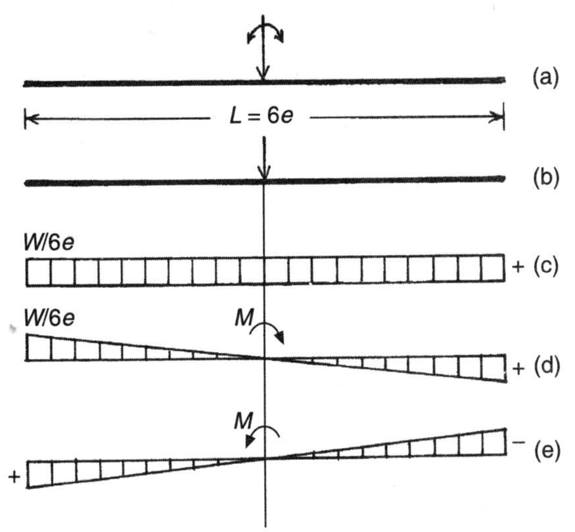

Figure 6.3 Case 3: Planning isolated footing with vertical load and varying moment which can act in the reverse direction also. (a) and (b) Layout with $L = 6e$. Place column at centre of footing, (c) Ground pressure due to vertical load only, (d) and (e) Ground pressure due to varying moments in reverse directions.

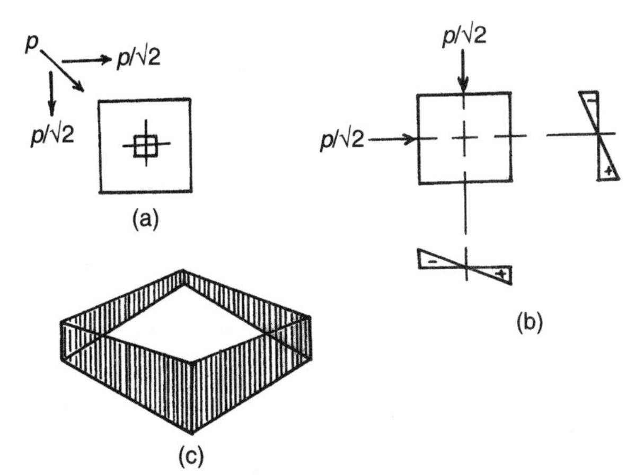

Figure 6.4 Isolated column footing with central vertical load *W* and moment *M* due to wind loading acting in any direction. (a) Resolution of diagonal load into two loads in *X-Y* directions, (b) Ground pressure due to moments, (c) Maximum pressure due to diagonal wind load.

6.5 SUMMARY

In this chapter we have delt with the design of isolated footings for columns subjected to vertical loads and the following cases of moments case (1) constant moment. Case (2) varying moment, acting only in one direction and case (3) varying moment acting in any direction.

EXAMPLE 6.1 (Alignment of footings under the column with *P* and *M* (acting always) to produce uniform ground pressure)

A square footing 2.6 × 2.6 m has to carry a vertical load of 600 kN with a moment due to eccentric load of 100 kNm along one of its axes. Determine the contact pressure if the column is placed at the centre of the footing. Plan the position of the column so that the bearing pressure on the footing will be uniform.

Reference	*Step*	*Calculation*
	1	*Determine the eccentricity of the load*
		$$e = \frac{M}{W} = \frac{100}{600} = 0.167 \text{ m}$$
		This should not exceed
		$$\frac{L}{6} = \frac{2.6}{6} = 0.43 \text{ m}$$
		0.167 < 0.43 m, which implies all pressures are positive.
	2	*Determine the pressures with the column at the centre*
		$$q = \frac{W}{B^2}\left(1 \pm \frac{6e}{B}\right) = \frac{600}{(2.6)^2}\left(1 \pm \frac{6 \times 0.167}{2.6}\right)$$
		q_{max} = 88.75 + 34.2 = 122.95 kN/m^2
		q_{max} = 88.75 − 34.2 = 54.55 kN/m^2
		Check average pressure $= \frac{122.95 + 54.55}{2} = 88.75 \text{ kN/m}^2$
		Total pressure = 88.75 × (2.6)^2 = 600 kN (Hence, it is okay)
Step 1	3	*Determine eccentricity of load to produce uniform pressure*
		Place the column with eccentricity of $e_2 = e_1 = 0.167$ towards the lower pressure side so as to produce a moment equal and opposite to that of the column moment. The resultant pressure will be uniform = 88.75 kN/m^2.

EXAMPLE 6.2 (Design of foundation for *W* and varying *M* in one direction)

Design a footing for a column loaded with W = 150 kN and a large varying moment in one direction of maximum value of 200 kNm. Assume the safe bearing capacity of foundation soil 120 kN/m^2 and a column size 400 × 300 mm.

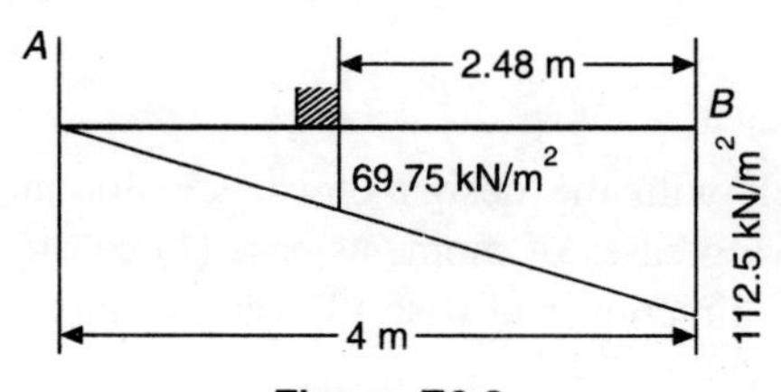

Figure E6.2

Reference	Step	Calculation
Figure 6.2	1	*Find area and L_x, length of footing taking service loads*
		Area required = $2W/q$
Eq. (6.7)		$A = \frac{2W}{q} = \frac{2 \times 150}{120} = 2.5 \text{ m}^2$
Sec. 6.3		$e_1 = \frac{M}{W} = \frac{200}{150} = \frac{4}{3} = 1.33$ m; Adopt $L = 3e_1$
		Adopt $L = 3 \times e_1 = 4$ m and $B = 2.5/4 = 0.63$ m
		Adopt 4 × 1 metre foundation ($L = 4$ m; $B = 1$ m)
		Place column with $e_2 = 1/2e_1 = L/6 = 4/6 = 0.67$ m
		Adopt $e_2 = 0.68$ m *on the opposite side of* e_1.
	2	*Calculate factored loads for design*
		$W = 1.5 \times 150 = 225$ kN
		$M = 1.5 \times 200 = 300$ kNm
	3	*Find maximum pressures*
		With W only at $e_2 = 4/6$ m and $Z = BL^2/6$
		$q_{max} = \frac{225}{1 \times 4} \pm 225 \times \frac{4}{6} \times \frac{6}{16} = 112.5$ and zero kN/m^2
		With M only on the footing
		$q_{max} = \pm (300 \times 6)/16 = \pm 112.5$ N/mm^2
		(These pressures oppose each other, see Figure E6.2)
		These pressures are less than 1.5 × 120 = 180 N/mm^2 allowed
	4	*Find depth for one-way shear with no moment*
		For a safe value, *assume* uniform pressure = W/BL over longer length.
		Larger length from the edge of the column (400 mm) to the edge of the footing = 2.0 + 0.68 – 0.20 = 2.48 m
		Total shear $= \frac{W}{BL}(2.48 - d) = \frac{225}{4}(2.48 - d)$
		Shear section = $Bd = 1 \times d$
Let		$\tau \times d = 56.25(2.48 - d)/4$
$\tau_c = 350$ kN/m^2		With $\tau = 350$ kN/m^2 required; $d = 0.343$ m
		$D = 343 + 10 + 75 = 428$ m
Use		For a conservative value, assume $D = 600$ mm
$\varphi = 20$ mm		$d = 600 - 75 - 10 = 515$ mm
	5	*Check for one-way shear with moment also* (As in step 3)
	6	*Check for punching shear at d/2 around column (as in other problems)*

Reference	Step	Calculation
	7	*Find maximum bending moment (Approx. method)* Maximum bending moment is produced from reaction from the longer side from the column, with W and M acting together. Pressure at the face of the column due to W $$=\frac{112.5}{4}\times 2.48 = 69.75\ \text{kN/m}^2$$
Example 2.7		BM on the face of the column on the RHS: x = 2.48 m $\text{B.M.} = (x^2/6)(2q_1 + q_2)$ $$=\frac{(2.48)^2(2\times 112.5+69.75)}{6} = 302\ \text{kNm}$$
	8	*Check capacity of section for BM* Capacity of section = $0.14 f_{ck} b d^2$ $= 0.14 \times 20 \times 1000 \times (515)^2 = 742$ kNm (For exact values, we may consider the weight of the footing separately as a UDL)
	9	*Determine longitudinal steel required* (d = 515 mm)
Table B.1		$$\frac{M}{bd^2}=\frac{132.5\times 10^6}{1000\times 515\times 515}=0.50$$ p with Fe 415 steel = 0.143% steel Adopt minimum main steel = 0.25% instead of 0.12% shrinkage steel prescribed. Find steel for full width = 1000 mm $$A_s=\frac{0.25\times 1000\times 515}{100}=1288\ \text{mm}^2/\text{m}$$ Use 6 Nos. 16 mm bars giving 1206 mm^2
	10	*Check for anchorage* $L_d = 47\phi = 47 \times 16 = 752$ mm is available
	11	*Find transverse moment (first method, Sec. 6.3.1)* Transverse offset from the column face = (1000 – 300) (1/2) = 350 mm Considering maximum pressure for the whole length of the slab, we have M_{max} = (112.5 – factored self-wt. pressure) × $(0.35)^2/2$ = 112.5 × $(0.35)^2/2$ = 6.86 kNm q for $\frac{M}{bd^2}=\frac{6.86\times 10^6}{1000\times 515\times 515}=0.025$. Thus, q is very low. Provide 0.2% steel as transverse steel.
	12	*Find area of steel (one metre width)*

Reference	Step	Calculation
		Let us provide 0.2% (instead of 0.12% for shrinkage)
		$A_s = \frac{0.20}{100} \times 4000 \times 515 = 4120 \text{ mm}^2$ (for full length)
		Use 21 Nos. of 16φ bars area = 4223 mm^2
		As $\frac{L}{B} = 4$, provide steel near or to the column in column band
		Column band = width of footing
IS 456 (34.3)		Reinforcement in col. band $\frac{2}{\beta+1} = \left(\frac{2}{4+1}\right) = \frac{2}{5}$ (β = Side ratio)
		The remaining $\frac{3}{5}$th area is distributed in the end bands. This should not be less than 0.12%. (We may also use the column band approach as explained in Sec. 6.3.1).
Table B.3		*Check for anchorage of transverse steel.* $L_d = 47\phi$
(See Sec. 2.4)		L_d for 16φ = 47 × 16 = 752 mm
		Length available for bond = 350 mm
		For satisfying the development length, the bar has to be bent up along the depth on either ends to provide L_d.
		(*Note:* As the amount of reinforcement supplied is much more than that required theoretically, the steel is stressed less and the average bond requirements need only to satisfy the stress in the steel. Thus, a reduction factor can be allowed and the upward bend can be avoided.)

EXAMPLE 6.3 (Foundation for a column with axial load and moment due to wind load from any direction)

A 400 × 400 mm column supports a circular water tank of 3 m diameter at its top. If calculated loads on the column are DL = 280 kN, water load = 170 kN and moment due to wind load in the *X*-direction is 30 kNm. Design a suitable footing for the column. Assume safe bearing capacity = 150 kN/m^2. (Refer Figure 6.4)

Reference	Step	Calculation
	1	*State the load combination for design*
		For serviceability – bearing capacity (three cases)
		1. DL + LL = 280 + 170 = 450 kN
		2. DL + WL = 280 kN + 30 kNm
IS 456		3. DL + 0.8 WL = 416 kN + 24 kNm
Table 18		For the above 3 cases, M_{max} = 30 kNm and W_{max} = 450 kN
(Sec. 3.2)		For limit state of collapse (four cases)

Reference	*Step*	*Calculation*
		1. 1.5(DL + LL) = 1.5(450) = 675 kN
		2. 1.5 DL + 1.5 WL = 420 kN + 45 kNm = 464 kN
		3. 0.9 DL + 1.5 WL = 252 kN + 45 kNmN
		4. 1.2(DL + LL) + 1.2 WL = 540 kN + 36 kNm
	2	*Find base area required*
		As wind can blow from any direction, the diagonal effect will be $\sqrt{2}$ or 1.41 times that in the *X*- and *Y*-directions. We will make a conservative estimate of the base required (SBC = 150 kN/m²).
		W = 450 kN and M = 30 kNm
		Area for axial load = 450/150 = 3 m²
		Max eccentricity e_1 = 30/280 = 0.10 m and $6e$ < 1 m
Sec. 6.4		Minimum length needed = $6e_1$ = 6 × 0.1 = 0.6 m
Step 2		Minimum area = 3 m²
		Adopt 2 × 2 m (4 m²) footing. $Z = BL^2/6 = 4/3$ m³
	3	*Check serviceability requirement ground pressures for three cases*
Step 1		*Case 1:* q = 450/4 = 112.5 kN/m²
		Case 2: q = 280/4 ± (1.41) × 30 × 3/4 (Diagonal wind)
		= 70 ± 31.7 = 101.7 and 38.3 kN/m²
		Case 3: q = 416/4 ± (1.41) (24 × 3/4) (Diagonal wind)
		= 104 ± 25.38 = 129.4 and 78.6 kN/m²
		All the pressures are positive and less than the safe bearing capacity.
	4	*Calculate pressures for limit states with wind in XX-direction:*
Step 1		1. q = 675/4 (No wind) = 168.8 + 0 = 168.8 kN/m²
		2. q = 420/4 ± 45(3/4)
		= 105 ± 33.75 = 138.8 to 71.3 kN/m²
		3. q = 252/4 ± 45(3/4) = 63 ± 33.8 = 96.8 to 29.2 kN/m²
		4. q = 540/4 ± 36(3/4) = 135 ± 27 = 162 to 108 kN/m²
	5	*Select pressures for design*
		Maximum is for Case 1.
	6	*Find depth for one-way shear* (τ_c = 350 kN/m²)
Eq. (2.7)		$d = \frac{q(L-b)}{700+2q} = \frac{168.8(2-0.4)}{700+337.6} = 0.26 \text{ m} = 260 \text{ mm}$
		Let D = 325 mm; d = 325 − 50 − 8 = 267 mm
	7	*Check for punching shear*
Eq. (2.11)		Section perimeter = 4(400 + 267) = 2668 mm
		τ_p = 1.12 N/mm²
		resistance = 2668 × 267 × 1.12

Reference	*Step*	*Calculation*
		= 797.84 kN > 675 (needed)
	8	*Find tension steel required for XX-direction*
		Case 1 is the worst case for flexure.
		Moment on the face of col. = 168.8 × $(0.8)^2/2$ /m width
		M = 54 kNm/m width
		$\frac{M}{bd^2} = \frac{54 \times 10^6}{1000 \times 267 \times 267} = 0.75$; Steel p = 0.22%
		Provide 0.25% steel
		A_s for 2 m width = (0.25/100) × 2000 × 267 = 1335 mm^2
		7 bars of 16 mm = 1407 mm^2
		Provide 7 bars of 16 mm with 2 m width with cover.
	9	*Check for average bond*
		$L_d = 47\phi = 47 \times 16 = 752$ mm
		This is just enough in the 800 mm cantilever part.
	10	*Steel in YY-direction*
		We provide the same steel in the *YY*-direction also.
	11	*Resistance against diagonal wind*
		As we provide for full wind in the *X*- and *Y*-directions separately, their combined action will be safe against the wind in the diagonal direction.

7

Combined Footings for Two Columns

7.1 INTRODUCTION

The term 'combined footing' is usually used for one footing which is designed to carry the loads of *two separate columns*. When three or more column loads in one line are supported on one long single footing, it is called a *column strip footing*. *Rafts* and *mats* support many columns located both in the *X*- and *Y*-axes.

There is also another type of footing called *balanced footing* involving two column loads, which is discussed in Chapter 8. In such footings, the external load is carried by cantilever action and, therefore, it is also called *cantilever footing*. Balanced footings are necessary where the external footing to be balanced is meant to exert *little* or *no pressure* on the ground. The beam used to connect two bases of columns is called a *strap beam*.

7.2 TYPES OF COMBINED FOOTINGS

In general, we can have the following three types of combined footings as shown in Figure 7.1. (Balanced footings and Cantilever footings are dealt with separately in Chapter 8.)

1. Combined *slab footing* with a continuous slab band
2. Combined *longitudinal beam and slab footing* with continuous slab base and a longitudinal *strap or spine beam* to the ends of the slab connecting the two columns

3. Combined *transverse beam and slab footing* with continuous slab base and separate transverse beam under columns.

The slab base can be rectangular or trapezoidal in shape, as shown in Figures 7.1 and 7.2.

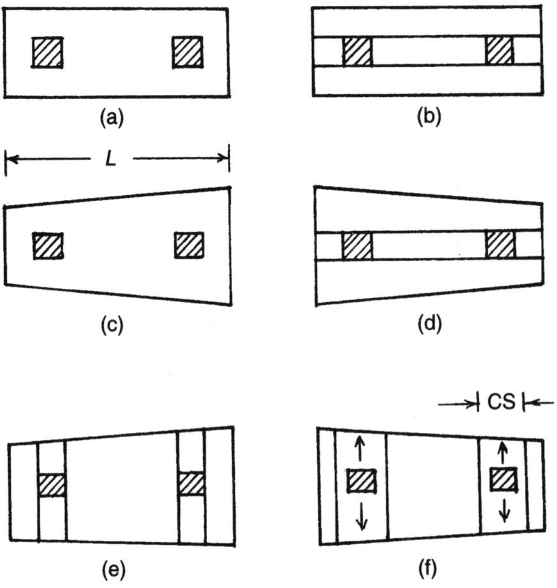

Figure 7.1 Five types of two-column combined footings, (a) Rectangular slab type, (b) Rectangular beam and slab type, (c) Trapezoidal slab type, (d) Trapezoidal beam and slab type, (e) Transverse beam and slab type, (f) Transverse distribution of column load on slab footing (CS—Column strip of width equal to column size + 0.75*d* on either side).

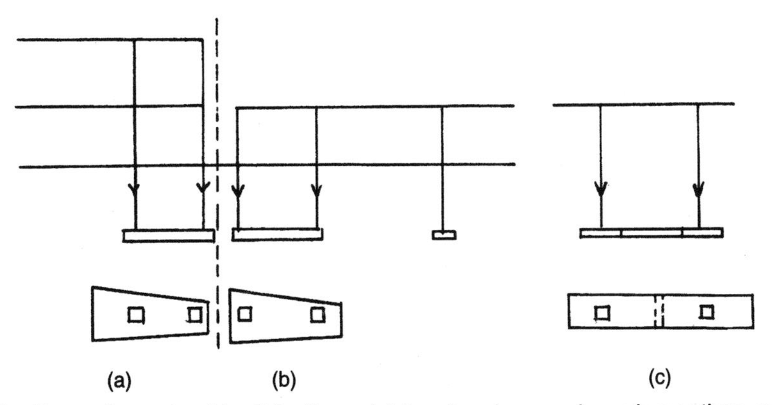

Figure 7.2 Two-column combined footings: (a) Load on inner column larger than on the outer, (b) Load on inner column smaller than on the outer, (c) Individual footing excavation overlap.

7.3 ACTION OF COMBINED FOOTINGS

The necessity to combine two footings to a single combined footing can arise under any one of the following conditions:

1. When the individual footings planned for the site overlap or come so near to each other that it will be more convenient to combine them.
2. If the exterior column of a structure is coming on or very close to the boundary line of the building site and there is space restriction to plan for a proper footing for the exterior column. We can *combine* the exterior and adjacent interior column foundations to a single foundation, with its centre of gravity coinciding with the centre of gravity of the column loads.
3. On the other hand, in the extreme case when the base of the exterior column *cannot be located at all*, we adopt a cantilever or *balanced footing*, which is described in Chapter 8. (This difference between combined footings and balanced or cantilever footings should be clearly understood.)

7.4 PLANNING LAYOUT OF COMBINED FOOTINGS

The principle to be followed in the layout of a combined footing for two columns is to have a rigid structure, where the Centre of Gravity (CG) of the foundation area coincides with the CG of the loads of the two columns. The foundation must have enough area to support the loads without exceeding the safe bearing capacity of the soil.

For designing such a footing, we first find which type of footing we should have rectangular or trapezoidal. The dimension can be derived from the following formulae:

$$\text{Total area required} = A = \frac{\text{Sum of column characteristic loads}}{\text{Safe bearing capacity}} \tag{7.1}$$

Secondly, the CG of loads and area should coincide. Let B_1 be the smaller, B_2 the larger breadth, and L be the length of the footing (Figure 7.3)

$$\text{Total area} = A = \left[\frac{B_1 + B_2}{2}\right] L \tag{7.2}$$

Let the distance of CG from B_2 be $\bar{x}$ given by

$$\bar{x} = \frac{L}{3}\left[\frac{2B_1 + B_2}{B_1 + B_2}\right] \tag{7.3}$$

From these conditions, the footing can be planned.

7.5 DISTRIBUTION OF COLUMN LOADS IN THE TRANSVERSE DIRECTION

The distribution aspect of design of combined footing should be very clearly noted. It is easy to visualize the design of the footing in the longitudinal direction. But it is also necessary in the combined slab footing (as different from the beams and slab footing) that the column loads be distributed transversely over the width of the footing (Figure 7.1(f)). For this purpose, we assume that (slightly different from what we had taken in the design of single rectangular footing) the column load is transmitted transversely, mainly over a *column strip* (transverse beam) of width equal to the column width plus 0.75 (effective depth) on either side of the edge of the column. If there is enough length of the slab over both sides of the column, this beam will be of $(b + 1.5d)$ in dimension as shown in Figure 7.1(f). This region has to be specially designed *as a beam* and provided with enough transverse steel. The rest of the transverse regions can be of minimum specified steel. (For foundation structure, it is better to specify a conservative minimum, and not the 0.12% shrinkage steel as in floor and roof slabs even though IS 456, Cl. 34.5.1 specifies that the minimum need be only as in roof slab.) A much more conservative provision of 0.2–0.25% should be provided for foundations. With ribs or strap beams, this problem does not arise. (This aspect is further explained in Sec. 7.8.1.)

7.6 ENHANCED SHEAR NEAR SUPPORTS

We have already seen in Chapter 2 that the allowable shear near supports is higher than that given in Table B.4. It is for this reason that we take sections for checking shear at a distance d away from the supports. In short cantilevers (e.g. corbels), the shear resistance is more by strut action and usually a much higher shear (τ_c') can be allowed. According to BS 110,

$$\tau_c' = \tau_c \left(1 + \frac{2d}{a_v}\right) \quad \text{but} \quad \not> 0.8\sqrt{f_{ck}}$$

where a_v is the distance of load from support.

Thus, in the above cases, we need to check only if the shear is less than $0.8\sqrt{f_{ck}} = 3.5$ N/mm^2 for M20 concrete.

7.7 COMBINED FOOTING WITH TRANSVERSE BEAMS UNDER COLUMN LOADS

The concept of providing a longitudinal beam to form a T beam with the slab as shown in Figure 7.1(e) is simple. We can also provide this transverse beam under the columns. This arrangement will require a very heavy arrangement of steel for the short transverse cantilevers under the column, but as the slabs are continuous slabs over two supports, the arrangement of steel in the slab is simple (ref. Example 7.6). Such beams are provided in those cases where space restrictions exist to provide a deep longitudinal beam between the columns.

7.8 STEPS IN DESIGN OF COMBINED SLAB FOOTINGS

The following are the step-by-step procedures in the design of *combined slab foundation*:

Step 1: Find the area of the footing to be used (from the characteristic loads plus self-load of footing and the safe bearing capacity of soil).

Step 2: Fix the dimension of the footing, depending on the layout restrictions and loads to be carried. We can have rectangular or trapezoidal slab footing with or without strap beams or transverse beams. The CG of the loads and the footings should coincide.

Step 3: Draw the shear force (SF) and bending moment (BM) diagrams in the *longitudinal directions* by using the "area method" or by integration (see Example 7.7).

Step 4: As we generally do not provide shear reinforcement *for slab footings*, find first the depth of the slab required in bending shear by assuming $\tau_c = 0.35$ (corresponding $p = 0.23\%$ of steel in Table B.2 of Chapter 2).

Step 5: Check the above depth for safety in punching shear also. $\tau_p = 0.25\sqrt{f_{ck}} = 1180$ kN/m^2 for $f_{ck} = 20$. Assume by any of the three methods shown in Chapter 2.

Step 6: Find the maximum bending moment where shear $V = 0$ and design longitudinal steel. (Even though IS specifies only 0.12% minimum shrinkage steel for foundations, it is advisable to provide at least 0.25% for important members such as foundations.)

Step 7: Check for average bond or development length for the diameter of steel chosen.

Step 8: Design for transfer of column loads in the transverse direction through a strip = column size + 0.75d on either side of the column. This strip is called the *column strip*. (This design is important in plane slab footing. For the rest of the slab in the transverse direction we need to provide minimum steel only.)

Step 9: Check for development length in the transverse direction also. In many cases, it may be necessary to bend up the steel as shown in Figure E7.5 (page 105), to get the necessary development length for tension steel.

Step 10: Check for bending shear also in the transverse direction in the column strip at d from the face of the column or the face of the beam.

Step 11: Detail the steel.

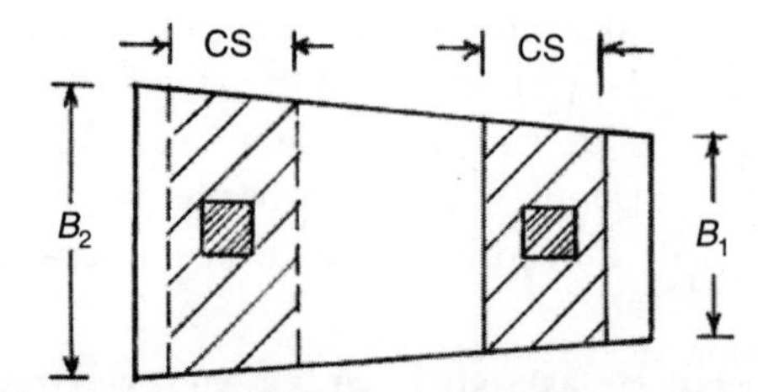

Figure 7.3 Transverse distribution of column loads in two-column combined slab footing through a column strip (CS = column size + 0.75 depth on either side).

7.8.1 Concept of Column Strip for Design of Transverse Steel in Combined Slab Footings

The mechanism of transfer of load from the column for determining the transverse steel has already been touched upon in Sec. 7.3. As this mechanism is important, it is further explained here.

In the design of isolated rectangular column footings (Chapters 3 and 4), we designed the transverse steel by assuming uniform pressure acting through the length of the footing in the transverse direction also. Hence, IS 456-2000 method is to find the total steel required on this assumption, but distribute more of the steel in a stipulated manner near the column part (see Sec. 3.7, detailing of steel).

In combined *slab footings*, as they are long along their length, we approach the problem in a different way. It is logical to assume that the column loads are transferred in the transverse direction through a *column strip* in the longitudinal direction of width equal to the length of the column + 0.75 times the effective depth of the slab on either side (if a column is at the end of the slab, the dispersion will be only on one side) as mentioned in Sec. 7.5. The transverse reinforcement of this column strip is designed as a cantilever slab or beam of width equal to the column strip cantilevering from the face of the column. The transverse steel is placed on top of the longitudinal steel. Nominal transverse reinforcement is provided for the regions beyond the column strips.

However, in combined beam and slab footing because of its rigidity, we assume that the ground pressure under the slab is uniform over the whole area of the footing.

7.9 STEPS IN DESIGN OF COMBINED BEAM AND SLAB FOOTING

We adopt a beam and slab footing when the depth required for a slab footing, especially from shear consideration, is large. The following steps can be followed for the design of a combined beam and slab footing:

Step 1: Find the area required based on service load and safe bearing capacity.

Step 2: Find CG of loads.

Step 3: Find the dimensions of the footing to suit the CG of loads to coincide with that of the area of the footing. If possible, provide cantilever to balance the moments.

Step 4: Find the factored ground pressure for design.

Step 5: Sketch the shear force and bending moment diagrams of the footing in the longitudinal direction and check whether a slab footing is possible.

Step 6: As we have to provide at least nominal shear reinforcements in the longitudinal beams, we need not restrict depth for shear. We find the required depth of beam from the maximum bending moment and to suit a reasonable percentage of steel. (Checking for punching shear is not required in T beams.)

Step 7: Calculate the tension steel required. (Unless there are restrictions about the depth of the beam, we choose a singly reinforced section.) We can use any method to design the T beam. An easy method is to assume a lever arm = $0.91d$ or (d – 1/2 slab) for T beam (instead of 0.84 for a rectangular beam).

Step 8: Check the anchorage requirement.

Step 9: Design a beam for shear and design shear reinforcement. In a beam we have to put at least nominal shear steel.

Step 10: Design the slab in the transverse direction. As we do not use shear reinforcements in slab portion, we must first find the depth of the slab for bending shear in transverse direction.

Step 11: Find the area of transverse (tension) steel in the slab to resist bending.

Step 12: Check for anchorage. If necessary, provide bends at the end of the rod for anchorage.

Step 13: Provide distribution steel in the slab along the length of the combined footing. This is placed on top of the transverse steel to get as large a value of d as possible for the longitudinal steel.

Step 14: Check bearing stresses of columns, if necessary.

Step 15: Detail the steel reinforcement.

7.10 SUMMARY

Combined footings for two columns can be planned in three ways, as shown in Figure 7.1, depending on the magnitude of the loads, the distance between the loads and the space available between the loads. This chapter dealt with their design. In the following examples, we assume the use of M20 concrete and Fe 415 steel.

EXAMPLE 7.1 (Layout of combined footings where footings overlap)

Two columns 300 × 300 mm carry 700 kN and 900 kN and are spaced at 3.0 m centres if the safe bearing capacity is 150 kN/m^2. Find a suitable layout for the foundation for the columns if the breadth is to be restricted to 2.4 m.

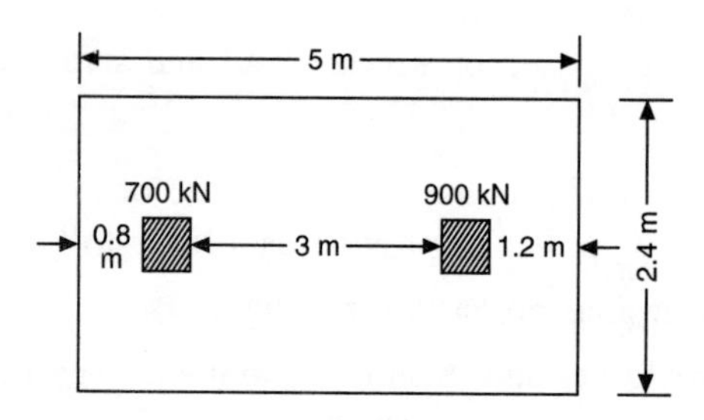

Figure E7.1

Reference	*Step*	*Calculation*
	1	*Choice of footing* Length required for the square footings For 700 kN col. = $\sqrt{700/150} = 2.16$ m For 900 kN col. = $\sqrt{900/150} = 2.45$ m

Reference	*Step*	*Calculation*
		As there is only 3 m between the columns, individual footings will be very near. Excavation is easier if they are combined.
		Calculate total load
		Load of columns = 700 + 900 = 1600 kN
		Wt. of footing = 160 kN (10%)
		Total = 1760 kN
	2	*Find area of footing required*
		$A = \frac{1760}{150} = 11.73\ \text{m}^2\ (\text{say}, 12\ \text{m}^2)$
		Length to be more than 3 m. Adopt 5 m with 1 m cantilever.
		Adopt a simple rectangular footing. L = 5 m;
		Breadth = 2.4 m; Area = 12 sq. m.
		Adjust the centre of gravity of footing to that of columns
Figure E7.1		CG of column loads $\bar{x} = \frac{700 \times 3}{1600} = 1.3$ m from 900 kN (point C)
		Distance of CG from 700 kN = 1.7 m
		Adjust the slab to have the same CG
		Offset at 700 kN AB = 5/2 – 17 = 0.8 m
		Offset at 900 kN CD = 2.5 – 1.3 = 1.2 m
		Total length = 0.8 + 3 + 1.2 = 5 m
		Find factored pressure for limit state design
		Factored load = 1.5 × 1760 = 2640 kN
		Upward pressure $\frac{2640}{12}$ = 220 kN/m^2
	3	*Draw the SF, punching shear and BM diagrams*
		We design the footing as shown in Example 7.2 for the following.
	4	*Design for one-way shear*
		One-way shear checked at effective depth from the face of the column. We can work out from fundamentals or use the formula used for footings [Eq. (2.7) in Example 2.3].
	5	*Design for punching shear*
		By any one of the three ways described in Example 2.5.
	6	*Design in longitudinal direction*
		Find the point where SF = 0 and find the maximum bending point. Find the depth by

Reference	Step	Calculation
		$\left(d = \sqrt{\dfrac{M}{0.14 f_{ck} b}}\right)$
		and A_s from M/bd^2 as described in Chapter 2.
	7	*Design in the transverse direction*
Sec 7.8.1		The column load on the slab is assumed to be carried by a column strip and is designed as a cantilever slab or beam of width equal to the column length +0.75d on both sides of the column as described in Example 7.2.

EXAMPLE 7.2 (Design of a rectangular combined slab footing with outer column directly on the property line)

Two columns 400 × 400 mm, one carrying 900 kN and another carrying 1600 kN, are having their centres 4.5 m apart. The outer face of the lighter column and the face of its footing have to be restricted along the property line. Design a suitable footing if the safe bearing capacity of the soil can be taken as 225 kN/m^2.

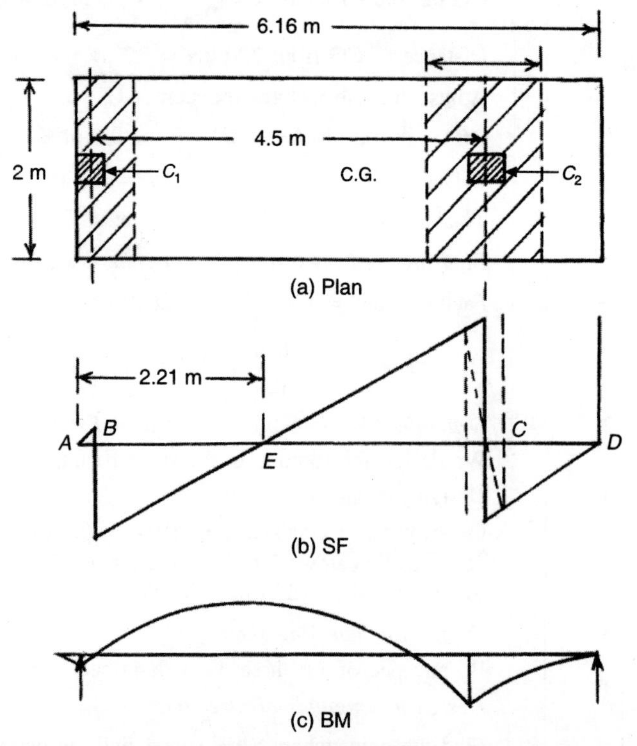

Figure E7.2 Two-column combined footing (Type 1. Rectangular slab type).

Reference	Step	Calculation
	1	*Find the area required for service loads*
Sec. 7.8		$A = \dfrac{900 + 1600 + \text{Self wt. } (250)}{225} = 12.2 \text{ m}^2$
		(10% of self-weight)
	2	*Find the dimensions of the footing*
		CG of loads; $x = \dfrac{1600 \times 4.5}{2500} = 2.88 \text{ m}$ from 900 kN
		2.88 m from the centre of 900 kN load from B; AB = 0.2 m
		Adopt a rectangular footing with CG at 2.88 m from column C_2.
		L = (2.88 + 0.2)2 = 6.16 m
		Footing extends 1.46 m beyond C_2.
		Breadth B = 12.2/6.16 = 1.98 m
		Adopt 6.16 × 2 m slab footing.
	3	*Draw the SF and BM diagrams*
		Ground pressure with factored load
		Load = (1.5 × 900) + 1.5(1600) = 1350 + 2400 = 3750 kN
		$q = \dfrac{1350 + 2400}{6.16 \times 2} = 304.8 \text{ kN/m}^2 \approx 305 \text{ kN/m}^2$
		$w = pr$; per unit length = 2 × 305 = 610 kN/m (width = 2 m)
		[Check load = 610 × 6.16 = 3757 kN]
		Shear at end = 0
		Shear at column point B = 1350 – 0.20 × 600 = 1230 kN
		Find x @ V = 0 = 1350 – 610x
		$x = AE$ = 2.21 m from the end.
	4	*Check for shear in the slab. Assuming* τ_c = 0.35 N/mm^2, *find the depth for shear at d from the face of the columns (one-way shear)*
		Col. 1 Method 1 (Left to right): Solving for d
		Shear at inner face of col. at B = 1350 – 0.4 × 610 = 1106 kN
		Shear at distance d from column
		= 1106 – 610d/1000 = 1106 – 0.610d
		Shear strength for depth d at τ = 0.35 N/mm^2
		= 2000 × d × 0.35/1000 = 0.70d
		Hence (1106 – 0.610d) = 0.70d; gives d = 844 mm

Reference	Step	Calculation
		Col. 1 Method 2: (by formula) (Eq. 2.7)
		Alternatively, use the analogy of rectangular footings, starting from the point of zero shear *E*.
Eq. (2.6)		Using the standard formula for d
Step 3		$q = 305$; $a/2 = 400/2 = 0.2$ m
		$L/2 = 2.21 - 0.2 = 2.01$
Figure E7.2		$d = \frac{q(L/2 - a/2)}{350 + q} = \frac{305 \times 1.81}{350 + 305} = 0.838 \text{ m} = 838 \text{ mm}$
		Col. 2 Method 1
		Shear at the inner face = 2400 – 1.66 × 610 = 1387 kN
		$(1387 - 0.610d) = 0.70d$ gives $d_2 = 1058$ mm (method 1)
		Col. 2 Method 2
		Starting from the point of zero shear using Eq. (2.6)
Eq. (2.6)		$q = 305$; $a/2 = 0.2$ m
		$L/2 = 2.49$
		$d = \frac{305\,(2.49 - 0.2)}{350 + 305}$ = 1.066 m. Adopt 1060 mm.
		Adopt depth 1060 mm (concrete or soil cover 75 mm)
		$D = 1060 + 75 + 10 = 1145$ mm
Section 7.6		[*Note: Check by BS Code at distance of 1.5d from the face*]
		Col. (1): Shear = $[1106 - (1.5d)\ 0.610] \times 10^3 = 0.70d$
		As $d_1 = 684$ mm, we need much less considering the slab from right to left
		Col. (2): We get $d = 858$ mm
		These are smaller than those we get by the IS method.
		Let us use the BS method for Shear Section 7.6.
		[Adopt $d = 885$ mm; $D = 885 + 10 + 16 + 75 \approx 986$ mm]
	5	*Check for the punching shear at d/2*
		Col: 400 × 400 and $d = 885$ mm, and col. (1) is on the edge of the footing. We will check if perimeter shear > col. load.
		$\tau_p = k \times 0.25\sqrt{f_{ck}} = 0.25\sqrt{20} = 1.12 \text{ N/mm}^2$
		Col. (1): Shear perimeter only on three sides for col. A
		$= (400 + 1060) + 2\left[400 + \frac{1060}{2}\right] = 3320 \text{ mm}$
		Capacity = 3320 × 1060 × 1.12 = 3941.5 kN > 1350. Hence safe.

Reference	Step	Calculation
		Col. (2): Shear perimeter on all sides = 4(400 + 885) = 5140 mm
Eq. (2.11)		Capacity = 5140 × 1060 × 1.12 = 6102 kN > 2400 kN
		[Though not necessary, we will calculate the actual loads at $d/2$ to be resisted, which will be less than the column loads.
		Punching shear at col. (1)
		= 1350 – 305 (0.842 × 1.285) = 1019 kN
		q_u = 305 kN/mm^2, which is less than capacity 3823 kN.
		Punching shear under col. (2)
		= 2400 – 305 (1.285 × 1.285) = 1896 kN
		which is less than the capacity 6102 kN.]
	6	*Design longitudinal steel for span support*
w = 610		Max BM, where SF = 0 @ x = 2.21 from A
		$$M_U = 1350 \times (2.21 - 0.20) - \frac{610 \times (2.21)^2}{2}$$
		= 1224 kNm
		Check depth for bending
		$M_U = 0.14 f_{ck} b d^2$
		$$d = \sqrt{\frac{1224 \times 10^6}{0.14 \times 20 \times 2000}} = 467 < 1060 \text{ mm}$$
		$$\frac{M_U}{bd^2} = \frac{1224 \times 10^6}{2000 \times (1060)^2} = 0.54$$
Table B.1		p_t = 0.16% [Slab steel > 0.23%]
		$$A_s = \frac{0.23 \times 2000 \times 885}{100} = 4071 \text{ mm}^2$$
		20 Nos. 16 mm bars gives A_s = 4022 mm^2
		Adopt 21 bars with 20 spacings.
	7	*Development length required* = 40φ
		= 47 × 16 = 752 mm
		is available from the point of maximum stress.
	8	*Design cantilever beyond col. (2)*
		w = 610 kN/m
		l = 1.26 m
		M_U = 610 × (1.26)^2/2 = 484.22 kNm
		$$\frac{M_U}{bd^2} = \frac{485 \times 10^6}{2000 \times 885 \times 885} = 0.30; \quad p_t = 0.085$$

Reference	Step	Calculation
		$$P_{t(\min)} = \frac{0.12 \times 2000 \times 885}{100} = 2124 \text{ mm}^2$$
		Provide 11 Nos. 16φ, giving 2212 mm^2
	8A	*Design transverse steel in a column strip as a transverse cantilever slab or beam*
B = 2m		The usual practice is to assume that the loads spread over a transverse *column strip beam in the footing of width* 0.75d *on either side of the outer edges of the column* (width b + 1.5d). If we place the transverse steel above longitudinal steel d = 900 mm (say)
		0.75d = 0.75 × 900 = 675 mm
		As the column is at the edge of footing,
		Distribution width under col. (1) 0.40 + 0.675 = 1.075 m width
		(*Note:* We may design this strip as a cantilever slab of 1.075 m with uniform pressure and find steel per metre width. Otherwise, we can design it as a cantilever beam of width 1.075 m and cantilever length to find the area of steel required for this width. We will adopt the first procedure.)
		Length of the column strip = 2 m
		Distribution width under col. (2) = 0.40 + (2 × 0.675) = 1.75 m
		q_2 = 2400/(2 × 1.75) = 686 kN/m^2, which is larger than q of step 3.
		Projection from the column = (2 – 0.4)/2 = 0.8 m
		$$M_u = \frac{686 \times (0.8)^2}{2} = 220 \text{ kNm/m}$$
		$$\frac{M_U}{bd^2} = \frac{220 \times 10^6}{1000 \times 900 \times 900} = 0.27$$
		p_t = 0.08% less than 0.12 (min)
		$$A_s \text{ per m width } = \frac{0.12 \times 1000 \times 900}{100} = 1080 \text{ mm}^2$$
		16 mm @ 175 mm spacing 1149 mm^2
		(Design for col. 1 similarly)
	9	*Check development length*
		Development length = 47ϕ = 762 mm
		$$\text{Length available } = \frac{2000 - 400}{2} = 800 \text{ mm}$$

Reference	Step	Calculation
		(Otherwise, we have to bind up the bars to get the development length.)
		(Similarly, design transverse steel under col. 1.)
	10	*Check the slab for shear in transverse direction at d*
		As the critical section at d is outside the footing, we omit this check.
		(Distance of footing = 1000 – 200 = 800 mm)
	11	*Check anchorage of the column to the footing*
		Limiting bearing stress (assume $\sqrt{A_1/A_2} = 1$)
		$= 0.45 f_{ck} \sqrt{A_1/A_2} = 0.45 \times 20 \times 1 = 9\ \text{N/mm}^2$
		Allowable bearing without dowel from the column
		= 400 × 400 × 9 = 1440 kN
		As col. (2) load is 2400 kN, it needs dowel anchorage, i.e. extension of column rods into the foundation which is the usual field practice.
	12	*Detail the steel*
		(a) Longitudinal steel for bending moment in *ABC* is placed at the top and *CD* at the bottom
		(b) Transverse steel is placed at the bottom on column strips under both columns and nominal transverse steel in the rest of the region at the bottom.
		(c) Check for nominal longitudinal steel at the bottom in *ABC* and at the top in *CD*.

EXAMPLE 7.3 (Design of a rectangular beam and slab combined footing)

Two columns 400 × 400 mm are spaced at 3.2 m centres carrying loads of 1000 kN each. If the width is restricted to 2 m and the safe bearing capacity is 200 kN/m^2, plan the layout of the footings.

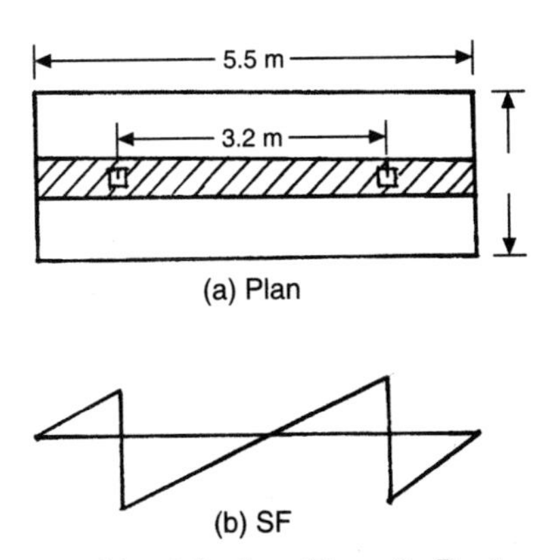

Figure E7.3A Two-column combined footing (Type 2. Rectangular beam and slab type).

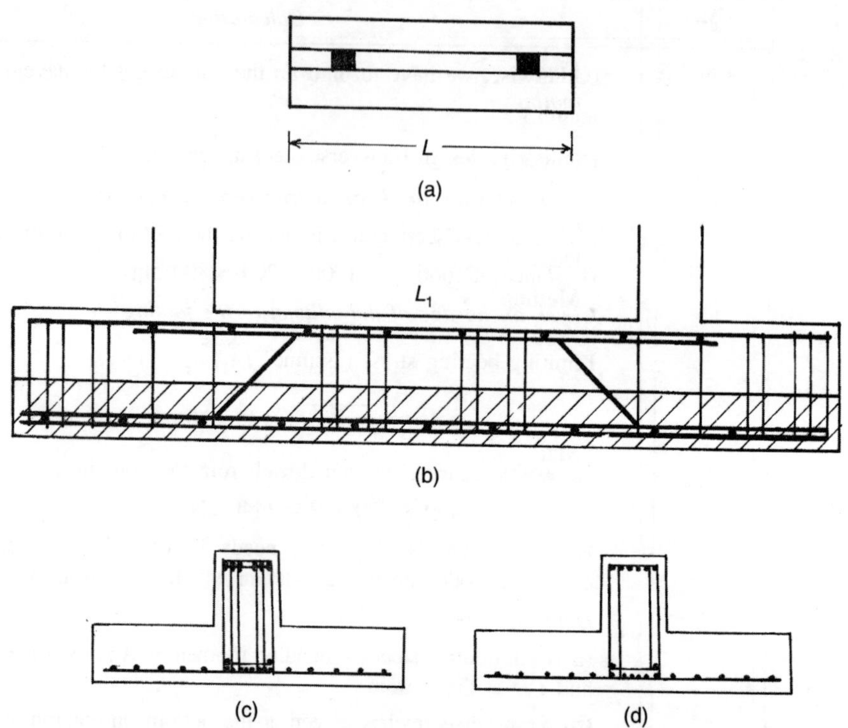

Figure E7.3B Detailing of beam and slab combined footing: (a) General plan, (b) Detailing of steel in beam, (c) Section of middle span near column, (d) Section of cantilever. (*Note:* Bent bars under column adds to shear resistance.)

Reference	*Step*	*Calculation*
Sec. 7.9	1	*Find the area required*
		$$\frac{\text{Total service load}}{\text{Safe BC}} = \frac{2000 + 200}{200} = 11\ \text{m}^2$$
	2 and 3	*Find the dimension of the footing*
		As the loads are equal to the CG at its middle, we can plan a symmetric layout. As the width is restricted to 2 m
		$L = \frac{11}{2} = 5.5$ m (As the loads are symmetric, plan for equal moments to occur at the centre and support.)
Figure 9.2 (Chapter 9)		For equal moments to occur at the centre and at the supports
		$$\text{Overhang} = \frac{L}{4.83} = \frac{5.5}{4.83} = 1.13\ \text{m}$$
		Total length = 3.2 + 1.13 + 1.13 = 5.46 m
		Adopt a 5.5 m × 2 m slab.

Reference	Step	Calculation
	4 and 5	*Find the magnitude of depth required for safety in shear if it is a slab footing*
		Upward pressure $q = \frac{W}{A} = \frac{1.5 \times 2000}{5.5 \times 2} = 273 \text{ kN/m}^2$
		For 2 m, width w = 546 kN/m for full width
Figure E7.3A		Depth for resisting shear of the order of, say, with $\tau_c = 0.35 \text{ N/mm}^2$
		Method 1: (From SF diagram)
		$(803 - 0.546d) \times 10^3 = 0.35 \times 2000 \times d$
		$d = \frac{800}{1.25} = 642 \text{ mm}$
		Method 2: (From point if zero shear and formula)
		$q = 273 \text{ kN/m}^2$ and $a/2 = 0.2$ m
		$L/2 = 1.6$ m (to the point of zero shear)
		$d = \frac{q(L/2 - a/2)}{350 + q}$
		$d = \frac{273(1.6 - 0.2)}{350 + 273} = 0.613 \text{ m} = 613 \text{ mm}$
		Note: We adopt a beam connecting the columns (inverted T) section for the footing to reduce the depth of the footing.
Chapter 9	6	*Design size of the rib beam connecting columns for bending as rectangular beams*
Figure 9.2		$q = 273 \text{ kN/m}^2$ and width 2 m (BM = $0.062\ wL_1^2$)
Balanced cantilever		$M = 0.062q(L_s)^2 = 0.062(273 \times 2)(3.2)^2 = 352$ kNm
		Considering the width of the beam equal to the size of the column.
(Figure 9.2)		Depth required, $d = \sqrt{\frac{M}{0.14 f_{ck} b}} = \sqrt{\frac{352 \times 10^6}{0.14 \times 20 \times 400}} = 560 \text{ mm}$
		Note: We will place transverse steel above longitudinal
		$D = 560 + 10 + 75 = 645$ mm
$\phi_1 = 20$ mm		Let us adopt D = 800 mm as otherwise shear steel requirement will
$\phi_2 = 16$ mm		be high.
Cover		Assume d = 700 mm
= 75 mm		We will design the beam to take the shear as in T beams. As regards the punching shear, it is not valid for beams with shear steel as we must liberally provide shear steel in the design of the beam.
	7	*Find the area of steel required for longitudinal steel as a T beam*
		In general, we should draw the shear force diagram and find M_{max} at

Reference	*Step*	*Calculation*
Step 4		shear is zero. However, as we have already planned a cantilever for the equal span moments, the span moment = 352 kNm (T beam). As the slab is in compression, we design it as T beam
		(Assume $z = 0.91d$).
		$$z = 0.91 \times 700 = 637 \text{ mm}$$
		$$A_s = \frac{M}{0.87 f_y z} = \frac{352 \times 10^6}{0.87 \times 415 \times 637}$$
		$= 1530 \text{ mm}^2$ (Use 5–20ϕ = 1571 mm^2)
		Percentage of steel = $\frac{1530 \times 100}{400 \times 700} = 0.54\%$
		(We may also adopt any other method for T beam design.)
		(*Note:* Here we have found the depth as a rectangular beam and steel as a T beam.)
	8	*Check for bond and anchorage*
		As we are using deformed bars, no check for local bond is to be considered.
Table B.3		Anchorage bond = 47ϕ = 47 × 20 = 940 mm, which can be supplied by these bars.
	9	*Design beam for shear q* per metre length is equal to
		2 × 273 = 546 kN
		Shear at d from the column (1.13 + 0.4 + 0.70) m from the edge of the foundation is given by
		1500 – 546(1.13 + 0.4 + 0.70) = 283 kN
		Shear stress = $\frac{283 \times 1000}{400 \times 700} = 1.01 \text{ N/mm}^2$
Table B.4		τ_c at 0.5% steel = 0.50 N/mm^2 (approx.) and $\tau_{c\,max}$ = 2.8 N/mm^2
Table B.6		Hence design for extra shear reinforcement.
		Design shear steel for the beam
		Factored shear = 283 kN
Step 8		The shear the concrete can take is obtained as follows
		0.5 × 400 × 700 = 140 kN
		Shear to be carried by steel = 143 kN
		Max spacing allowed = 0.75d = 525 mm
Table B.7		V/d in kN/m = 143/70 = 2.04 mm
		Fe 415 using 12 mm ϕ two-legged stirrups spacing = 400 mm
		(*Note:* We now proceed to design along the width)

Reference	Step	Calculation
	10	*Check for shear of cantilever slab in transverse direction* (for τ_c, $p = 0.36\%$) Find the necessary depth to resist the above value at d from the face of the beam per metre length. Cantilever slab length = 800 mm Shear at d from beam = shear strength, i.e. $\left(\frac{800-d}{1000}\right)(273)\times 10^3 = 1000\times d\times 0.42$ $d = 315$ mm; $D = 315 + 8 + 75 = 398$ mm Adopt 400 mm depth and make the final adjustments of steel. $d = 400 - 75 - 8 = 317$ mm
	11	*Design slab in bending in transverse direction* (*Note*: As the transmission of load is through a rigid beam (unlike in a slab), we assume the ground reaction is uniform. We do not use the column strip concept for beam and slab footing. Minimum depth = 150 mm.) Cantilever part of slab T beam = (2 – 0.4)/2 = 0.8 m $\frac{BM}{m} = \left(\frac{1}{2}\right)273\times(0.8)^2 = 87.4$ kNm *Find percentage of steel required* $\frac{M}{bd^2} = \frac{165\times 10^6}{1000\times 315\times 315} = 1.66$ $p = 0.25\%$ (The maximum steel percentage usually allowed in slabs is 0.75%.) $A_s = \frac{0.25}{100}\times 1000\times 315$ per m $= 788$ mm²/m Adopt 16 mm @ 250 mm = 840 mm
	12	*Check for anchorage steel* $L_p = 47\phi = 752$ mm < 800 mm (available)
	13	*Provide distribution steel* Provide 12 mm @ 250 mm centres
	14	*Check for column bearing as explained in Chapter 4* (*Note:* Detailing of a beam and slab combined footing is shown in Figure E7.3(B).

EXAMPLE 7.4 (Design of trapezoidal combined slab footing)

Two columns of 500 × 500 mm are spaced at centres, with the external column carrying 700 kN and the internal column 1100 kN. The external edge of the footing is not to be farther than 0.50 m from the centre of the external column. Design the footing neglecting self-weight of the footing assuming an SBC = 150 kN/m².

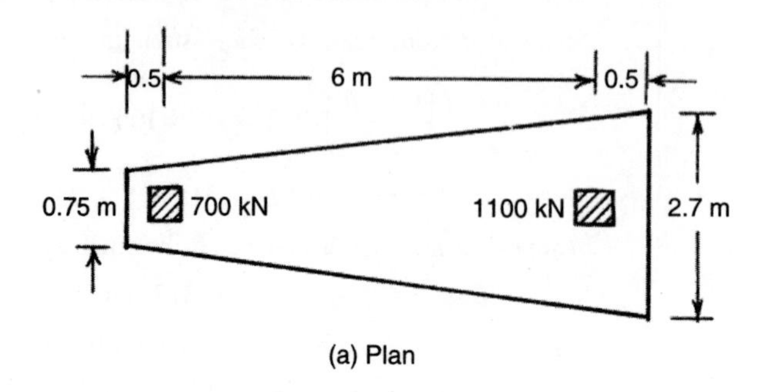

(a) Plan

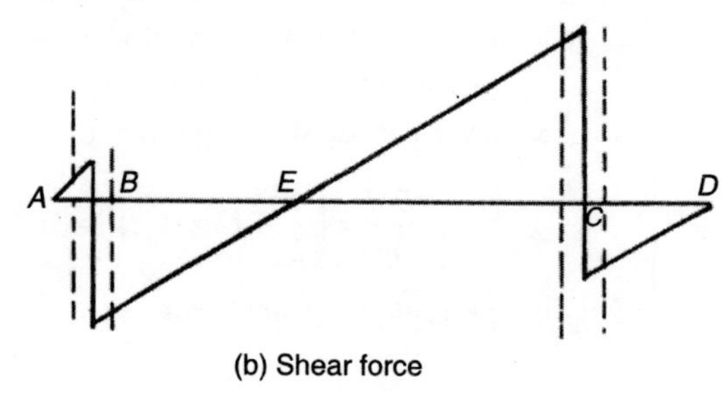

(b) Shear force

Figure E7.4 Two-column combined footing (Type 3. Trapezoidal slab type).

Reference	Step	Calculation
	1	*Find CG of loads from internal column*
		$$\bar{x} = \frac{700 \times 6}{(1100 + 700)} = 2.33 \text{ m} \text{ from the centre of } C_2$$
		If we need a rectangular footing, then
		$L = 2(2.33 + 0.5) = 5.76$ m only
		This is less than 6.0 m between the columns. Hence, adopt a trapezoidal slab footing 0.5 m beyond C_2 also. Total length $L = 7.0$ m and 0.5 m beyond C_1.
	2	*Find dimensions of footing*
		$$A = \frac{1800}{150} = 12 \text{ m}^2$$
		Let the breadth be B_1 and B_2.

Reference	*Step*	*Calculation*
		CG should be @ 2.33 + 0.5 = 2.83 m from col. C_1 $2.83 = \frac{7}{3}\frac{(2B_1 + B_2)}{B_1 + B_2}$ (i) $\left(\frac{B_1 + B_2}{2}\right) \times 7 = 12 \text{ m}^2$ (ii) Solving B_1 = 0.75 m and B_2 = 2.70 m, we get $A = \frac{1}{2}(0.75 + 2.70) \times 7 = 12 \text{ m}^2$
	3	*Find variation of width with x from the left end (A)* Width at end A = 0.75. Width at x from $A = 0.75 + \left(\frac{2.70 - 0.75}{7}\right)x = 0.75 + 0.28x$ Col. $C_1 = x$ = 0.75 m and B = 0.75 + (0.28 × 0.75) = 0.96 m @ Col. C_2. At C = 0.75 + 0.28(6.5) = 2.57 m
	4	*Find q_u with factored load* Factored loads = 1.5 × 700 = 1050 kN = 1.5 × 1100 = 1650 kN $q = \frac{1050 + 1650}{12} = 225 \text{ kN/m}^2$
	5	*(a) Draw SF and BM diagrams* *[We will proceed on the area basis. See also step 5(2)]* SF left of B = (area between A & B) × q $= \left[\frac{0.75 + 0.96}{2}\right] \times 0.5 \times 225 = 96 \text{ kN}$ SF right of B = 1050 – 96 = 954 kN SF right of C = $\left[\frac{2.70 + 2.57}{2}\right] \times 0.5 \times 225 = 296 \text{ kN}$ SF left of C = 1650 – 296 = 1354 kN *(b) Find the position of zero shear at x from A (point E)* Breadth = 0.75 + 0.28x Load on strip A to E = x (mean width) (225) kN $= x\left[0.75 + \frac{0.28x}{2}\right] \times 225 = 1050$ $0.75x + 0.14x^2 - 4.67 = 0$ $x^2 + 5.35x - 33.3 = 0$ x = 3.69 m from end A.

Reference	*Step*	*Calculation*
		Draw SF diagram.
		(c) Find maximum BM
		M_{max} @ SF = 0, i.e. 3.69 m from A
		Width at 3.69 m = 0.75 + (0.28 + 3.69) = 1.78 m
		Pressure from A to E = Same as col. load = 1050 m
		CG of the area from end $A = \dfrac{3.69}{3}\dfrac{[(2\times 0.75)+1.78]}{(0.75+1.78)} = 1.59$ m
		Distance of the edge of C_1 from E = 3.69 – 0.75 = 2.94 m
		BM = 1050(2.94 – 1.59) = 1417 kNm
		Width at 3.69 m = 0.75 + 0.28x = 1.783 m
	5.1	*Alternatively, we can proceed using integration of variation of loading to find SF and BM*
		At end A = 225 × 0.75 = 168.75 kN/m
		At end D = 225 × 2.70 = 607.50 kN/m
		Variation = $168.75 + \dfrac{607.50 - 168.75}{7}x$
		= 168.75 + 62.68x
		Calculation of shear by integration
		Shear force = $Q = \int_0^x (168.75 + 62.68x)\,dx$
		$Q = 168.75x + 31.34x^2 + c_1$
		At $x = 0$, $Q = 0$, Hence $c_1 = 0$
		Q left of B @ $x = 0.75$: Q = 144.3 kN
		Q right of B @ $x = 0.75$: Q = 1050 – 144.3 = 905.7 kN
		At x = 6.5 left of C = (168.75 × 6.5) + [31.34 × $(6.5)^2$] – 1050
		= 1371 kN
		Right of C = 1650 – 1371 = 279 kN
		Calculation of x, where SF = 0
		$31.34x^2 + 168.75x = 1050$
		$x^2 + 5.38x - 33.50 = 0$
		x = 3.69 m (as before)
		Calculation of BM
		We know $M = \int Q\,dx$
		$M - \int_0^x (168.75x + 31.34x^2)\,dx$
		$= 84.37x^2 + 10.45x^3 + c_2$

Reference	Step	Calculation
		As $M = 0$ @ $x = 0$; $c_2 = 0$
		BM at $x = 3.69 = M_1$ due to uniformly distributed load – M_2 due to c_1
		Column load = 1148.8 + 525.0 – 1050 (3.69 – 0.75)
		= 1413 kNm
	6	*Find depth for slab footing for one-way shear*
		Distances from zero shear to centre C_1 = 3.69 – 0.5
		= 3.19 = $L/2$
Example 2.3		Consider as col. footing a = 0.50 m
Eq. (2.7)		$d = \dfrac{(l/2 - \alpha/2)q}{350 + q} = \dfrac{225(3.19 - 0.25)}{350 + 225} = 1.15\text{ m} = 1150\text{ mm}$
		This depth required, without shear reinforcement is very large. *We may adopt a lesser depth with shear steel, or alternately the design as a beam and slab combined trapezoidal footing*
		Let us adopt d = 720 mm = 72 cm and provide shear steel.
		Design for shear at d from col.
		Shear at 3.19 m from col. centre = 0
Example 2.3		V at d from col. edge = $(l/2 - a/2 - d)q$ per metre width
		= (3.19 – 0.25 – 0.72) × 225 = 500 kN/m
		$\tau = \dfrac{V}{bd} = \dfrac{500 \times 10^3}{1000 \times 720} = 0.69\text{ N/mm}^2 > 0.35\text{ N/mm}^2$
		Excess shear = $(\tau - \tau_c)\, b \times d$ per metre width
		$= \dfrac{(0.69 - 0.35) \times 1000 \times 720}{1000}\text{ kN} = 245\text{ kN/m}$
		Width of the slab at d from the edge of the col. C_2
		Distance from A = 6.25 – 0.97 = 5.28 m
		= 0.75 + (0.28 × 5.28) = 2.23 m
		Excess shear V for full width = 245 × 2.23 = 546.1 kN
		$\dfrac{V}{d}\left(\dfrac{\text{kN}}{\text{cm}}\right) = \dfrac{546}{72} = 7.6$
		(V/d for 2-legged 10 mm stirrups @ 30 cm = 1.89)
Table B.7		10 mm 8-legged stirrup @ 30 cm gives 4 × 1.89 V/d = 7.56
	6(a)	*Check for punching shear* (Take col. of 1100 kN)
		Punching shear = V_p at $d/2$ from the column
		$V = 1650 - (0.50 + 0.72)^2 \times 225 = 1315$ kN
τ_p for M_{20}		$V_p = \tau_p\,(2a + 2b + 4d) \times d$
= 1180 kN/m^2		Shear resistance = 1180 × 4 × (0.5 + 0.72) × 0.72

Reference	Step	Calculation
		= 4146 kN > 765 kN
		(Capacity is also greater than 1650 col. load)
	7	*Tension steel for* M_{max} (width = 1783 mm)
Step 6 Table B.1		$\frac{M}{bd^2} = \frac{1417 \times 10^6}{1783 \times (720)^2} = 1.53 < 2.76$
		Single reinforced slab: p = 0.477% steel
		(Calculate longitudinal steel required)
		$\frac{0.477}{100} \times 1783 \times 720 = 6123 \text{ mm}^2$
		No. of 20 mm dia bars = $\frac{6123}{314.2} = 20$ bars
		(Check for development length)
	8	*Design transverse steel*
		Column strip = a = 1.5d
		Take internal column, 1100 kN
		Width col. strip = 0.50 + 1.5(0.72) = 1.58 m
		Cantilever length on both sides of the column $= \frac{2.75}{2} = 1.37$ m
		For a quick design, we assume $W/2$ is carried by cantilevers.
Table B.1		$M = \left(\frac{W}{2}\right)\left(\frac{l^2}{2}\right) = \frac{1650}{2} \times \frac{(1.37)^2}{2} = 774$ kNm
		This is resisted by a beam of 1.58 m width and 720 mm depth.
		Find $M/bd^2 = \frac{774 \times 10^6}{1580 \times 720 \times 720} = 0.94$
		p = 0.28%
		$A_s = \frac{0.28}{100} \times 1580 \times 720 = 3185 \text{ mm}^2$
		No. of 16 mm rods $= \frac{3185}{201} = 16$ rods
		Check development length. If necessary, bend the rods.

EXAMPLE 7.5 (Design of trapezoidal beam and slab combined footing)

The foundations of an exterior column 250 × 250 mm, 0.6 m from the boundary and a carrying characteristic load of 600 kN are to be combined with those of an interior column 450 × 450 mm carrying a characteristic load of 1200 kN. The columns are spaced 4.5 m apart. The foundation can extend 1.5 m towards the inside from the interior column. Design a trapezoidal footing with spine beam. Assume the safe bearing capacity is 180 kN/m^2.

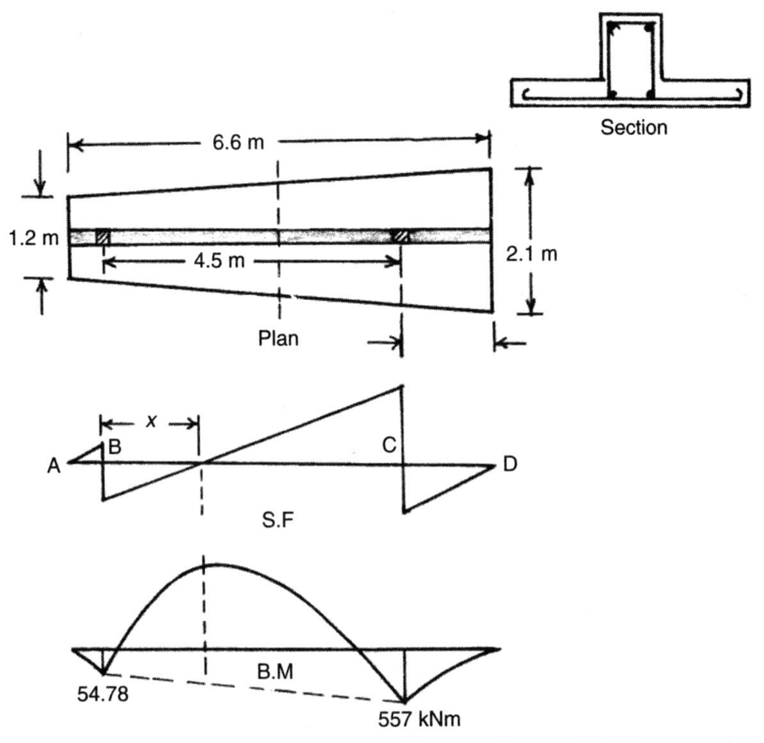

Figure E7.5 Two-column combined footing (Type 4. Trapezoidal beam and slab type).

Reference	*Step*	*Calculation*
	1	*Find the area required with characteristic load*
		$$\frac{\text{Total load}}{\text{SBC}} = \frac{600 + 1200 + 200}{180} = 11.1\ \text{m}^2$$
		Assume 11 m^2
	2	*Find CG of loads from the end of the inner footing*
		$$\bar{x} = \frac{(1.5 \times 1200) + (6 \times 600)}{1800} = 3\ \text{m from } C_2$$
	3	*Find the dimension of the footing to suit CG of loads*
		As the interior load is much heavier than the exterior load, the shape will be trapezoidal.
		Distance between col. = 4.5 m
		End from exterior col. = 0.6 m

Reference	*Step*	*Calculation*
		End from interior col. = 1.5 m (given)
		Total length = 6.6 m. Let B_1 = 1.2 m
		Find B_2 (larger end) for CG at 3 m from B_2.
		$\bar{x} = \frac{L}{3}\left(\frac{2B_1 + B_2}{B_1 + B_2}\right)$
		$3 = \frac{6.6}{3}\frac{(2 \times 1.2 + B_2)}{(1.2 + B_2)}$
		B_2 = 2.1 m
		Adopt L = 6.6 m; B_1 = 1.2 m; B_2 = 2.1 m
		As the larger column is 450 × 450 mm, we choose the rib size = 450 mm.
	4	*Factored ground pressure (with factored loads)*
		Ultimate soil pressure (excluding weight of base)
		= 1.5(600 + 1200) = 900 + 1800 = 2700 kN
		Factored soil pressure q_v = 2700/11 = 245.5 kN/m^2
	5	*Draw the SF and BM diagrams in longitudinal direction*
		If we design the footing as a T beam, because of the trapezoidal shape, the load from the ground pressure increases from the exterior to the interior end.
		W_1 at the left end = 245.5 × 1.2 = 295 kN/m
		W_2 at the right end = 245.5 × 2.1 = 516 kN/m
Figure E7.5		However, as an approximation, we will take the average width in each zone AB, BC and CD and assume they act as UDL.
		Width at A = 1.20 m
		Width at B = (0.6 m from A) $1.2 + \frac{0.9 \times 0.6}{6.6} = 1.28$ m
		Width at C = $1.2 + \frac{0.9}{6.6} \times 5.1 = 1.89$ m
		Average width of AB = $\frac{1}{2}(1.20 + 1.28) = 1.24$ m (a)
		Average width of BC = $\frac{1}{2}(1.28 + 1.89) = 1.59$ m
		Average width of CD = $\frac{1}{2}(1.89 + 2.1) = 1.99$ m (b)
		Factored load W_1 = 900 kN
		Load left of B = 0.6 × 1.24 × 245.5 = 182.6
		Shear at B = 900 − 182.6 = 717.4 kN

Reference	*Step*	*Calculation*
		Find shear at right of C and then left of C
		Load on CD = $\left(\frac{1.89+2.1}{2}\right)\times 1.5\times 245.5 = 734.7$ kN
		Shear left of C = 1800 – 734.7 = 1065.3 kN
		Let the point of zero shear be x
		$\frac{x}{717}=\frac{4.5}{1783}$ or x = 1.8 m from B or 2.7 from C
		Draw the SF diagram
		Draw the BM diagram
		$M_A = 0$
		$M_{BA} = 182.6 \times \left(\frac{0.6}{2}\right) = 54.78$ kNm (Cantilever)
		$M_{CD} = 734.7 \times \left(\frac{1.5}{2}\right) = 551$ kNm (Cantilever)
		SSBM on span BC = $\frac{WL}{8}=\frac{(1.59\times 245.5\times 4.5)\times 4.5}{8}$
		= 988 kNm
		max MB = 988 – $\left[54.8+\frac{(551-54.8)}{4.5}\times 1.27\right]$
		= (988 – 194.8) = 793.2 kNm
	6	*Find the minimum depth of the singly reinforced beam*
		Even though the beam is a T beam, we design it as a rectangular beam $M = 0.14 f_{ck} bd^2$ for Fe 415 steel. Thus
		b = 450 mm of larger column
		$d=\sqrt{\frac{793\times 10^6}{0.14\times 20\times 450}}$ = 793 mm; using ϕ = 25 mm
		D = 793 + 75 + 25/2 = 880, say 900 mm
		Assume d = 810 mm
	7	*Find the tension steel required*
		$\frac{M}{bd^2}=\frac{793\times 10^6}{450\times 810\times 810}=2.7$
Table B.1		p = 0.946% (This is the limit for singly reinforced beams. We may reduce the depth and can use a doubly reinforced beam also.)
		$A_s=\frac{0.946}{100}\times 450\times 810 = 3448\ \text{mm}^2$
		Use 7 no 25 mm giving 3436 mm^2

Reference	Step	Calculation
	8	*Check for average load*
		We need anchorage = $47\phi = 47 \times 25 = 1175$ mm. This can be achieved, if necessary, by providing bends at the end of rods.
	9	*Design for shear and shear steel*
		V_{max} @ d from the column edge (225 + 900) = 1125 mm from the centre line of the column
		$V = 1065.3 - (1.125 \times 1.62 \times 245.5) = 618$ kN
Table B.4		$\tau = \frac{618 \times 10^3}{450 \times 810} = 1.68\ (\text{N/mm}^2);\ \tau_c = 0.60\ \text{N/mm}^2$ for $p = 0.9\%$
		Requires shear reinforcement. Design the shear steel.
	10	*Design a slab in the transverse direction (Find the depth required without shear steel)*
		As width varies, we find the depth for the maximum cantilever span at the far end of C.
		Average ½ width of CD = 2 m (approx)
		Assuming $\tau_c = 0.35$ N/mm^2 – Depth for shear = d
Eq. (3.7)		$d = \frac{q(L/2 - a/2)}{350 + q}$ in m and kN; $L = 2$ m, $a = 0.45$ m
		$= \frac{245.5\,(1 - 0.225)}{350 + 245.5} = 0.32$ m
		$D = 320 + 75 + 10/2 = 400$ mm
	11	*Find area of transverse steel required per* metre *length*
		$M_{max} = \frac{wl^2}{2} = \frac{245.5 \times (0.775)^2}{2} = 73.73$ kNm
		$\frac{M}{bd^2} = \frac{73.73 \times 10^6}{1000 \times 320 \times 320} = 0.72$
		$p = 0.22\%$ (As this is a low percentage, provide same steel throughout.)
		$A_s = \frac{0.22 \times 1000 \times 320}{100} = 704\ \text{mm}^2/\text{m}$
		Provide 16 mm @ 275 mm (A_s = 731)
	12	*Check anchorage length*
		$47\phi = 47 \times 16 = 752$ mm. In smaller widths, the bars have to be bent up for this anchorage.

EXAMPLE 7.6 (Combined trapezoidal footing with transverse beams and longitudinal slab)

In a factory, an exterior column 250 × 250 carrying characteristic load of 600 kN is to be combined with an interior column 450 × 450 mm carrying a characteristic load of 1200 kN. The columns are spaced at 4.5 m, and the foundation can extend to 1.5 m from the interior column. The depth of the foundation between the columns should be kept to a minimum due to the interference of pipelines. Hence, a combined footing with transverse beam is planned. Design a trapezoidal footing with the *column on a transverse beam and longitudinal slab* between these beams if the safe bearing capacity is 180 kN/m^2.

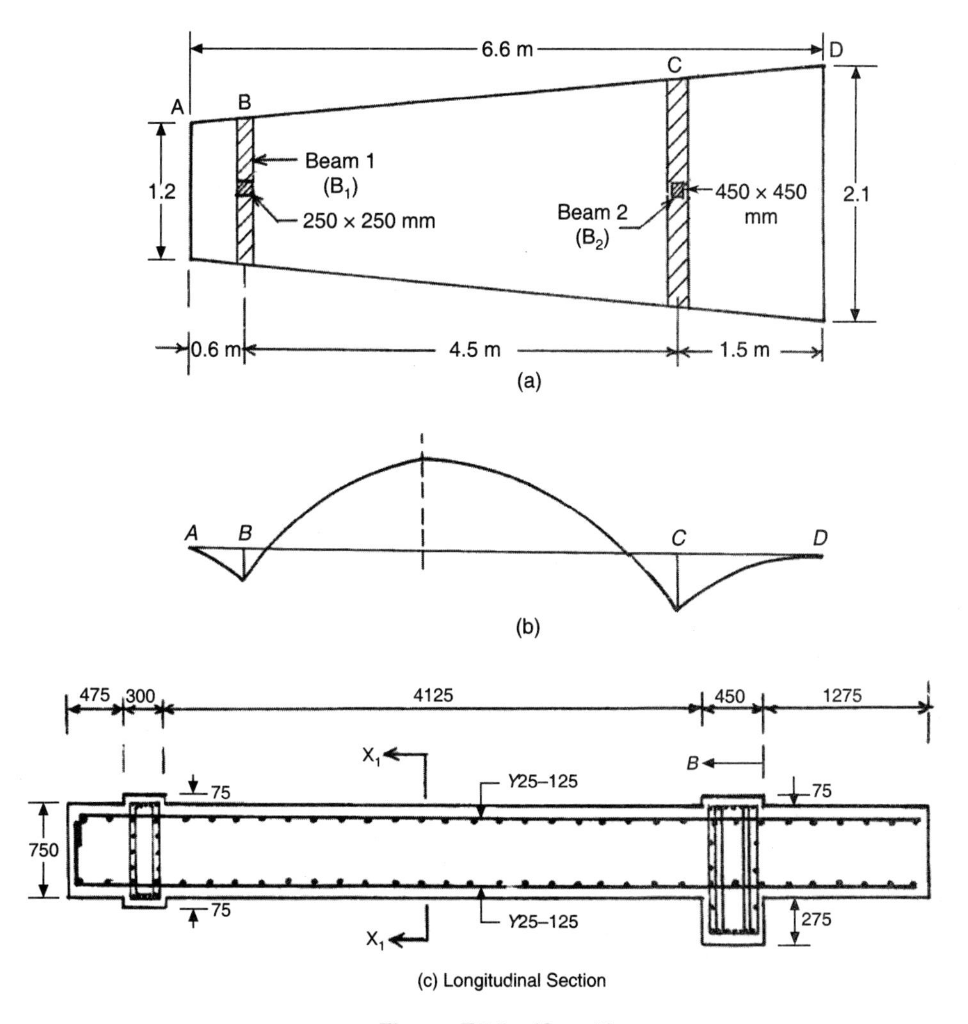

Figure E7.6 (Contd.)

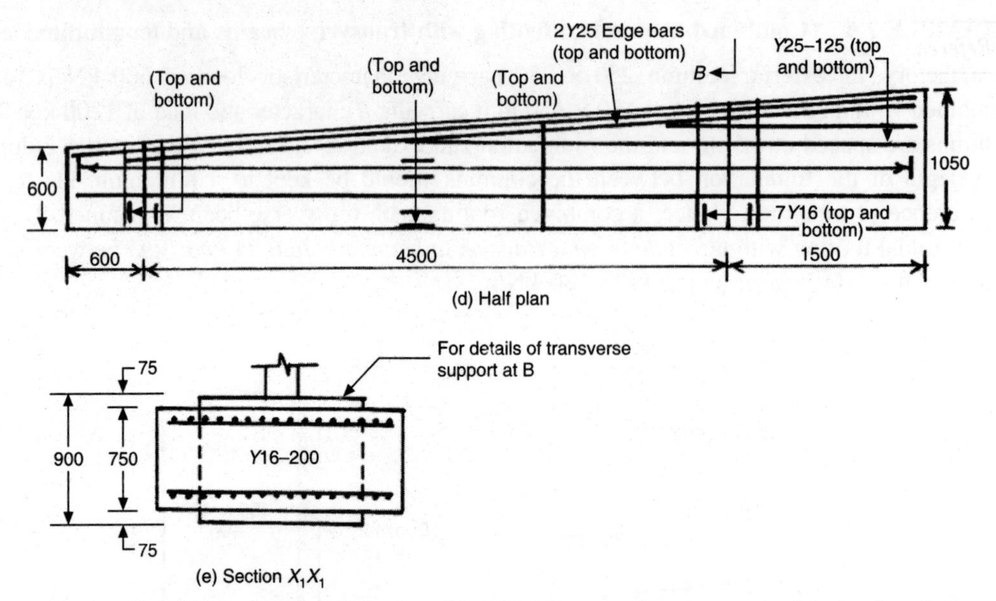

Figure E7.6 Detailing of two-column combined transverse beams and slab footing (Type 5): (a) Plan showing arrangement, (b) Bending moment in slab, (c) Longitudinal section, (d) Half plan—detail of steel, (e) Section A–A.

Reference	*Step*	*Calculation*
Example 7.5	1	*Find the area required with characteristic load*
		$\frac{\text{Total load}}{\text{SBC}} = \frac{600 + 1200 + 200}{180} = 11.1\ \text{m}^2$
		Assume 11 m^2; $\bar{x}$ = 3 m from B_2
		B_1 = 1.2 m; B_2 = 2.1 m; L = 6.6 m
		Factored loads
		@ B = 1.5 × 600 = 900 kN
		@ C = 1.5 × 1200 = 1800 kN
		Ultimate soil pressure = 245.5 kN/m^2
		(Structural action—Assume transverse beams of (300 mm) width and the required depth below column at B (250 × 250 mm) and beam of (500 mm depth required) below column at C (450 × 450 mm). These beams are to be designed as cantilever beams. The slab spans between these beams with uniformly distributed loads. These types of footings will require heavy cantilever support beams but the slab will be shallow and its reinforcement will be simple, continuous over two supports, the two beams.) The required footing is shown in Figure E7.6(a).
	2	*Find the design moments of the transverse cantilever beams*
		Breadth of footing at $\text{B} = 1.2 + \frac{(2.1 - 1.2)}{6.6} \times 0.6 = 1.28\ \text{m}$
		Cantilever at col. $\text{B} = \frac{(1.28 - \text{col. size})}{2} = 0.515\ \text{m}$

Reference	*Step*	*Calculation*
		Breadth of the footing at col. at $C = 1.2 + \left(\dfrac{0.9 \times 5.1}{6.6}\right) = 1.9$ m
		Cantilever at col. $C = \left(\dfrac{1.9 - 0.45}{2}\right) = 0.725$ m
		Load on cantilever at B = 900/2 = 450 kN (approx.)
		Load on cantilever at C = 1800/2 = 900 kN (approx.)
		(*Note:* The exact load will be slightly less due to reduced reaction under the space occupied by the columns)
	3	*Design the transverse beam for shear*
		As these are short beams, shear will be more critical than bending moment. These are to be designed with shear at the face of the support with increased shear at supports as in corbels. However,. for a safe design we design for shear at support with increased shear (see Sec. 7.7).
		At col. B (Total width = 1.28 m),
		(Cantilever = 0.64 from the centre = 0.515 from the edge of the column)
		Shear at the edge of the column at B = $\dfrac{450 \times 0.515}{0.64} = 362$ kN
Table B.6		Shear at the edge of the column at C = $\dfrac{900 \times 0.725}{0.95} = 687$ kN
		As we have to provide shear reinforcement, max shear allowed for M20 is 2.8 N/mm^2, assume $\tau_c = 1.5$ N/mm^2; $\tau = \dfrac{V}{bd}$
		Breadths of end beams are 300 and 500 mm; $\tau = \dfrac{V}{bd}$
		d_1 = Depth required for column at B $= \dfrac{362 \times 10^3}{300 \times 1.5} = 804$ mm
		d_2 = Depth required for column at C $= \dfrac{687 \times 10^3}{500 \times 1.5} = 916$ mm
	4	*Design shear steel in transverse beam*
		As in Examples.
	5	*Design the transverse beam for bending as a cantilever beam* 450 × 916 mm
Step 2		For column C,
		$M = \dfrac{900 \times (0.725)^2}{2} = 236.5$ kNm

Reference	*Step*	*Calculation*
Table B.1		$\dfrac{M}{bd^2} = \dfrac{236.5 \times 10^6}{450 \times (916)^2} = 0.62 < 0.95$
		p = 0.18%; provide 0.72% as required for shear
		(*Note:* As the beam is short, anchorage requirements will also be satisfied only with high percentage of steel subjected to lower stress and also bends at the ends. Thus, at B even though only 2 of 16 mm may be ample, it is better to provide 5 of 16 mm to reduce the tension value and thus adjust bend requirements. Similarly provide at C.)
	6	*Design the longitudinal slab*
		Width at A = 1.2 m
		Width at D = 2.1 m; L = 6.6 m
		Width at A = 1.2 m
		Variation = $1.2 + \dfrac{0.9}{6.6}x = 1.2 + 0.136x$
		Width at B = 1.2 + (0.136 × 0.6) = 1.28 m
Step 1		Width at C = 1.2 + (0.136 × 5.1) = 1.89 m
		W_1 pressure at A = 1.2 × q = 1.2 × 245.5 = 294.6
		W_2 pressure at D = 2.1 × 245.5 = 515.5
		Variation of pressure = $294.6 + \left(\dfrac{515.6 - 294.6}{6.6}\right)x$
		$w = 294.6 + 33.5x$ as distributed load per metre length
		$\text{SF} = \int w = 294.6x + 16.75x^2$
		$\text{BM} = \int \text{SF} = 147.3x^2 + 5.58x^3$
		(We have also to take moment of the column load for the final values.)
	7	*Draw SF longitudinally taking x_1 from A*
		Point where SF = 0 be x_1 (where reaction = col. load)
		$16.75x_1^2 + 294.6x_1 = 900$
		$x_1^2 + 17.59x_1 - 53.73 = 0$
		x_1 = 2.65 m from A (2.05 from B) and 2.45 m from C
	8	*Find max BM where SF = 0*
		M_{max} @ 2.05 m from B column
		= (900 × 2.05)—due to varying load from end A
		= 1845 − [147.3 × $(2.65)^2$ + 5.58$(2.65)^3$]
		= 1845 − 1138 = 707 kNm
		Width at x = 1.27 m = 1.2 + 0.136 × 1.27 = 1.37 m
		Moment per metre width = 707/1.37 = 516 kNm/m
		Design the slab for this BM per metre width with respect to the depth required and steel required as in other cases.

8

Balanced Footings

8.1 INTRODUCTION

Balanced footings (also known as strap footings) and *cantilever footings* form a special group of footings. Balanced footings consist of *two separate footings* for two columns connected by a strap beam as shown in Figure 8.1(a). As briefly explained in Chapter 7 Section 7.1. A cantilever footing always becomes necessary when the foundation of a column cannot be built directly under the column or when the column should not exert any pressure below. It is then necessary to balance it by a *cantilever arm* rotating about a fulcrum and balanced by an adjacent column (or a mass of concrete or by piles) where footings cannot be built. The situations under which these footings become necessary are the following:

Case 1: The column load comes on the property line and the necessary foundation cannot be built on the property line. It can also happen when the soil is not very good for its foundation or there are restriction placed for the construction of a footing by the adjacent building.

Case 2: The column load comes up over an existing structure and no pressure needs to be exerted over the structure.

Case 3: The distance between columns to be combined is very large and the combined footing becomes very narrow with high bending moment.

As the name implies, the planning of these footings essentially consists of a balancing element and making use of the principle of levers. We should be aware that a cantilever element will be subjected to high tension at the top and it will also have to transmit heavy shear. It should *be also rigid* as excessive deflection will cause movement of the superstructure above.

8.2 TYPES OF BALANCING USED

Depending on the field situation [e.g. the magnitude of the load on the balancing column (internal column) and the column to be balanced (external column), the nature of soil, and so on], the following three base types are commonly met with. (Let us call the column to be balanced the *external column* and the column used for counterbalance the *internal column*).

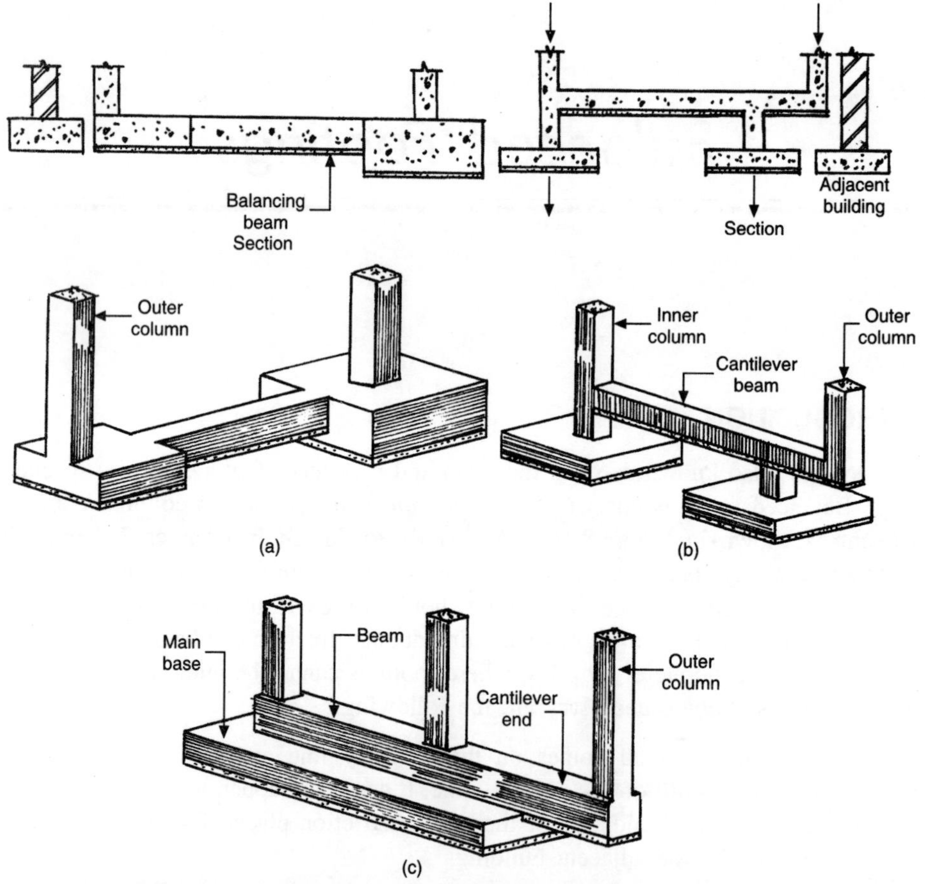

Figure 8.1 Balanced and cantilever footings: (a) Balanced footing, (b) and (c) Cantilever footings. (*Note:* In balanced footings we make centre of gravities of loads and footings to coincide so that foundation pressure will be uniform. In cantilever footings as the centre of gravities of loads and footings may not coincide the foundation pressure may not be uniform).

Type 1: *Balanced footing with fulcrum near external column* Refer to Figure 8.2(a). If the footing can be built near the external column, we assume a "*fulcrum base*" *footing near the external column* C_1 of load W_1 and balancing it by an *internal column or a block of concrete* C_2 of load W_2. A factor of safety of not less than 1.5, should be provided when balancing against rotation. The beam connecting W_1 and W_2 is called the *strap beam* and it should be cast integral with the footing.

$$W_2L_2/W_1L_1 > 1.5 \tag{8.1}$$

Type 2: *Balanced footing with fulcrum near internal column* Refer to Figure 8.2(b). The external column is balanced by an eccentrically placed internal column so that *the fulcrum is near the internal column* C_2. Such a case occurs when we have a heavy internal column, or when no foundation can be built near the external column.

$$W_1(L - e) = W_2e$$

$$\text{or} \quad e = W_1L/(W_1 + W_2) \tag{8.2}$$

Type 3: *Cantilever footings* Refer to Figure 8.2(c). One or both the columns on the opposite sides are not to exert any weight below it. In such a case, the columns can be connected by a stiff beam on a base with their ends cantilevered over a fulcrum base. The ends of the cantilever parts should be free or placed over a compressible base so that the columns at the ends do not exert any pressure under the soil below the cantilever.

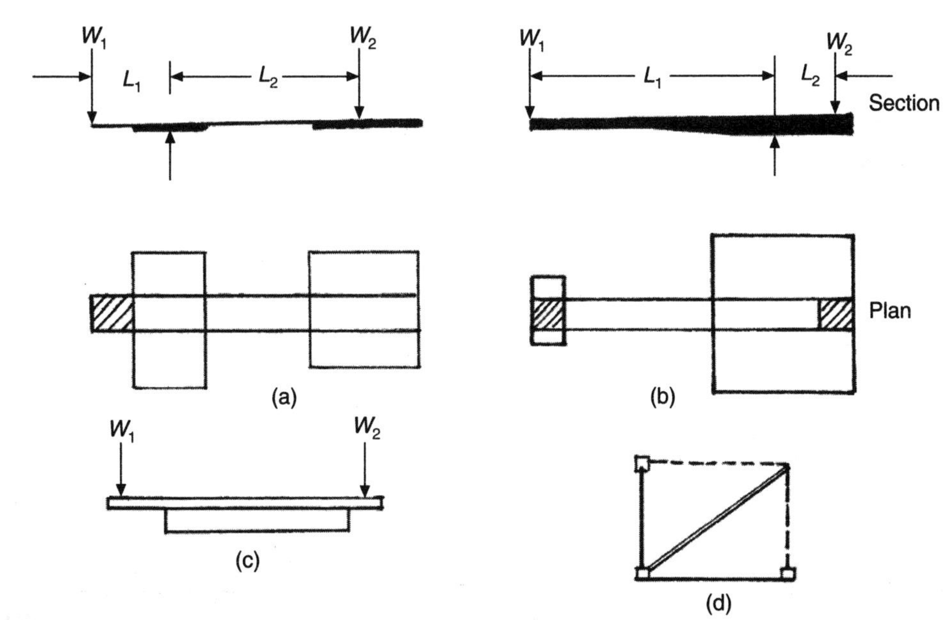

Figure 8.2 Types of balanced and cantilever footings: (a) Type 1. Fulcrum base near external footing balanced by an internal column or a dead weight, (b) Type 2. Fulcrum base near the heavy internal column, (c) Type 3. Cantilever footing with ground beam, (d) Balancing of corner columns.

Type 4: *Balanced cantilever foundation on piles* This occurs when the soil at the site needs a pile foundation (see Chapter 17).

[*Note:* A corner column has to be balanced it has to be balanced in two directions, or we may use a diagonal cantilever as shown in Figure 8.2(d).]

8.3 LOADS TO BE TAKEN FOR CALCULATION

For the calculation of the layout of the footings, we take it as serviceability condition and use only the characteristic loads. For the design of the structure, we use limit state design and the factored loads.

8.4 BASIS OF DESIGN

As we have seen, balanced footings consist of two separate footings and a strap beam connecting them. If we can make the centre of the areas of the system coincide with the centre of gravity of the loads, we can assume uniform distribution of loads under the footings. If these two centres do not coincide, we have the vertical load and moment acting on the system due to the eccentricity so that the distribution of base pressure will not be uniform but can be assumed to be linearly varying. The difference between balanced footings and cantilever footings can be discribed as follows.

In a balanced footing, we make the centre of gravity of the loads and the centre of the areas to coincide. Hence, the ground pressure will be uniform. In cantilever footings, in general the two centres may not coincide, so we have a moment in addition to the vertical loads. Hence, the ground pressure will be varying.

8.5 SUMMARY

Cantilever footings are also combined footings used in special situations. In a balanced footing, we make the centre of gravity of the loads and the centre of the footings to coincide so that the ground pressure is uniform. In a cantilever, as the centres may not coincide, the ground pressure will vary along the length of the footings. The examples show how they can be designed.

EXAMPLE 8.1 (Design of Type 1 balanced footing—External column balanced by an internal footing with a fulcrum near external column)

An outer column 300 mm square with a characteristic load of 600 kN is built on the boundary and it has to be balanced by an internal column 400 mm square carrying 900 kN. The centre lines of the columns are spaced 5 m apart and the safe bearing capacity (SBC) of the soil can be taken as 120 kN/m^2. Give a layout of the system.

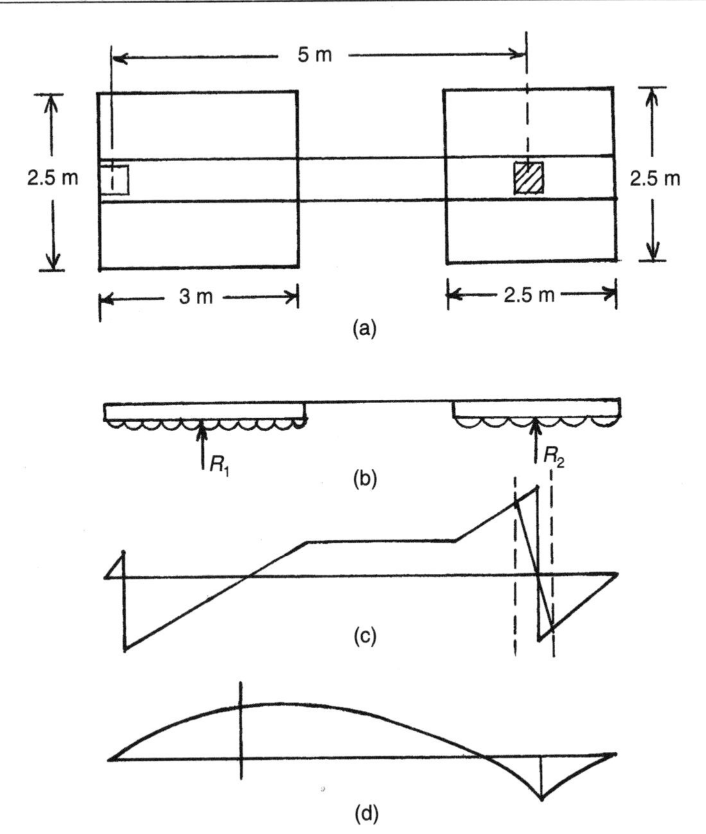

Figure E8.1 Type 1. Balanced footing: (a) Plan, (b) Section, (c) Shear force diagram, (d) Bending moment diagram.

Reference	*Step*	*Calculation*
		As the weight of the internal column is not very large, we adopt type 1 footing with fulcrum near the external column so that the lever arm is long.
		We can either dimension each of the footings so that the CG of the areas and loads coincide, resulting in uniform pressure, or adopt suitable base dimensions and then check the resulting ground pressure taking the unit as a whole, which may not be uniform, and design for non-uniform pressure. We will adopt the first method.
	1	*Preliminary dimensioning of footing* (F_1)
		Area needed for col. $C_1 = \dfrac{600}{120} = 5\ \text{m}^2$
		Provide 1.5 times the area needed = 7.5 m^2
		2.5 m wide and 3.0 m along the centre line

Reference	Step	Calculation
		Hence, we fix the fulcrum at 3.0/2 = 1.5 m from the end near W_1, and so
		Distance of R_1 from $W_1 = L_1 = 1.5 - 0.3 = 1.2$ m from W_1
		Distance of W_2 from $R_1 = L_2 = 5 - 1.2 = 3.8$ m
		Take moments about or W_2. Then,
		$3.8R_1 = 600 \times 5$ or $R_1 = 780$ kN ($> W_1$)
		$R_2 = 600 + 900 - 780 = 720$ kN ($< W_2$)
	2	*Find FS against overturning using characteristic loads*
Eq. (8.1)		$\text{FS} = \dfrac{W_2L_2}{W_1L_1} = \dfrac{900 \times 3.8}{600 \times 1.2} = 4.75 > 1.5$
	3	*Find dimension of footing for R_2 so that CG of loads and areas coincides*
		The size of the square foundation F_2 for R_2 is given by
		$R_2 = \sqrt{\dfrac{R_2}{\text{SBC}}} = \sqrt{\dfrac{720}{120}} = 2.5 \text{ m}$
		Adopt $F_2 = 2.5 \text{ m} \times 2.5 \text{ m}$
Sec. 8.4		*Recalculate necessary breadth of footing F_1 so that CG of loads and areas of footings coincides.*
		$x_1 = \text{CG of loads} = \dfrac{W_1 \times 5}{W_1 + W_2} = \dfrac{600 \times 5}{600 + 900} = 2$ m from W_2
		Let us find the CG of areas we have assumed.
		Area of $F_2 = A_2 = 2.5 \times 2.5 = 6.25$ m^2. Find A_1 required for CG to be same as that of loads.
		$x_2 = \dfrac{A_1 \times 3.8}{A_1 + A_2} = \dfrac{A_1 \times 3.8}{A_1 + 6.25}$ should be 2 which gives $A_1 = 6.94$
		Breadth for length 3 m $= \dfrac{6.94}{3} = 2.3$ (2.5 m × 3 m is OK)
	4	*Calculate uniform pressure for factored load*
		$q = \dfrac{1.5 \times 600 + 1.5 \times 900}{6.94 + 7.5} = 156 \text{ kN/m}^2 < 1.5 \times 120 \text{ (SBC)}$
		Assume a strap beam wide rigid strap beam run on top of the footings connecting them.
	5	*Design of footing F_1 . 3.0 × 2.5 m in size*
		The depth of the footing slab required depends on the cantilever BM of the slab on both sides of the rigid strap beam.
		Cantilever overhang of the foundation slab $= \dfrac{2.5 - 0.50}{2} = 1$ m on both sides of the beam.

Reference	Step	Calculation
		$$\text{B.M.} = \frac{156 \times 1 \times 1}{2} = 78 \text{ kNm/m}$$
		$$\text{Depth } d = \left(\frac{78 \times 10^6}{0.14 \times 20 \times 1000}\right)^{1/2} = 167 \text{ mm}$$
		We can slope the slab D = 350 mm near the beam to at the edges. (Find steel D = 350 mm.)
		$$d = 350 - 75 = 275 \text{ mm}$$
Table B.1		$$\frac{M}{bd^2} = \frac{78 \times 10^6}{1000 \times 275 \times 275} = 1.03; \; p = 0.5\% \text{ steel}$$
		Check for one way shear
		d at 275 mm from the face of the strap beam
		$$= 275 - \frac{125 \times 275}{1000} = 241 \text{ mm}$$
		$$V = 156 \times (1 - 0.30) = 109 \text{ kN/m}$$
		$$\tau = \frac{109 \times 10^3}{1000 \times 241} = 0.45 < \tau_c, \; \text{for } 0.5\% = 0.48$$
	6	*Design of strap beam*
		Draw SF along the length. From the SF diagram of the beam, it is clear that the max design BM is at the junction between the beam and where only the beam slab of F_1 section resists bending.
		$$M = 900 \times 2.85 - \frac{(2.5 \times 156) \times (3)^2}{2} = 810 \text{ kN/m}$$
		Let width of the beam be 500 mm
		$$d = \sqrt{\frac{810 \times 10^6}{0.14 \times 20 \times 500}} = 760 \text{ mm}$$
		$$D = 750 + 75 + 10 = 835. \text{ Adopt } 835 \text{ mm}$$
		Find the percentage of steel as in other cases and detail the beam.
	7	*Design of footing F_2*
		Design the footing F_2 as we designed F_1

EXAMPLE 8.2 (Design of Type 1 balanced footing with fulcrum near the external column using a concrete counter weight)

An external column 250 × 250 mm carrying 120 kN is to be balanced by a counter weight 4.4 m from the edge of the column as there are no adjacent internal columns. Assuming bearing capacity = 140 kN/m^2, give a layout of the system.

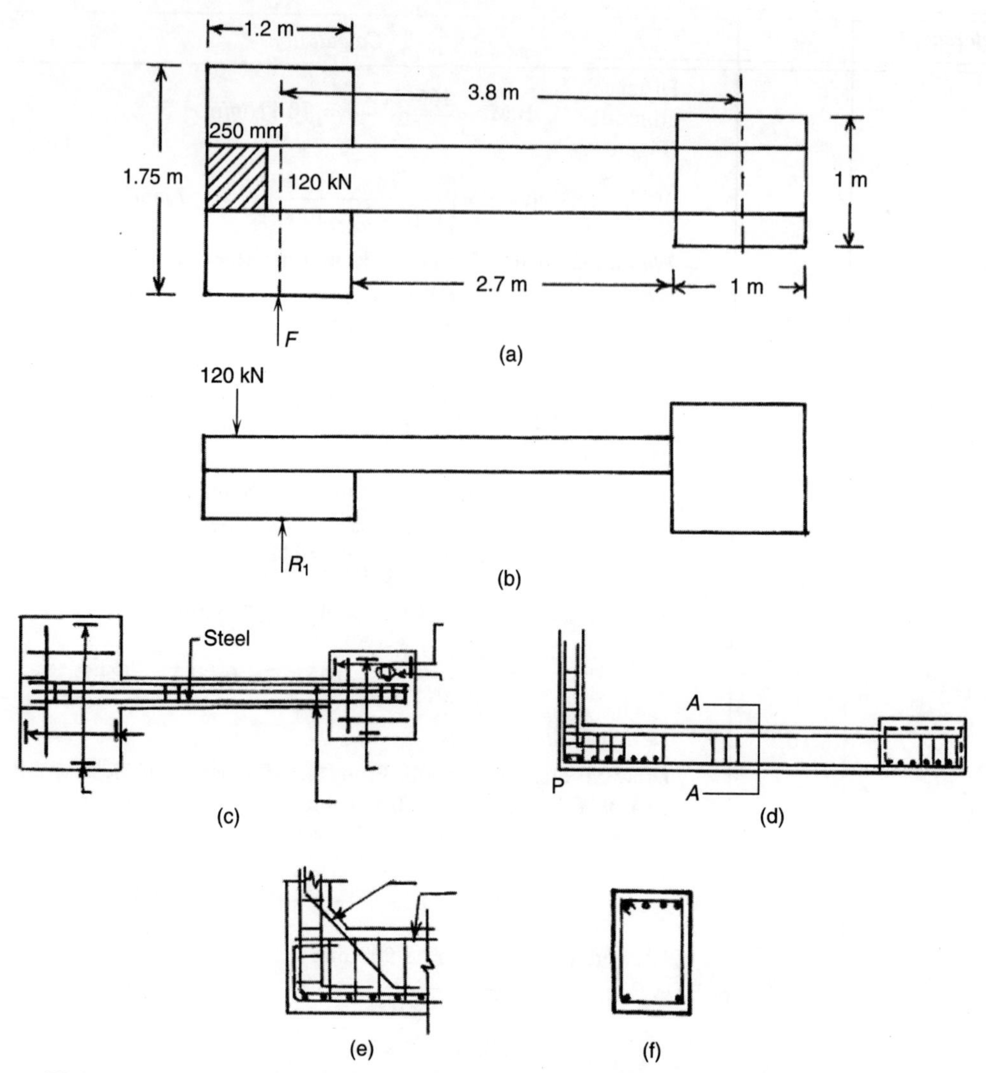

Figure E8.2 Type 1. Balanced footing balanced by a dead weight: (a) Plan, (b) Action of lever, (c) Details of steel-plan, (d) Details of steel section, (e) Detail at P, (f) Section A-A.

Reference	*Step*	*Calculation*
	1	*Estimate dimension of footing F_1* Area needed $= \dfrac{120}{140} = 0.85\ \text{m}^2$ Assume thrice the area = 3 × 0.85 = 2.55 m^2 Assume 1.2 × 1.75 m = 2.1 m^2 (1.2 m lengthwise)

Reference	Step	Calculation
		Distance of the fulcrum = 0.6 m from the end. (We can assume this arbitrarily also.) Reaction R_1 has to carry all the vertical loads (Figure E8.2)
		Value of lever arm length,
		L_1 for $R_1 = 0.6 - 0.125 = 0.475$ m
	2	*Find a suitable lever arm length L_2 for equilibrium in terms of L_1*
		For 15 kN, a reasonable value for the counter weight
		Find the lever arm length L_2 for 15 kN:
		$L_2 = \frac{120 \times 0.475}{15} = 3.8$ m from R_1
		Note: FS against rotation is given by
		FS = $\frac{3.8 \times 15}{120 \times 0.475} = 1$ (only). [We can also increase L_2 or W_2 to give a higher FS]
	3	*Find base resistance*
		Assuming 1 m × 1 m in plan F_2
		Depth of the balancing base $= \frac{15}{1 \times 1 \times 24} = 0.625$ m
		Self wt. of the footing at F_1 assuming 500 mm depth of the footing (0.5 m) is obtained as
		Self wt. of F_1 = vol × 25 = (2.1 × 0.5) × 24 = 25 kN
		Self wt. of the beam assuming 300 mm wide 450 mm deep and length equal to 2.7 m = 0.3 × 0.45 × 2.7 × 24 = 8.7 kN
		The total wt. of load and foundation is
		120 + 25 + 8.7 + 15 = 168.7 kN
		Pressure = $\frac{168.7}{2.1}$ = 80 kN/m^2 < 140 kN/m^2
	4	*Design for footing F_1 the fulcrum (with factored loads)*
		Forces acting on F_1 are the following, except wt. of F_1
Step 2		Factored vertical load = 1.5 (120 + 8.7 + 15) = 216 kN
		The only unbalanced moment about R_1 is due to the wt. of the beam
		= (1.5 × 8.7) 1.90 = 24.8 kNm
		Design the footing for these loads by finding the ground pressures under the above loads and moment as in Example 8.1.
	5	*Design of strap beam*
		Max moment to be transmitted will be that at the junction between the footing and the beam and is moments on the right hand side

Reference	Step	Calculation
		$M = (1.5 \times 15) \times 3.2 + (1.5 \times 8.7) \times (2.7/2) = 89.6$ kNm
		Find the depth required
		Find also the area of steel required and check for anchorage length.
		Check for shear
		$V = 22.5 + 13 = 35.5$ kN
Notes Table B.7		Provide shear reinforcement for beam. (All beams should be designed for shear and at least nominal stirrups should be provided even if the shear stress is low)
	6	*Construction of counter weight*
		The mass of concrete is to be built as the construction proceeds. Mesh reinforcement should be provided to take care of homogeneity and shrinkage.

EXAMPLE 8.3 (Design of Type 2 balanced footing with the fulcrum near the internal column using an eccentric internal footing)

An outer column 250 × 250 mm carrying a characteristic load of 120 kN has to be balanced with an inner column 300 × 300 carrying 700 kN. The columns are spaced 5 m apart. If the safe bearing capacity of the soil is 140 kN/m^2, give a layout of the system and indicate how to design the whole system.

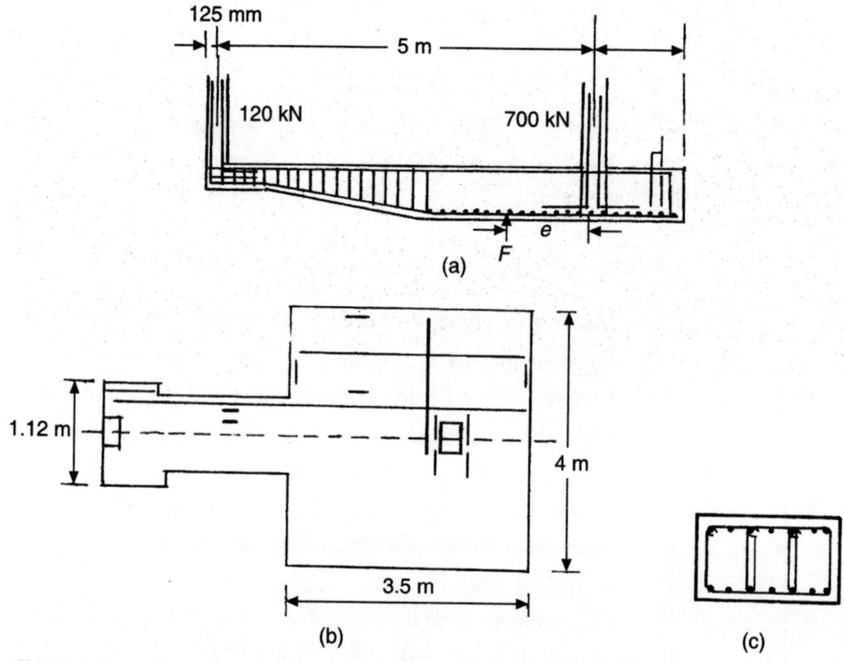

Figure E8.3 Type 2. Balanced type footing: (a) Longitudinal section showing fulcrum F, (b) Plan, (c) Section of cantilever.

Reference	Step	Calculation
		(As the inner column is a heavy column compared to the outer, we can balance the outer footing by eccentric placement of the inner footing.) Type 2 design.
	1	*Find eccentricity required and size of inner footing under* C_2
		$W_1 \times 5 = (W_1 \times W_2)\ e$ and $W_1 + W_2 = 120 + 700 = 820$ kN
		$e = \dfrac{120 \times 5}{820} = 0.73$ m
		The eccentricity to be $< L/6$. Hence the total length
		$L = 6 \times 0.73 = 4.38$ m
		As the dead load of the footing acts through the CG of the footing and it also contributes to the stability, the needed length of the footings can be less than 4.38, depending on the dead weight of the footing. Let us *assume* that due to this reduction, the breadth required is 3.5 m only with $e = 0.73$ m = 730 mm
		In this cantilever footing, the pressure under the footing will not be uniform because of the eccentricities. The base area required for ($W_1 + W_2$ + self-weight) will be much more than that required under uniform pressure. Assume area required as 2.5 times than that required with ($W_1 + W_2$) only.
		$A = \dfrac{820 \times 2.5}{140} = 14.6 \approx 14\ \text{m}^2$
		Assume 3.5 × 4 m eccentrically loaded footing for W_2
	2	*Layout of footing under* C_2 *interior column footing for* F_2
	2.1	*Calculate design pressure assuming uniform pressure*
		The whole load of the system also acts on the footing of size 3.5 × 4 m = 14 m^2
		(a) DL of the footing = (3.5 × 4 × 0.6 (assumed) × 24) = 200 kN
		(b) Load from cantilever beam = 30 kN (assumed)
		(c) Total load at R_2 = 700 + 230 = 930 kN = W_3
		(d) Load from W_1 = 120 kN
		Factored load = 1.5 (930 + 120) = 1395 + 180 = 1575 kN
		$q = \dfrac{1575}{14} = 112.5\ \text{kN/m}^2$
	2.2	*Find depth from bending shear with all loads* (τ_c = 350 kN/m^2)
Eq. (2.7)		$d = \dfrac{q(L-b)}{700+2q} = \dfrac{112.5\,(4-0.3)}{700+(2\times 112.5)} = 0.45\ \text{m}$ = Adopt D = 600 mm
		$d = 600 - 75 - 10 = 515$ mm

Reference	Step	Calculation
	2.3	*Find depth for punching shear for* C_2 (t_p = 1.12 N/mm^2)
Example 2.5		Punching shear for W_2 only
		Shear resistance = $(4a + 4d)\ t_p = 4(300 + 515) \times 1.12 = 3651 > 1395$
Step 2.1 above		Hence, safe. (Note: $R_2 \times 1.5 = 930 \times 1.5 = 1395$ kN)
	2.4	*Find soil pressures with characteristic loads with eccentricity*
		Direct load $\frac{D}{A} = \frac{930 + 120}{14} = 75 \text{ kN/m}^2$
Step 1		Eccentricity = 730 mm (with $z = bd^2/6$)
		Bending stress due to W_2 only $= \frac{(700 \times 0.73)6}{4 \times (3.5)^2} = \pm 63 \text{ kN/m}^2$
		max pressure = 75 + 63 = 138 < 140 kN/m^2
		min pressure = 75 – 63 = 12 kN/m^2
	3	*Layout of footing* F_1 *under* C_1 *to carry load to the beam*
	3.1	*Estimate dimensions*
		Column = 250 × 250 mm — load 120 kN
		Generally, as a rule, we place the column on a footing of the same bearing capacity.
		Area needed $= \frac{120}{140} = 0.9 \text{ m}^2$ approx.
		Adopt a footing over the cantilever beam 750 mm wide along the beam, and depth 300 mm.
		Adopt a beam 1000 mm long.
	3.2	*Determine the various weights to find* R_1 and R_2 *with characteristic loads*
		(a) Weight of the exterior footing = Area × depth × unit wt.
		= 0.9 × 0.3 × 24 = 6.5 kN; CG at the centre of the footing
		(b) Wt. of the beam $= \left(\frac{0.6 + 0.3}{2}\right) \times 1 \times 24 = 10.8 \text{ kN}$
Figure E8.3		CG from $B_2 = \frac{L}{3}\left(\frac{2B_2 + B_1}{B_1 + B_2}\right) = \frac{1895}{3}\left(\frac{0.6 + 0.6}{0.3 + 0.6}\right) = 842 \text{ mm}$
		(c) Wt. of the footing F_2 = 3.5 × 4.0 × 0.6 × 24 = 202 kN
		(d) Col. W_1 = 120 kN at (375 – 125 = 250 mm from R_1)
		(e) Col. W_2 = 700 kN at 730 mm from R_2
		Taking moments about R_2
		$R_1 = \frac{(6.5 \times 4.02) + (10.8 \times 2.59) + (120 \times 4.27) - (700 \times 0.73)}{4.02}$

Reference	Step	Calculation
		= 14 kN (approx.) which is small. So also the pressure.
		ΣW = 120 + 6.58 + 10.8 + 202 + 700 = 1039.3 kN
		$R_2 = (\Sigma W - 14)$ = 1039.3 − 14 = 1025 kN
		For factored load condition
		R_1 = 1.5 × 14 = 21 kN (propping force)
		R_2 = 1.5 × 1025 = 1538 kN
	4	*Structural design of the cantilever beam*
		The cantilever is fixed on to F_2 and the maximum factored moment produced by (a) & (b) & (c)
		(a) Wt. of the footing = (1.5 × 6.5) × 2.27 = 22.13
		(b) Wt. of the beam = (1.5 × 10.8) × 0.842 = 2.27
		(c) Column load = (1.5 × 120) × 2.520 = 453.60
		Total = 478.00 kNm
		Design the cantilever for this moment
	5	*Design of footing* F_2
		Take F_2 as a support for the cantilever with
		(a) Maximum moment due to DL and LL = 478 kNm (= M) (anticlockwise)
		Minimum moment due to DL only (to be calculated)
		(b) In addition, we have self wt of F_2 = 1.5 × 202 = 303 kN
		(c) Wt of eccentric column C_2 = 1.5 × 700 = 1050 kN
		Total = 1353 kN(= P)
		(d) Moment due to C_2 = 1050 × 0.730 = 766 kN (clockwise)
		Net moment with (DL & LL) = 766 − 478
		= 288 kNm clockwise
		Also, FS against rotation $= \frac{766}{478} = 1.6$ (OK)
		Design footing F_2 as a footing subjected to load and moment as in Chapter 6.
	6	*Check the whole system as a single footing*
		The whole structure can be taken as a single footing with the external loads. Check the base pressures by using
		$q = \frac{P}{A} \pm \frac{M}{I} y$, where $I = \sum Ar^3$
		$M = P \cdot e$
		where e is the eccentricity between the centroid of the whole area and the CG of all loads acting on the structure.
		Detail the steel
		(See Figure E8.3)

EXAMPLE 8.4 (Design of Type 3 balanced footings where both columns should not exert any pressure below)

Figure E8.4 shows the section of a car shed to be constructed between two buildings and for which excavation cannot be carried out under the outer walls. Give a layout for the foundation for construction of the building.

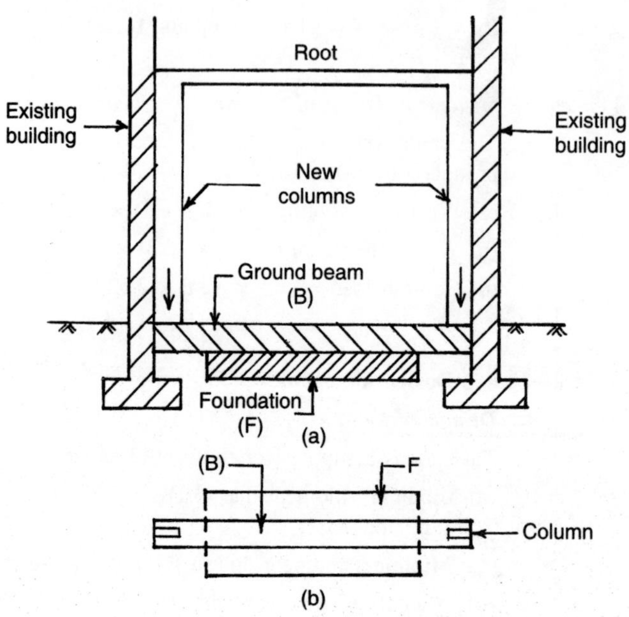

Figure E8.4 Type 3. Cantilever type of footing: (a) Section, (b) Plan.

The layout is shown in Figure E8.4. Design can be carried out by using the principles explained in this chapter.

EXAMPLE 8.5 (Design of a balanced cantilever foundation on piles)

Refer Example 17.5 dealing with pile caps.

Strip Footings under Several Columns

9.1 INTRODUCTION

In Chapter 5, we dealt with wall footings which are strip footings under continuous walls. In this chapter, we will deal first with strip footings under equally loaded and equally spaced columns (producing uniform ground pressure) and also strip footings under unequal column loads.

In places where we expect large settlement, it may be better to provide individual footings even if it touches each other as otherwise the reinforcement needed for continuity will be large. However, in good soils continuous footings are useful as they can later be used as support for connecting beams for foundations of internal partition walls inside the building (see Sec. 5.6). They are also useful to bridge over soft spots in the foundation. These footings can be of rectangular section with or without shear steel or more often of T-beam or U-beam section as shown in Figure 9.1. (Elastic settlement of raft foundation is given in Sec. 14.5)

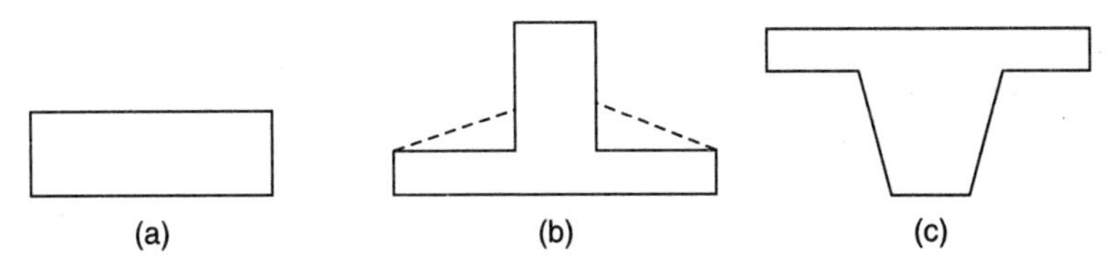

Figure 9.1 Types of strip foundations: (a) Slab, (b) T beam, (c) U beam.

9.2 DESIGN PROCEDURE FOR EQUALLY LOADED AND EQUALLY SPACED COLUMNS

A rectangular slab can be used as a strip foundation under these columns. If the strip is for equally spaced and equally loaded columns, the contact pressure can be assumed as uniform if the CGs of the load and the footing coincide. Usually, an end overhang will reduce the internal moments. A cantilever equal to span/$\sqrt{8}$ (or 0.354 times the span), if provided, can reduce the end moments for a two-column footing equal to the span moment. (This principle is usually adopted for many other arrangements in structures).

The bending moment values for cases up to 5 column loadings (4 spans), as given by the BS Code and as shown in Figure 9.2, can be used for design. Alternatively, moment distribution with 30% redistribution of moments to even out the positive and negative moment can be also used.

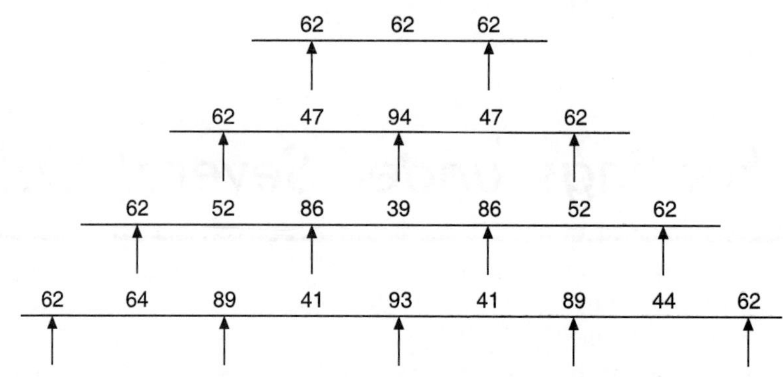

Figure 9.2 Elastic bending moment coefficients (× 10^3) for UDL (w), for equal spans with cantilever length = 0.354 × inner span. (BM = (coeff.) (w) l^2 without redistribution). [*Note:* For a single span with cantilever on both sides the cantilever length will be (1/4.83) × total length]

9.3 ANALYSIS OF CONTINUOUS STRIP FOOTING FOR UNSYMMETRIC LOADING

There are cases where a number of unequally loaded columns have to be founded on a reinforced slab or slab and beam foundation. A slab with an upstanding T beam or downstanding U beam strip foundation is very suitable for column loads with irregular column spacing or varying column loads. The ribs will make the foundation more rigid and will be able to distribute the loads to the foundation more efficiently.

9.3.1 Analysis of Strip Footing with Unsymmetrical Loads

The following three methods are commonly used *for an approximate and quick analysis only* with unequal column loads. But the best and exact method is to treat it as a beam on an elastic foundation as described in Chapters 22 and 23.

Method 1: *Conventional method using varying pressures:* The conventional method is to treat the foundation as a rigid member and find the maximum and minimum pressure due to eccentricity of the CG of loads and CG of the strip foundation along the length from Eq. (9.1).

Pressure $$q = \frac{P}{A} \pm \frac{Pe}{Z}$$

where $Z = bL^2/6$
(b = breadth; L = length)

$$q = \frac{P}{b \times L}\left(1 \pm \frac{6e}{L}\right) \tag{9.1}$$

Having found the pressures at the extreme ends, assuming linear distribution of pressures, the bending moment and shear force diagrams can then be drawn by simple statics. This often leads to high BM to be provided for the strip. The method also requires the strip to be *very rigid* to produce the uniformly varying contact pressure.

Method 2: *Conventional method of adjusting CG of strip to obtain uniform pressure:* In this method, we plan the length of the cantilever at both ends so that the CGs of the loads and the strip coincide. Then we assume uniform pressure below the slab. The beam is analyzed as a continuous beam. In many cases, this may not be economical as this leads to large cantilever moments at the end of heavily loaded columns.

Method 3: *Alternate method:*[1] The bending moment diagrams obtained by methods 1 and 2 are far from that obtained by other exact methods such as the theory of beams on elastic foundations, as described in Chapters 22 and 23. Theoretically the ground pressures are distributed symmetrically about both sides of the load and the radius of influence extends only to four times the "elastic length", L_e. Hence we proceed as follows to get an approximate distribution of the ground pressures. This method has no theoretical basis but it is reported that it produces good results when used with good practice of detailing of steel.

First, we assume the loads are distributed uniformly in the immediate vicinity of the column loads, say, to both sides of the load up to the centre line of adjacent spans. (This assumes that the columns are spaced $>4L_e$.)

Secondly, we redistribute the above distribution so that *in each span* (between the column loads) it is uniform, i.e. in the spans the loads are averaged and made uniform as in Figure 9.3.

Thirdly, we analyze the strip as a continuous beam, each span being loaded with an averaged uniformly distributed load. We can use moment distribution or any other method. (As the whole method is an approximation, a rough value of $wl^2/10$ for both positive and negative moments is also a good approximation.)

[*Comments:* A theoretically exact method of analysis is to use the theory of beams on elastic foundation. It is, however, logical to imagine that there will be more pressure immediately below the load and the pressures will even out away from the load point. The above alternate method of design has been reported to have been used in soft clay of safe bearing capacity of about 75 kN/m^2 in and around Chennai and the performance of these foundations has been reported as satisfactory.]

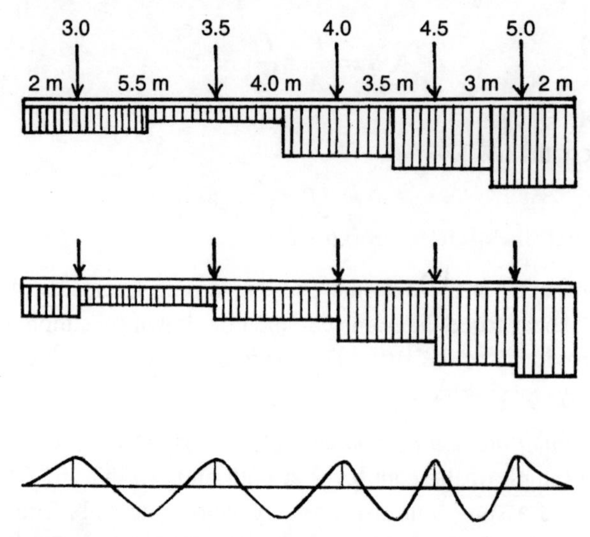

Figure 9.3 Approximate method of analysis of strip foundation under unsymmetric column loads (Method 3).

9.4 DETAILING OF MEMBERS

The general principles of detailing of the strip foundations as a slab, T beam or U beam are shown in Figure 9.4. Other details are to be incorporated as in ordinary continuous beams.

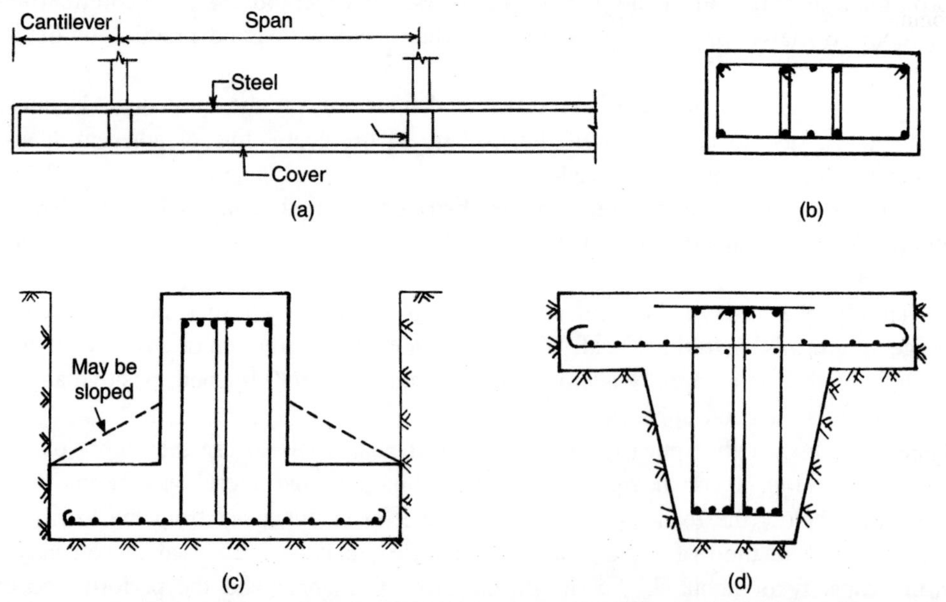

Figure 9.4 Strip foundation under many columns: (a) Arrangement with cantilever to reduce moments, (b) Rectangular slab section, (c) T beam section, (d) U beam section.

9.5 SUMMARY

Strip footings under a number of columns have many advantages. The design of these foundations for equally spaced and equally loaded columns is simple. An approximate method for varying loads and spans has been described in this chapter. Strip footings can be rectangular slabs or T beams or U beams.

EXAMPLE 9.1 (Analysis of a footing for a series of equally spaced, equally loaded columns)

A series of four 250 × 250 mm columns each carrying (DL + LL) of 500 kN are spaced at 4 m, as shown in Figure E9.1. Design a suitable continuous strip footing if the safe bearing capacity is 100 kN/m².

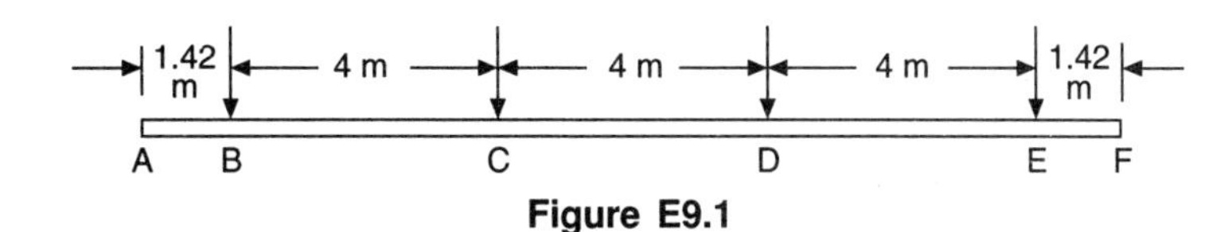

Figure E9.1

Reference	*Step*	*Calculation*
	1	*Plan the footing*
		Provide cantilever on both sides = 0.354L
Figure 9.2		Cantilever = 0.354 × 4 = 1.42 m (to balance BM)
(comments)		Total length = 14.84 m (Assume self wt. 10%)
		Breadth required $= \dfrac{2000+200}{100 \times 14.84} = 1.48$ m (BC = 100 kN/m²)
		Adopt 1.5 m width. As the CGs coincide, q is uniform.
	2	*Find design pressure with factored load*
		Factored soil pressure $q = \dfrac{1.5 \times 2000}{14.84 \times 1.5} = 134.8\ \text{kN/m}^2$
		Pressure for 1.5 m width = w = 134.8 × 1.5 = 202.2 kN/m
	3	*Find BM from Figure 9.2 for 4 supports*
		Ending moment M = coeff. (wL^2)
		$wL^2 = 202.2(4)^2 = 3235.2$ kNm for 1.5 m width
Figure 9.2		Using coefficients BM per metre width
		(Support) M_B = 0.0625 × 3235.2 = 202.2 kNm/m
		(Span) M_{BC} = 0.051 × 3235.2 = 165 kNm/m
		(Support) M_C = 0.086 × 3235.2 = 237.8 kNm/m
		(Span) M_{CD} = 0.039 × 3235.2 = 126.0 kNm/m
		We will determine steel after finding the depth necessary for one-way shear.

Reference	Step	Calculation
	4	*Find depth for one-way shear*
		Draw SF diagram assuming zero shear at the centre of the strip.
		Load in one panel = 202.4 × 4 = 808.8 kN
		With the loading of the column 750 kN, and the loading of 808.8 kN in the span, the SF diagram can be easily drawn (Figure E9.1).
		Max shear at end support = 750 – (202.2 × 1.42) = 463 kN
		V_c shear at d from the face of the support = 463 – 202.2(0.125 + d) kN
		Assuming τ_c = 0.35 N/mm² = 350 kN/m²
		B = breadth = 1.5 m; $Bd\tau_c = V_c$
		$(1.5 \times d) \times 350 = 438 - 202.2d$
		$d = \frac{438}{727.4} = 0.60 \text{ m} = 600 \text{ mm}$
		[By the formula from the point where $V = 0$ (at the extreme end)
		V_{max} = 463 kN. Assume τ = 0.35 N/mm² (350 kN/m²).
Step 2		$q = 134.8 \text{ kN/m}^2, \ \frac{a}{2} = 0.125, \ \frac{L}{2} = \frac{463}{202.2} = 2.29 \text{ m}$
		$d = \frac{q(L/2 - a/2)}{350 + q} = \frac{134.8(2.25 - 0.125)}{350 + 134.8} = 0.59 \text{ m}$
		Choose d = 600 mm]
	5	*Choosing the structure*
		We can design this foundation as a slab without shear steel. We may also proceed for design as a T beam with shear reinforcement as follows. The final choice will depend on which will be cheaper.
	6	*Layout as a T beam*
		Beam width to accommodate 250 mm, col. width = 300 mm
		max shear stress allowed for M_{20} concrete = 2.8 N/mm²
		For beam, let us adopt τ_c = 1.5 N/mm² = 1500 kN/m²
		Adopt the width of the rib as 300 mm = 0.3 m
		Taking shear at d of the beam from the face of the first column.
		463.2 – 202.2(0.125 + d) = 0.3 × d × 1500
		652.2d = 438 or d = 0.67 m
		Let D = 750 mm, d = 750 – 50 – 8 = 692 mm
		(Alternatively, we can choose d and design for shear in beams.)
		We design the footing as a T beam

Reference	Step	Calculation
		300 × 750 with 300 × 150 slab. The projection of the slab on either side is 600 mm < 3 times the thickness of the slab. We can design the beam as a U beam also.
		[*Note:* For design as slab only, we proceed with checking for punching shear and design of steel as in other examples. We do not check for punching shear in beams as we have to provide minimum shear steel in beams. The design of an inverted T or U beam is also carried out as in other examples.]

EXAMPLE 9.2 (Strip footing for columns with non-uniform loads)

A series of five columns is to be supported on a 20 m × 2 m strip foundation. Determine the SF and BM for design. Assume safe bearing capacity as 100 kN/m². The loads are 300, 350, 400, 450, and 500 kN at 2, 7.5, 11.5, 15 and 18 m from one end.

Reference	Step	Calculation
	1	*Find eccentricity of load*
		Taking moments about the left-hand end
		$\bar{x}$ of load $= \dfrac{(3\times 2)+(3.5\times 7.5)+(4\times 11.5)+(4.5\times 15)+(5\times 18)}{20}$
		$= 11.78$ m from G
		$\bar{x}$ of strip $= \dfrac{20}{2} = 10$ m from G
		$e = 11.78 - 10.0 = 1.78$ m to the left
	2	***Method 1: Analysis with eccentricity***
	2.1	*Find pressure variation assuming the strip is rigid*
		Assume varying pressure due to eccentricity to find maximum and minimum pressures from characteristic loads.
Eq. (9.1)		Ground pressure $= \dfrac{2000}{20\times 2}\left(1+\dfrac{6\times 1.78}{20}\right)$
		$= 76.81$ and 23.19 kN/m²
		Maximum pressure is less than the bearing capacity 100 kN/m² and the pressure is not negative anywhere in the base (Assume self-weight of slab is ≈ 20 kN/m²). We calculate loads and moments per metre width and draw BM and SF diagrams.
		Factored design pressure = 1.5 × (service loads) = 1.5 × 76.81
		= 115.22 and
		1.5 × 23.19 = 34.79 kN/m² with 2 m width.
		q_1, q_2 = 230 and 70 kN/m per metre length.

Reference	Step	Calculation
	2.2	*Find the SF diagram due to factored load*
		We find SF by summation on forces from the left to the right. We proceed from the left end to the right end and then from the right end to the left end and check whether these totals tally.
	2.3	*Find BM diagram due to factored load*
		By taking moment of forces to the left of section till the right end and from the right of section to the left end, we check whether they also tally.
	3	***Method 2:* By adjusting the footings to have zero eccentricity**
	3.1	*Find cantilever necessary at ends to make CGs coincide*
		CG of loads = 11.78 – 2.0 = 9.78 m from RHS load
		CG of total length of strip between loads $= \dfrac{16}{2} = 8$ m from G
		Eccentricity = 9.78 – 8 = 1.78 m to the left
		We have to give a cantilever of at least 0.5 m to the right of F. Hence, we have to give a cantilever of (2 × 1.78) + 0.5 = 4.06 m to the left of B.
		Now CG of strip = (0.5 + 16 + 4.06)/2 = 10.28 from load E
		= (20.56)/2 – 0.5 = 9.78 from load E
		As the CGs of loads and foundation coincide, we can assume uniform ground pressure.
	3.2	*Find uniform pressure*
		$w = \dfrac{2000}{2 \times 20.56} = 48.63 \text{ kN/m}^2$
		Factored load = 1.5 × 48.63 = 72.9 kN/m
	3.3	*Find SF and BM due to the uniformly distributed load for 2 m width*
		Note: There is a cantilever length of 4 m, which will give a large bending moment.
	4	***Method 3:* Alternate Empirical Method**
	4.1	*Distribute the column load symmetrically to both sides of the load on to the strip (2 m wide)*
		Column A: $w_1 = \dfrac{3}{2 + 2.5} = 0.67 \text{ kN/m}$
		Column B: $w_2 = \dfrac{3.5}{(2.5 + 4.25)} = 0.52 \text{ kN/m}$
		Similarly, divide all the loads between the distances between centre lines of adjustment spans.

Reference	Step	Calculation
	4.2	*Convert these into uniformly distributed load in various spans by finding the averages* Cantilever load = 0.67 kN/m $\text{Load span AB} = \dfrac{0.67 \times 2.5 + 0.52 \times 2.5}{5.0} = 0.59 \text{ kN/m}$ Proceed to all the spans similarly. Check also whether ground reaction = downward load (0.67 × 2) + (0.65 × 5) + (0.81 × 4.5) + (1.19 × 3.5) + (1.52 × 3.0) + (1.67 × 2.0) = 20 kN (characteristic loads)
	4.3	*Analyze the beam for factored load by moment distribution* Find the shear force diagram and BM diagram and design for shear and bending moment.
	4.4	*Alternate approximate method of analysis* Assume the BM as ± $w(l^2/10)$ at support and midspan without distribution for each span and find the area of steel.
	5	***Design of strip footing from BM and SF diagrams***
	5.1	*Design for BM* The bending moments in the above calculations are along the centre line of the column. Design the reinforcements based on the moments at the face of the column.
	5.2	*Design for one-way shear* If the depth obtained for the shear corresponding to the percentage of tension steel is large, reduce the depth and provide the necessary shear steel. Otherwise, adopt T or U beam.
	5.3	*Design for punching shear in slab foundation* Find the punching shear around the most heavily loaded column. The depth provided should be in safe punching shear.
Sec. 7.5	5.4	*Design transverse reinforcement for column strip around columns* If a slab is adopted, the distribution of the column load transversely through the column strip (distance = column + 0.75d) on either side of the column should be designed as explained in combined footings. [*Note:* The strip foundation can be in the form of a slab or a T or U beam with slab. If it is a T or U beam, it should be designed as a beam and transverse reinforcement in the slab should be designed for cantilever action.]

REFERENCE

[1] Veerappan, A. and A. Pragadeeswaran (Eds.), *Design of Foundations and Detailing,* Association of Engineers, Tamil Nadu Public Works Department, 1991.

Raft Foundations

10.1 INTRODUCTION

Raft foundations are mostly used on soils of low bearing capacity, where the foundation pressures are to be spread over a large area. They are also used in places where the foundation soils are of varying compressibility and the foundation has to bridge over them. The geotechnical aspects for selection of the type of raft to be used are described in books on soil mechanics and foundation engineering [1][2]. However, the following points which are very important in the layout of raft foundations must be remembered by all designers:

1. First, the *differential settlement of raft foundations in a given noncohesive soil is found to be only half that of footings in the same soil designed with same intensity of loading*, so that from consideration of allowable settlement the bearing capacity allowed from settlement consideration for raft foundation is double the value allowed in footings. (We may also remember the thumb rule that the safe bearing capacity of *footings* in kN/m^2 for total settlement of 25 mm in sandy soils is roughly 10 times the N value of the soils. Also, for footings designed for maximum settlement of 25 mm, the *expected differential settlement* is ¾ of 25 = 18 mm. Hence, for the safe bearing capacity of rafts equal to twice that of footings, the differential settlement will be of the order of 18 mm incohesionless soils. In cohesive soils the above thumb rule refers only to the safe bearing capacity and not the allowable bearing capacity which has to take settlements also into consideration.)

2. Secondly, the design load to be considered on rafts is only equal to the characteristic load minus the weight of the soil excavated. *This is a very important conception in raft design.* Thus, every metre of soil excavation reduces the load on the raft by about 18 kN/m^2 (the density of the soil). [We should also note that the total characteristic load (DL + LL) from each floor of a multi-storeyed residential flat is only about 10 kN/m^2, so that one metre excavation is nearly equivalent to load from two floors.]
3. Thirdly, rafts have large capacity to bridge over soft spots in a non-homogeneous foundation. So, in non-homogeneous soils such as fills it gives better results than any other shallow foundation.
4. The expected maximum settlement of rafts in clay soils should be calculated by the theory of consolidation explained in books on soil mechanics.

10.2 RIGID AND FLEXIBLE FOUNDATIONS

We have already seen in Chapter 1 that rigid foundation (e.g. a footing) is one where we can assume to settle uniformly under a load, the settlement of all the parts being the same. If we assume soil as a set of independent springs (Winkler Model), then the contact pressure will also be uniform. (It can be found from books on soil mechanics that this is not true. The pressures at the periphery will be found to be larger than at the centre in elastic soils such as clay. With time and increasing loads, it may even out to some extent.)

On the other hand, in a flexible beam there will be more settlement under the loads, so pressures will be varying under the foundation. If we have a raft on solid rock, the load will be transferred to the rock by dispersion at 45 degrees and the effects will be felt only around the loads. We will examine this problem in more detail in Chapter 22 on Beams on Elastic Foundation. According to the above behaviour, there are two models that we can use for the design of raft foundations, namely

1. The classical rigid model assuming uniform contact pressure under a raft when its CG coincides with the CG of loads and linear distribution in eccentrically loaded rafts.
2. Modern flexible models using the theory of soil structure interaction, depending on the rigidity of the raft.

Design by both methods is only an approximation of the real behaviour. Any raft cannot be considered as fully rigid. Also, any theoretical soil structure interaction in the absence of exact soil data (which is very difficult to get for soils covering a large area) is an approximation. However, the second method, by all means, can be considered more exact than the first. It was because of these uncertainties regarding magnitude and sign of bending moment at various points that we had a provision in earlier National Building Code of India as follows:

1. Rafts designed on empirical methods should be reinforced with equal amounts of the calculated steel on top and bottom (as the sign of the moment as calculated may not be the same in the field). Thus, we use two times the area of steel ($2A_s$) calculated.
2. Rafts designed by more exact flexible methods can be reinforced with total of $1.5A_s$, with 50% steel provided on the side opposite to that required by calculation.

These requirements have been discontinued in the code [3] but many practise them in their designs even today for safety.

10.3 COMMON TYPES OF RAFTS

The commonly used types of rafts are the following:

1. Simple plain slab rafts (Flat plate rafts for light weight buildings)
2. Flat slab rafts for framed buildings
3. Beam and slab rafts for framed buildings
4. Cellular rafts
5. Piled rafts
6. Annular rafts
7. Strip rafts forming grid rafts (Grid foundations)

The basic principles involved in their selection are discussed in the following sections.

10.3.1 Plain Slab Rafts for Lightly Loaded Buildings

In *lightly loaded buildings* in poor soils such as fills for light residential buildings, plain reinforced slabs can be used to spread the superload to a large area. In most cases, the slab is used as the ground floor of the building also. They can be as shown in Figure 10.1.

(a) R.C. slab only
(b) R.C. slab with edge beam [edge beam can be above or below the raft]
(c) R.C. slab with edge beams and cross beams below wall loads

Thickening the slab or providing a beam at edges adds to its stiffness and if planned downstanding, it also helps in preventing erosion of soil below the raft.

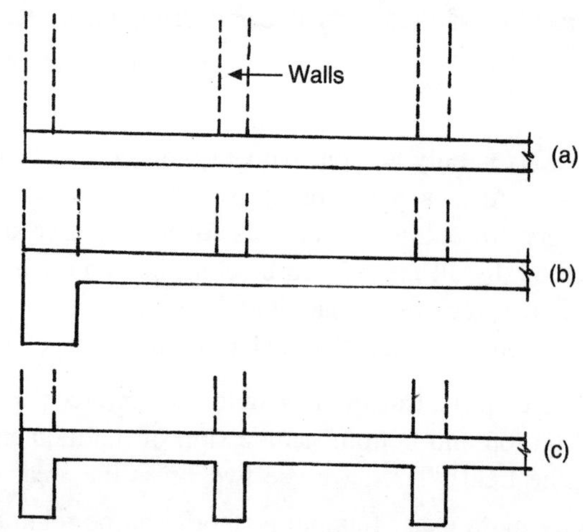

Figure 10.1 Simple reinforced concrete mats for light buildings: (a) Simple mat, (b) Mat with side beams, (c) Raft with beams under walls.

10.3.2 Flat Slab Rafts for Framed Buildings—Mat Foundation (Figure 10.2(a))

Flat slab rafts consist of concrete slab with constant thickness over the plan area. They are also called *mat foundations*. Flat slab rafts are used in framed buildings *where the column spacings are not large and are uniform*. These are no beams as in "beam and slab rafts" in them *except the edge beams*. (We should always provide edge beams all around the periphery of these types of rafts for reasons explained in Sec. 10.7 and Chapter 11.) The soil under these mats should not be very compressible. Larger stresses are produced in deeper depths in large rafts so that in clay soils consolidation settlement also increases with the size of the raft. In ideal flat slab rafts, internal column spacings should be uniform and the loading should also be symmetrical.

10.3.3 Beam and Slab Rafts (Figure 10.2(b))

We choose beam and slab rafts under the following conditions:

- When the slab depth in flat slab construction is excessive as it happens when the spans are large.
- When the arrangement of the column loads is irregular.
- When we expect large variation in the nature of the foundation soil under the plan area.

These rafts can be considered more rigid than the flat slab rafts and they are the most commonly used types of raft foundation in practice.

10.3.4 Cellular Rafts (Figure 10.2(c), (d))

In very weak soils, if the depth of soil excavation to the bottom of the raft is large enough to offset the superload, then no pressure will be exerted on the foundation. Such foundations can be partly compensated or fully compensated. Fully compensated foundations are called *floating foundations*. We need special rigid rafts of cellular construction or other types of rigid construction to carry the large loads to these foundations and also withstand the side earth pressures. Cellular rafts are used in these situations.

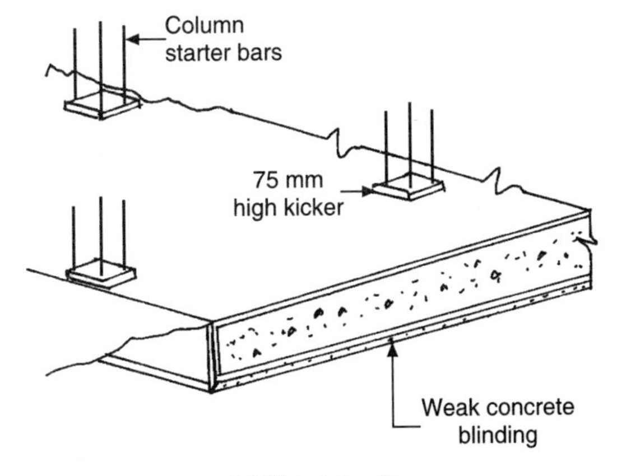

(a) Flat slab raft

Figure 10.2 (Contd.)

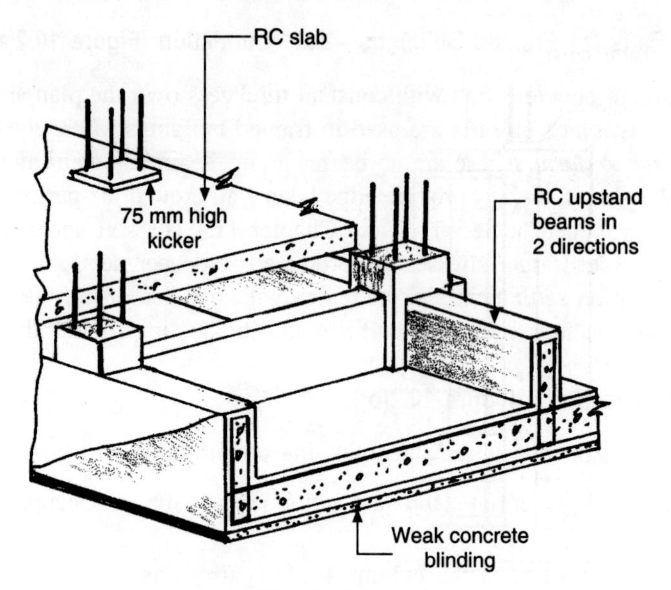

(b) Beam and slab raft

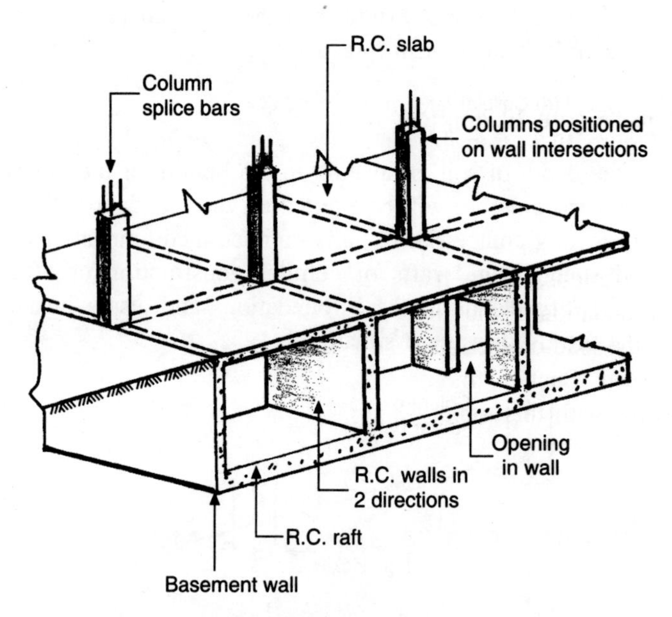

(c) Cellular raft (general view)

Figure 10.2 (Contd.)

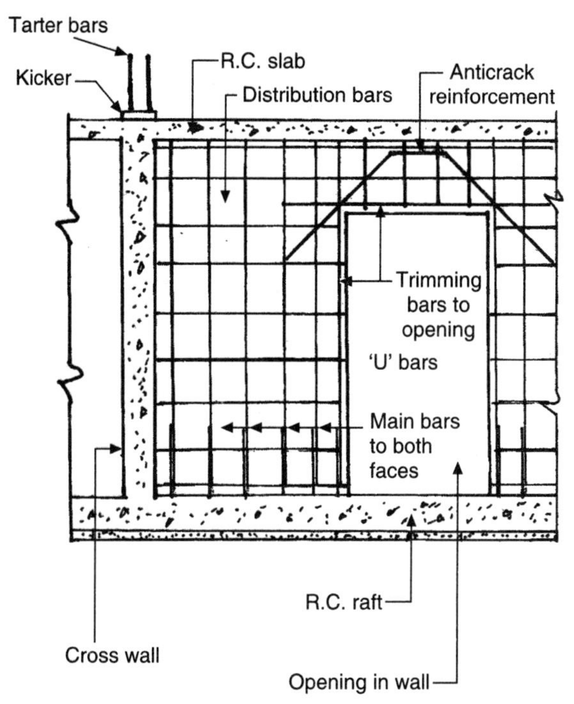

(d) Cellular raft showing detailing of wall

Figure 10.2 Conventional types of raft foundations.

10.3.5 Piled Rafts

In conventional pile foundations, all the loads are assumed to be taken by the piles. However, when designing a raft foundation, if the raft has enough bearing capacity but there is excessive settlement, then part of the load on the raft can be relieved by installing a few piles so that raft settlement can be reduced. Such a foundation is called a *combined piled raft foundation* (CPRF). Chapter 14 deals with this topic.

10.3.6 Annular Rafts

In some cases such as tall towers, we prefer to have annular rafts. Chapter 15 gives a brief summary of it.

10.3.7 Grid Foundation

When the bearing capacity of the soil is good and there is no need to cover the whole plan area, we use strip rafts which usually form a grid foundation in the *X*- and *Y*-directions. We deal with these types of foundations in Chapter 25.

We will discuss about each of these types in subsequent chapters.

10.4 DEFLECTION REQUIREMENTS OF BEAMS AND SLABS IN RAFTS

IS 456, Cl. 32.2 defines effective span as follows:

1. For simply supported beam or slab, it is taken as the clear span plus effective depth or centre to centre of support, whichever is less.
2. For a continuous beam or slab where the width of support is less than 1/12th span, the effective span is taken as the clear span.

IS 456, Cl. 23.2.1 gives the deflection requirements as length/effective depth or *L/d* ratio for beams for roofs and floors of buildings. They can be assumed to be applicable for beams for spans up to 10 m. The following span/effective depth ratios are to be maintained for beams *in floors and roofs of buildings* (beams usually have 1% of steel):

Cantilever: 7
Simply supported members: 20
Continuous members: 26

However, as roof and floor slabs usually contain low percentage of steel (less than 0.5%), these ratios for slabs can be multiplied by a correction factor. Hence, IS 456, Cl. 24 allows the following ratios for slabs subjected to LL *loadings up to* 3 kN/m^2 and spans up to 3.5 m using Fe 415 steel for floors and roofs:

Simply supported slabs: 28
Continuous slabs: 32

For two-way slabs, the shorter span is taken for calculating the above ratios. For flat slab roofs, these ratios are to be multiplied by a factor 0.8; thus for continuous flat slabs, the span depth-ratio should be $32 \times 0.8 = 25.6$ only.

However, caution should be taken in using these ratios to rafts. These span effective depth ratios are not applicable to rafts. It should be clearly noted that these ratios are for roof slabs, where the characteristic loading is only of the order or less than 10 kN/m^2 (5 kN/m^2 DL, 2 kN/m^2 for finish and 3 kN/m^2 LL). The total load *of each floor* (including slabs, beams and columns) will be of the order of 15 to 25 kN/m^2. As an ordinary raft has to support 3 to 4 floors at least, the contact pressures under rafts will be of the order of 75 to 100 kN/m^2. Hence, care should be taken to choose proper span/depth ratios for raft members. We will have to choose larger depths and hence much lesser span/depth ratios as explained in Chapters 11 and 12.

10.5 GENERAL CONSIDERATIONS IN DESIGN OF RIGID RAFTS

The following principles are used in the design of rigid rafts by conventional methods:

1. The raft plan dimensions should be made in such a way that, if possible, the centre of gravities of the raft and the loads should coincide (Figure 10.3). In such cases, we assume that the contact pressure is uniform. Otherwise, the pressures under the foundation will vary under the raft.

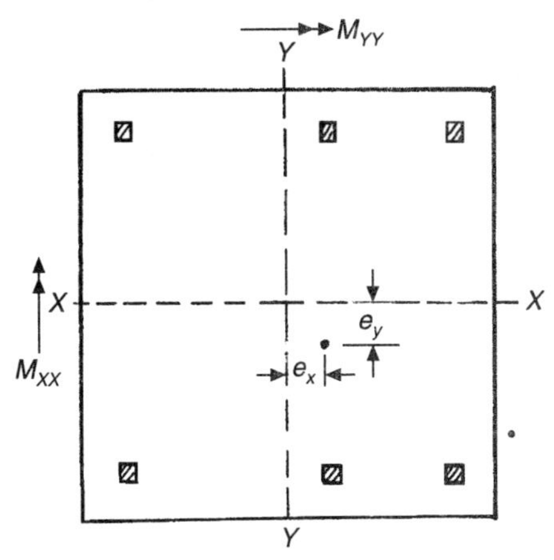

Figure 10.3 In layout of mats (and rafts) their centre of gravity should coincide with the centre of gravity of loads or it should be minimal. [By convention e_x produces M_{XX} about YY axis and e_y produces M_{YY} about XX axis. M_{XX} is moment for steel in X direction and M_{YY} is the moment for steel in the Y direction. See Sec. 4.6]

2. The characteristic loading on the soils should not exceed the allowable bearing capacity of the soil allowed for rafts. In granular soils, the N values can be used for the determination of the bearing capacity as explained in Sec. 10.1. Allowable bearing capacity for allowable settlement in rafts will be twice that of footings.
3. The loading in clayey soils should be such that the settlement due to long-term effects of consolidation should be within permissible limits.
4. The probable effects of uplift forces should be taken into account if they can be present with the variation of ground water level at the given place.
5. If parts of a building have large difference in loading intensity (due to varying heights or loading) or if there are large areas of weak spots within the raft area, it is advisable to separate the foundations of the heavily loaded part from the lightly loaded part and the weak soil part from the stronger parts. Similarly, joints should be introduced when there is a change in the directions or intensity of loading of the raft. For large rafts, construction joints have to be incorporated to work in stages. They should be properly planned at points of minimum shear.

We shall deal with the design of each type of raft in subsequent chapters.

10.6 TYPES OF LOADINGS AND CHOICE OF RAFTS

The following are the general rules for planning the types of rafts to be used in a given situation:

1. For symmetrical arrangement of spacing of columns and loadings (where the ground pressure is uniform), we adopt a *flat slab raft* and analyze it by the ACI empirical method called *direct design method* (DDM). The layout should satisfy the conditions specified for DDM.

2. With uniform ground pressure but with unequal spans which do not satisfy the conditions for DDM, we can adopt a flat slab, but it has to be analyzed by the *equivalent frame method* (EFM). However, beam and slab rafts are generally preferred in such cases.
3. With symmetrical arrangement of *unequally loaded columns* but with uniform ground pressure, we adopt a *beam and slab raft* construction.
4. On soils of very low bearing capacity, we compensate the loads on the raft by excavation of soil and design the raft as a *cellular raft*.
5. With non-symmetrical loading with the centre of gravity of loads not coinciding with the centre of gravity of the area, the ground pressures will vary form place to place. It is advisable to adopt a beam and slab construction for such cases. To some extent, the space around the loading area should be adjusted to give the least eccentricity. The slab should then be designed for the maximum pressures. The beam loading must be for the maximum values of the varying pressure in its location.

10.7 RECORD OF CONTACT PRESSURES MEASURED UNDER RAFTS

There are a number of records of measurement of contact pressures under raft foundations. Most of them indicate large contact pressures (up to three times) at the edges compared to the contact pressures near the centre. The edges seem to be more stressed than the central parts. This is true, especially during and immediately after the construction of the raft. As the time progresses, the contact pressure at the centre tends to increase. This non-uniform distribution is more noted in large rafts (30 m and over in size) than in small rafts. We must be aware of this behaviour. So, it is advisable to provide a peripheral beam in all rafts, especially flat slab rafts, to bridge over these large edge stresses.

10.8 MODERN THEORETICAL ANALYSIS

For an exhaustive treatment of elastic analysis of various types of raft foundations by computer methods see reference [4]. This book can be consulted for design of rafts for special layout of loads.

10.9 SUMMARY

Raft foundations have several advantages such as higher allowable bearing capacity and lower settlement over footing foundations. There are different types of rafts that are used in practice. Their choice depends on the loading, contact pressure, and the site conditions. The intensity of loadings of members of the raft is much more than the loadings on the members of roofs and floors. A raft has to support many more storeys and hence the span/depth ratios of raft members will be much different from those of roof members specified in the codes of practice for floors and roofs. The principles of design of different types of rafts are described in Chapters 11 to 15.

REFERENCES

[1] Thomlinson, M.J., *Foundation, Design and Construction*, ELBS, Longman, Singapore, 1995.

[2] Varghese, P.C., *Foundation Engineering*, Prentice-Hall of India, New Delhi, 2005.

[3] *National Building Code 2005*, The Bureau of Indian Standards, New Delhi, 2005.

[4] Hemsley, J.A., *Elastic Analysis of Raft Foundations*, Thomson Telford, London, 1998.

Design of Flat Slab Rafts—Mat Foundations

11.1 INTRODUCTION

A flat slab raft consists of a main slab of uniform thickness supporting a number of columns without any beams. In American practice, it is also called a *mat foundation*. The columns can have enlargement or splay at their ends. (They are called column heads.) Slabs with plain columns without column heads are called *flat plates* and those with enlargement of columns are called *flat slabs*. Flat slabs can be further given increased thickness of slab around the columns, called *drop panels*, for increased resistance against punching shear. One of the great advantages of a flat plate is that its top surface itself can be used as the ground floor without any filling.

Flat slab rafts are commonly used with equally spaced columns, with symmetrical column loadings, with not too large spacing and uniform soil conditions. Such slabs can be analyzed by the simple direct design method (DDM). Flat slab rafts can also be used for irregular layout of columns. Such layout has to be designed by more exact methods, such as the equivalent frame method (EFM). Beam and slab rafts are more suited for irregular layout instead of flat slab rafts. In this chapter, we deal with the analysis of flat slab rafts by DDM in detail and also briefly examine the EFM method. We will follow the IS and ACI Codes for the design [1, 2].

11.2 COMPONENTS OF FLAT SLABS

Flat slab rafts usually consist of the following elements:

- Main raft slab
- Columns with or without capital and drop panels
- Edge beam along the periphery of the raft (flat slabs should always be provided with edge beams along the periphery along the discontinuity for better stability (see Sec. 10.7).

In flat slabs, there are no internal beams. Columns have to carry all the loads directly from the roof and floors to the slab. There are four types of column connections to the slab as shown in Figure 11.1.

1. Column ending without any enlargement. Such flat slabs are called *flat plates.* In extreme cases of punching shear, shear head reinforcement can be used to strengthen them in punching shear.
2. Columns thickened at the base ending with only enlarged base are called *column capitals*. This enlargement helps the columns to transfer the load from the slab more efficiently and also reduce the punching shear.
3. Columns ending with column capitals and thickening of the slab below the column capital called *drop panel.* If the effect of these drop panels is to be taken in design, its length should not be less than 1/6th the span on the corresponding side of the column (IS 456, Cl. 31.2). These column capitals and drop panel help the column to distribute their loads to the raft slab more efficiently, especially when the loads are very heavy.
4. In the fourth type of flat slab foundations, the thickening can be provided under the slab without any capitals.

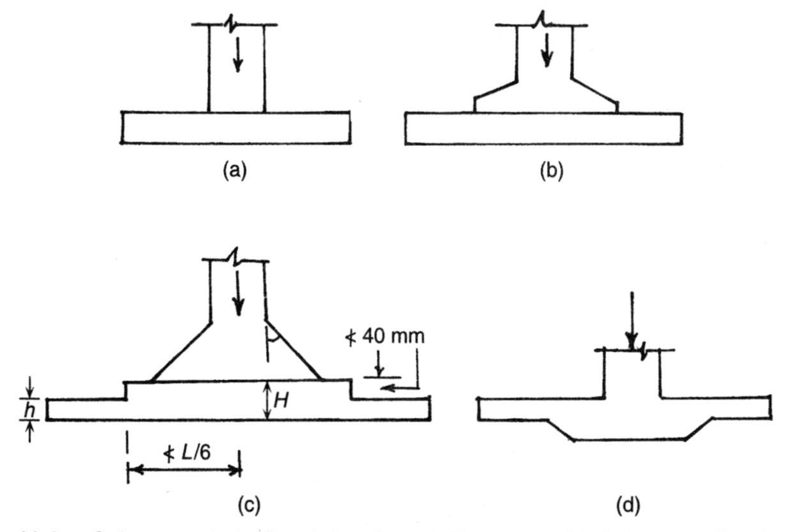

Figure 11.1 Column ends in flat slab rafts: (a) Flat plate, (b) Column ending in capital, (c) Column ending in capital and drop panel, (d) Thickening slab into column bases.

11.3 PRELIMINARY PLANNING OF FLAT SLAB RAFTS

The rough dimensions that can be used for the preliminary planning of flat slab rafts are discussed below.

11.3.1 Columns

1. ***Main column.*** These columns must be always short columns, usually square or circular in shape. A diameter equal to 1/8 to 1/10th the storey height and not less than 1/16th the larger span is usually adopted. We must remember that because there are no beams between columns, any unbalanced moments at supports (near columns) between adjacent spans have to be balanced (or transferred) through the columns only so that columns can be subjected to additional torsional shear and bending moment. Hence, columns should be made short and strong. *Long columns should not be used in flat slab rafts.*
2. ***Column capital.*** Usually a dimension equal to 1/5 but not more than 1/4 the shorter span length is used as the size of the column capital. It is usually sloped at 45° to join the column.
3. ***Drop panel.*** The slab above the capital can be thickened to 1.25 to 1.5 times that of the main slab [H = 1.25 to 1.5h] to form a drop panel around the column capital. This drop panel if used should *be at least* 1/6 the span on each side so that the total width will not be less than 1/3 of the smaller span of the surrounding panels. The effective thickness for the calculation of the steel area is to be taken not more than 1/4 the projection of the panel beyond the capital. In foundations, the drop panel can also be provided below the slab directly under the column.

11.3.2 Main Slab

Even though we provide a minimum span effective depth ratio of 26 to 30, in roof design for foundations which carry more load, the minimum depth to be adopted should be 250 mm. (The minimum thickness of the slab specified for flat slabs for roofs and floors given in IS 456, Cl. 31.2 is only 125 mm.) With two layers of 12 mm steel and 75 mm cover for soil, this will give an effective depth of [250 – (75 + 6 + 12)] = 157 mm only. The cover may be reduced to 50 mm when blinding concrete base is used. We also usually provide not less than 0.25 to 0.5% steel as reinforcement both ways. The calculated depths may be adjusted according to these empirical data. A span depth ratio of 12 to 14 will be more appropriate for depth of the slab.

11.3.3 Edge Beams

Even though flat slabs and flat plates have no internal beams, we must always provide a peripheral edge beam and an apron in flat slab foundation for better distribution of exterior moments that occur around external columns (see Table 11.2). The *height of these edge beams* should not be less than 1.5 times the thickness of the main slab. The details of the usual edge beam are shown in Figure 11.2.

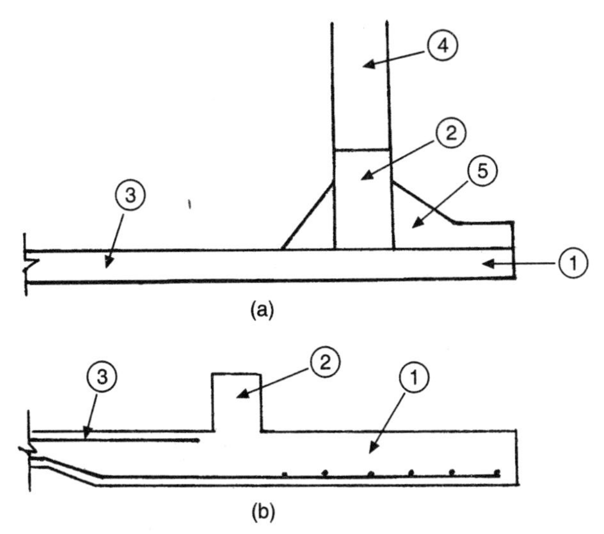

Figure 11.2 Detailing of apron slab: (1) Apron, (2) Edge beam, (3) Mat foundation, (4) Column, (5) Column capital.

11.4 ANALYSIS OF FLAT SLAB BY DIRECT DESIGN METHOD

The regular layout of flat slabs can be analyzed by an empirical method called *direct design method.* This method was first proposed in the ACI code, which has been adopted by IS 456 also.

IS 456–2000, Cl. 31.4.1 gives the limitations of the use of *DDM to flat roofs*. In roofs, we separately analyze the system for DL, LL for pattern loading. In foundation analysis, we have the dead load effects much more than the live loads so that all analyses can be carried out with the total dead load plus live load put together. Pattern loading will have very little effect in the analysis of foundations.

The conditions under which all flat slab raft layout can be analyzed by DDM are the following:

1. The loads on the raft slab should be uniformly distributed. For a raft, this means that the centre of gravities of the loads and the area of the raft should coincide. The flat slab should be also rigid enough to distribute the column loads to the foundation so that we can assume the foundation pressure as uniform.
2. The settlement of the foundation should not be more than permissible (see Table 14.3) so that the differential settlement is within allowable limits. In raft foundations, we allow more settlement than in footings as its differential settlement is taken as 1/2 that in footings.
3. The panels should be rectangular and the ratio of the longer to the shorter sides should not be more than two so that the slabs will have two way actions. (Some restrict the ratio to 4 : 3.)
4. There should *be a minimum of three spans* in each direction (along *X*- and *Y*-axis) and the adjacent spans should not differ in their spans by more than 1/3 the larger span.

5. The offset of the column in the X- and Y-directions should not exceed 10% of the span in that direction. This means that the length (and width) of adjacent panels should not differ by more than 10% of the greater length (or width).
6. The end spans should not be longer than the interior spans.
7. There should be an edge beam along the periphery.

11.5 METHOD OF ANALYSIS [1] [2] [3]

The procedure for the analysis of the flat slab is as follows. We analyze the slab as a two-way slab spanning both in X- and Y-directions, *carrying the whole load on the slab in both directions in the same way independent of each other.* Let L_1 and L_2 be the spans and $L_1 > L_2$.

The analysis of the slab for spanning in the YY-direction is shown in Figure 11.3. The foundation pressure is acting upwards are the flat slab raft. We first cut the slab along the YY-direction through the middle of the slab in each span. The portion between these cuts, together with the column above, is considered as frames in the YY-direction. We will have thus internal frames consisting of internal columns and slabs on both sides and external frames with external column the slab only on one side.

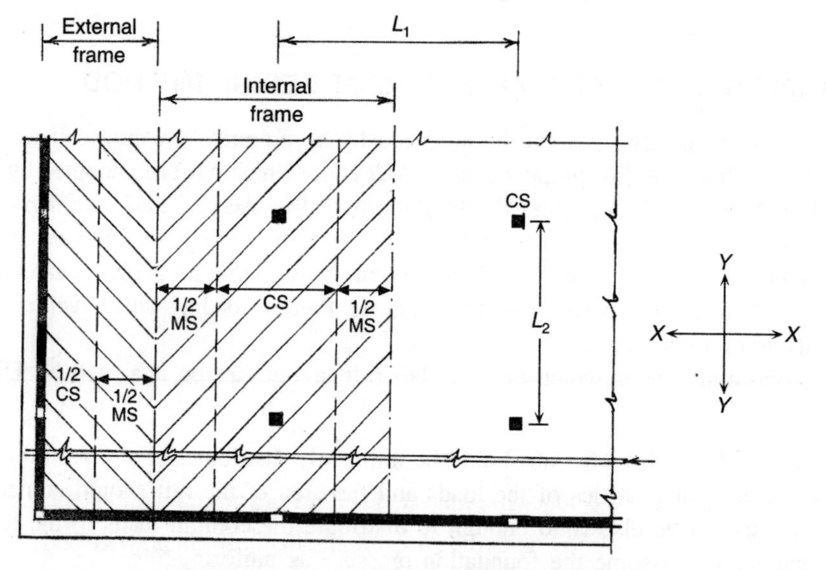

Figure 11.3 Analysis of flat slab rafts.

We again cut the slabs attached to the columns along a line $L_2/4$ from the centre on the two sides of the column. These are called 1/2 column strips (1/2 CS). Thus, the internal column line will have 1/2 column strip on both sides. The portion between our earlier cuts along the centre line and the 1/2 column strips is called 1/2 middle strips (1/2 MS) as shown in Figure 11.3.

Thus, for an internal column line, there will be 2 half column strips on either side of the line of columns. Together, it is called the *column strip*. Then these will be 1/2 middle strips on either side. The column strips are considered to be more rigid than the middle strips.

We next consider the span of the slabs as the *distance between the edges of the equivalent rectangular column capitals and designate it as* L_n, the effective span for analysis. If we take a panel as a whole, the load on the panel for analysis will be load on the area equal to L_2L_n, the load on span $W = wL_nL_2$, where w is uniformly distributed load. We calculate free bending moment as

$$\text{Free BM} = \frac{WL_n}{8} = \frac{wL_2L_n^2}{8} \tag{11.1}$$

We can assign or distribute a part of it as –ve moment at support and the rest as +ve moment in the middle. This is called longitudinal distribution. This distribution, according to the IS Code, is given in Figure 11.4 and Table 11.1.

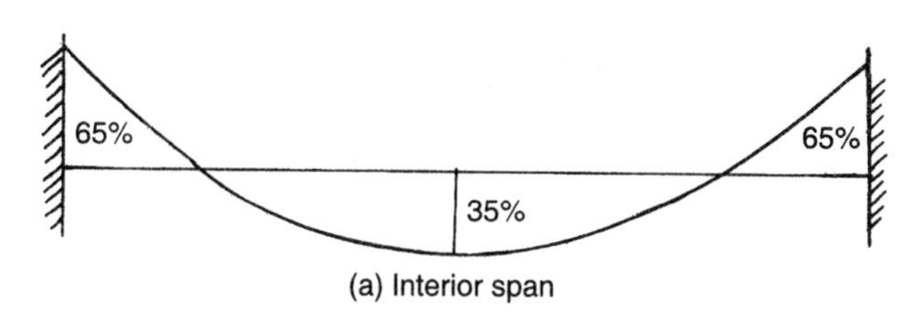

(a) Interior span

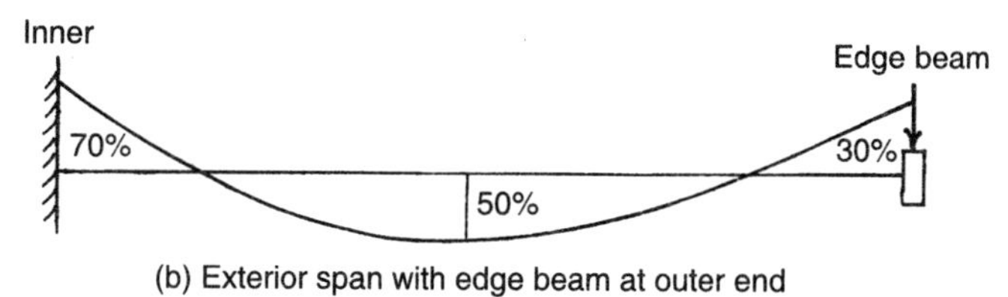

(b) Exterior span with edge beam at outer end

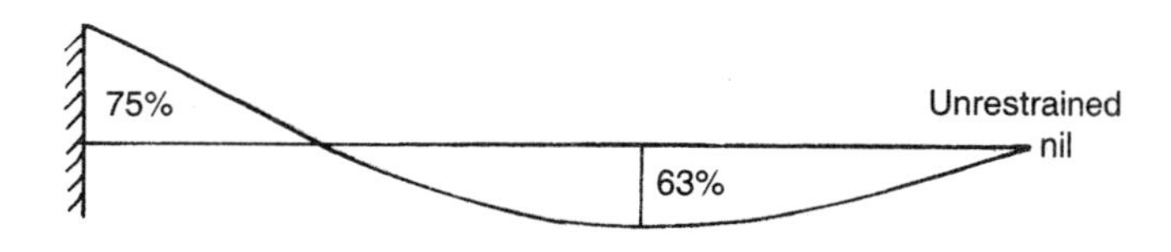

(c) Exterior span with end on wall (unrestrained)

Figure 11.4 Table 11.1. Longitudinal distribution of M_0 in analysis of flat slab by Direct Design Method (DDM). (Transverse distribution is shown in Table 11.2).

TABLE 11.1 Longitudinal Distribution of M_0 in Flat Slabs (Figure 11.4)

No.	*Description of the spanning of the slab*	*% Distribution of M_0*		
		Interior support	*Centre of span*	*Exterior support*
1.	Interior spans (fixed at ends)	65	35	65
2.	Exterior span with edge beams	70	50	30
3.	Exterior span without edge beam	70	52	26
4.	Exterior span on wall (end free)	75	63	0

Then we can redistribute the –ve and +ve moments transversely to the column strip and middle strip. The column strip being more rigid will naturally take more moments than the middle strip. This distribution, according to the IS Code, is given in Table 11.2. ACI Code and IS have given the same coefficients for longitudinal distribution (as –ve and +ve moments) and transverse redistribution (between the column strip and the middle strip). These empirical values can be directly used in a design. [In equivalent frame analysis, which we have to use for irregular layout of columns, the span moment is found by moment distribution, but for redistribution we use the same coefficients as used in DDM.]

11.5.1 Values for Longitudinal Distribution and Transverse Redistribution

As already stated, Table 11.1 and Figure 11.4 give us the values according to the ACI Code for distribution of the free bending moments into the support –ve moments and the span +ve moments. Table 11.2 gives us the transverse redistribution of these positive and negative moments to column strips. The balance of the moment is assigned to the middle strips.

TABLE 11.2 Transverse Distribution of Moments into Column Strips (Balance Distributed to Middle Strips)

No.	*Description*	*Percentage in column strip*
1.	Positive moment in all spans	60
2.	Negative moment at interior support	75
3.	Negative moment at exterior support in end span with edge beam	75*
4.	Negative moment at exterior support without edge beam	100* (Balance assigned to MS)

[*Note:* In the exterior span with edge beam, we get a better distribution of the end moments, and this is one of the advantages in providing an edge beam.]

11.5.2 Shear in Flat Slabs

Flat slabs have to be designed for one-way shear (also called beam shear) as well as punching shear as explained in Chapter 2. In addition, when the ratio between adjacent spans is more than 1.25, there will be a significant difference between the moments at the span junction. This difference has to be transferred by bending and torsion through the column-slab connection. Hence, the punching shear will be magnified by a factor [1]. However, ACI allows this to be neglected if adjacent spans are not different by 30%.

11.5.3 Bending of Columns in Flat Slabs

For the design of roofs, IS 456, Cl. 31.4.5 requires design of columns, especially end columns, for moments produced by differential loading or non-loading of adjacent spans. As we design rafts for full live and dead loads, we need not consider pattern loading in foundation design. It is always better we *provide an edge beam around the periphery* which will enable the whole width of the slab (i.e. the column and middle strips instead of the column strip only) acting in load transfer in the external span.

The value of the moment for the design of the external column can be derived from the IS 456, Cl. 31.4.5.2 by assuming that the outer span does not exist. Hence,

$$\Delta M = 0.08\, wL_2L_n^2$$

This will give a conservative design value as the full value of the live and dead loads is taken into account in this case.

11.6 LIMITATIONS OF DIRECT DESIGN METHOD FOR MATS

First of all, we should remember that in *flat slab roofs*, the loading on each floor can be assumed to be uniform. On the other hand, in mats we transfer column loads into the ground through the flat slab raft and *assume it is uniformly distributed.* In most cases, the loadings of the middle columns of a building will be different from the peripherial columns. Hence, the raft should be stiff enough to distribute the load evenly on the ground. Hence the equilibrium of forces (downward load and ground reaction) has to be always checked in the analysis of strips.

Secondly, as already pointed out, for DDM differences, the lengths of adjacent spans must be within the specified limits. Otherwise, transfer of moments has to be provided and we have to use the equivalent frame method for analysis. If the difference in the adjacent spans is within the limits specified by DDM (i.e. not more than 10% of larger span), we use the greater of the –ve moments obtained by the larger span for design.

11.7 EQUIVALENT FRAME METHOD OF ANALYSIS FOR IRREGULAR FLAT SLABS

Direct design method cannot be considered applicable to flat slabs when the adjacent spans differ by more than 10% of the larger span, or where the longer span to the shorter span (length to width) exceeds 4:3, or where there are less than 3 spans in the two directions. For these flat slabs, the equivalent frame analysis is applicable.

In equivalent frame method, which is described in advanced books on reinforced concrete [3], the system we got by cutting through the midspan is to be modelled to an equivalent frame and analyzed by moment distribution to find the total –ve and +ve moments, which is similar to the longitudinal distribution in DDM. The transverse distribution is again made, as indicated in Table 11.2, by arbitrary coefficients.

11.7.1 Method of Equivalent Frame Analysis

As described in IS 456 (2000), Cl. 31.5, in equivalent frame method, the BM and SF are determined by an analysis of the structure as a continuous frame with the following assumptions:

1. The structure is considered to be made up of equivalent frame on column lines taken once longitudinally and again transversely through the building. Each frame consists of a row of equivalent columns or supports bounded laterally by the *centre line of the column or supports.*
2. For determining the relative stiffness of the members, the MI of the slabs and the columns may be taken as that of the gross cross-section of the concrete alone.

3. Each frame is analyzed for the uniform contact pressure due to the total load on the raft (analysis for LL and DL need not be carried out separately).
4. The analysis can be done by moment distribution or any other method.
5. Redistribution of moments is also allowed.

11.7.2 Transverse Distribution of Moments along Panel Width in EFM

After the analysis by moment distribution to find the total negative and positive moments, the distribution of +ve and –ve moments for the column and middle strips can be made as in DDM, given in Table 11.2.

11.7.3 Approximate Method for Eccentrically Loaded Raft

An approximate method for the analysis and design of an eccentrically loaded but otherwise regular in layout flat slab raft is shown in Example 11.3.

11.7.4 Approximate Design of Flat Slab Rafts (Calculation of BM and SF from Statics)

In roof design, the superloads are known and can be assumed to be uniformly distributed in an indeterminate structure. In foundation design, only the column loads are known and *we assume* that the ground pressures are uniformly distributed. Hence, the frame we take for analysis is statically determinate. Thus, we first balance the loads by correcting the ground pressures and draw the BM and SF by simple statics. Having known the support moments and span moments, we can distribute them transversely to the midlle strip and column strip according to Table 11.2 and detail the steel as recommended.

11.8 DETAILING OF STEEL

The rules which are used for detailing flat slab roofs can be followed for the rafts also. These are given in Figure 16 and Cl. 31.7 IS 456. Detailing of steel can be done with straight bars or bent bars. Detailing with straight bars is shown in Figure 11.5. It can be summarized as follows.

General condition

1. The spacing of reinforcement in flat slabs should not exceed *two times the slab thickness, except* when the slab is ribbed or cellular. (In ordinary slabs such as floor slabs, it can be three times.)

Column strip

2. *Positive steel in column strip:* In column strips, 50% of the bottom or +ve steel should extend to $0.125L$ from the centre of the next column line and 50% should come very near to the centre line of the column (to a maximum gap of 75 mm from the centre line of columns) or made continuous.
3. *Negative steel in internal column strip:* In internal column strips, 50% of the top or –ve steel should extend $0.2L$ from the face of the column, and the rest of it should

extend to $0.3L$ from the face of the column for flat slabs without drops and $0.33L$ for those with drops.

4. *Negative steel in the external column strips:* The –ve steel should extend beyond the centre line of the column and the anchorage should not be less than 150 mm.

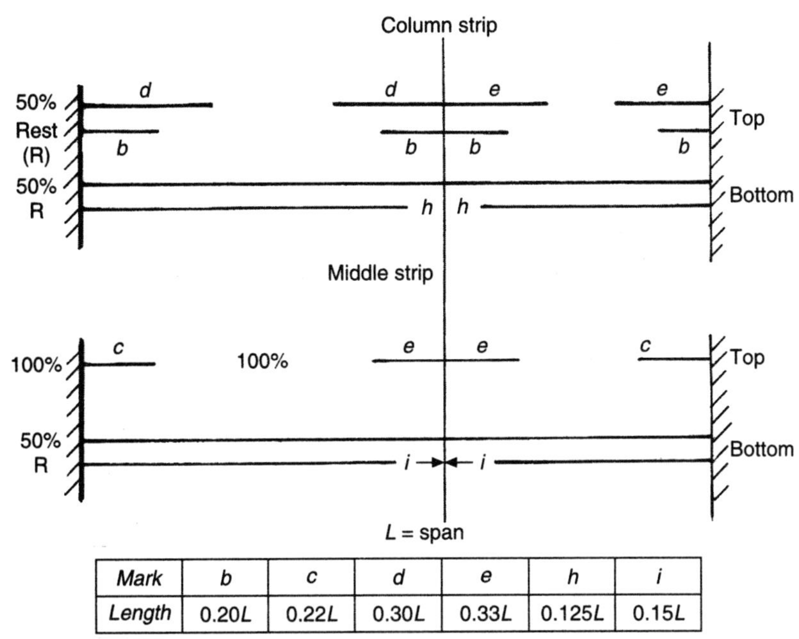

Mark	b	c	d	e	h	i
Length	0.20L	0.22L	0.30L	0.33L	0.125L	0.15L

Figure 11.5 Detailing of steel in flat slabs (IS 456, Figure 16).

Middle strip

5. *Positive steel in the middle strip:* 50% of the bottom or +ve steel should extend to $0.15L$ from the centre line of the support and the rest should extend very near to the centre line of the column (to a maximum gap of 75 mm from the centre line of columns).

6. *Negative steel in the middle strip:* 100% of the top or –ve steel should extend at least $0.22L$ from the face of the support.

Integrity steel

7. *Integrity steel:* According to the revised ACI code, it is mandatory to place at least two of the bottom bars in the column strip as *continuous reinforcement properly spliced* in each direction in the column core and properly anchored at the ends. These are called *integrity bars*. (This is provided because near the column, with negative moment tension acts at the top and cracking may occur there. Without any bottom steel sudden failure can take place due to unforeseen causes.)

11.9 DESIGN OF EDGE BEAM IN FLAT SLABS

It has been pointed out that in that slab foundation the outer edges of the raft should be stiffened by adopting an edge beam along the perimeter as shown in Figure 11.2. The depth of these beams should not be less than 1.5 times the slab thickness; proper *L*/*d* ratios should also be maintained. It can have its rib as upstanding or downstanding. An apron beyond the beam is also usually provided for mats as in the case of beam and slab rafts.

The loads and bending moments on the edge beams will be as shown in Figure 11.6. These beams are designed like regular beams—upstanding beams can be treated as T beams in the spans and downstanding beams as T beams at the supports.

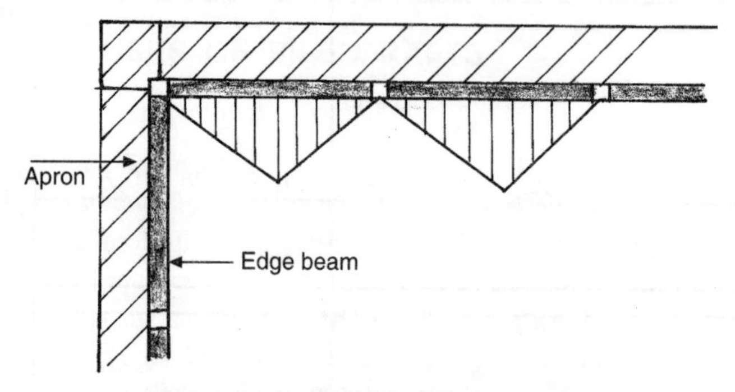

Figure 11.6 Design of edge beams.

11.9.1 Design of Slab around Edge Beam and Its Corners

The peripheral slab that is provided beyond the edge beam is designed as the cantilever slab with the corner portions designed on the principles explained in Chapter 12 for beam and slab rafts.

11.10 USE OF FLAT SLAB IN IRREGULAR LAYOUT OF COLUMNS

It is always preferable to use a slab and beam raft instead of a flat slab for irregular layouts of columns unless the latter works to be much more economical. This is especially true in compressible soils. The slab and beam rafts have better capacity to distribute the load evenly on the foundation. They are more rigid, so the assumption of uniform contact pressure can be assumed to be more valid.

As already mentioned, the analysis of irregular flat slab rafts is to be carried out in a way similar to the analysis of an equivalent frame described in RC textbooks [3]. The bending moments are to be found by analysis, and the transverse redistribution can be carried out as shown in Table 11.2.

11.11 SUMMARY

Regular flat slab rafts are used in fairly incompressible soil with regular layout of columns with symmetric loadings. They give an even surface on top without upstanding beams as in beam and slab rafts. They should always be provided with an edge beam. The analysis and design of regular flat slab by DDM is done in this chapter. The analysis of irregular flat slab by equilibrium frame method is similar to the analysis of such roofs.

EXAMPLE 11.1 (Preliminary planning of flat slab rafts with column capital and drop panel)

Plan a flat slab raft for a column layout at 3 m centres in the X- and Y-directions as in Figure E11.1. If the column loads are symmetrical so that CG of the column loads falls in line with CG of the raft slab, arrive at a preliminary dimension of the raft.

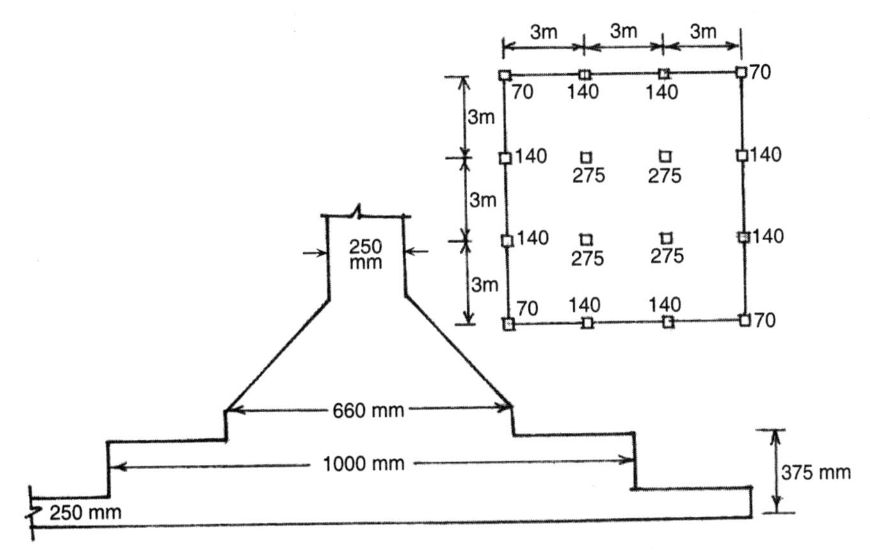

Figure E11.1

[Number beside columns indicate loads in kN]

Reference	*Step*	*Calculation*
	1	*Check if DDM is applicable*
Sec. 11.4		The spans are equal and loading is symmetrical. CG of loads and raft coincides (Conditions of Sec. 11.4 OK).
	2	*Estimate size of column*
Sec. 11.3 (IS 456, Cl.31)		We will adopt a square short square column L/B = 12 (both ends fixed).
		Size of col. = $\dfrac{3000}{12}$ = 250 mm square
	3	*Estimate size of column capital*
Sec. 11.3		$D = \dfrac{1}{5}$ to $\dfrac{1}{4}$ span, i.e. $\dfrac{3000}{5}$ to $\dfrac{3000}{4}$ = 600 to 750 mm
Figure E11.1		Adopt 660 mm square splayed at 45° as column capital. Adopt bottom offset of 40 mm.
	4	*Estimate slab thickness*

Reference	Step	Calculation
Sec. 11.3		$h_{min} = L_n/12 = (3000 - 660)/12 = 195$ mm
		Adopt a minimum h = 250 mm; 50 mm cover over mud mat concrete with two layers of steel.
Sec. 11.3		$d = 250 - 12 - 6 - 50 = 182$ mm
	5	*Estimate size of drop panel*
Sec. 11.3		Drop thickness H = (1.25 but not more than 1.5)h
		$H = 1.5 \times 250$; Adopt 375 mm
	6	*Estimate length of drop*
		They should be rectangular and not less than 1/3 span or 1/6 span in each direction (not less than L/6 of the shorter span)
		$\frac{L}{3} = \frac{3000}{3} = 1000$ mm total length
	7	*Estimate the size of the edge beam*
		Height $\nless 1.5$ slab thickness = $1.5 \times 250 = 375$ mm
		Adopt 500 × 300 edge beam.

EXAMPLE 11.2 (Design of a flat slab raft by DDM)

Design a flat slab raft with edge beam for a layout of column loads as shown in Figure E11.2. Assume the safe bearing capacity from settlement considerations as 50 kN/m^2. Assume columns are 300 × 300 mm enlarged to 600 × 600 mm as capital.

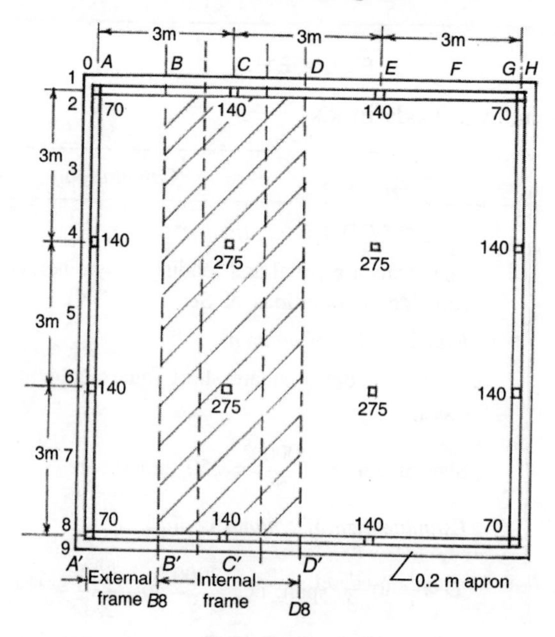

Figure E11.2

(*Note:* 1. Numbers beside columns represent column load in kN. 2. The slab is divided by coordinates *A*, *B*, *C*, *D* and 1, 2, 3, 4, ...)

Reference	*Step*	*Calculation*
	1	*Check whether DDM is applicable*
Sec. 11.4		Yes, satisfies all the conditions. (There are 3 spans.)
	2	*Find CG of loads*
		As the loads and spans are symmetrical, the CG of loads and raft is the same.
	3	*Find ground pressure and check BC*
		Let us provide 200 mm wide apron around to facilitate an edge beam. Plan of raft will be 9.4 × 9.4 m = 88.36 m^2
		Ground pressure with characteristic loads
		$$q_c = \frac{(4 \times 275) + (8 \times 140) + 4(70)}{88.36} = 28.3 \text{ kN/m}^2$$
		It is less than BC = 50 kN/m^2. Hence, safe.
	4	*Factored ground pressure for limit state*
		q = 1.5 × 28.3 = 42.5 kN/m^2
	5	*Design of flat slabs*
		Divide into strips in the *XX*- and *YY*-directions by cutting through midspans. Consider the resulting slab structure as frames in the column. We get *internal frames BD–D′B′* and *external frames AB–B′A′* to be analyzed. Let us consider a *YY*-internal frame *BD–D′B′*. This frame will have internal span and external span at the ends and column strip and middle strip as shown in Figure E11.2.
		We will analyse one of the interior frames.
	6	*Consider equilibrium of strip BCD–D′C′B′*
		Factored downward col. load = 1.5 (140 + 275 + 275 + 140)
		= 1245 kN = P_1
		Total upward reaction = 42.5 × 3 × 9.4 = 1198.5 kN = P_2
		$$\text{Unbalanced load} = \frac{1245 - 1198.5}{3 \times 9.4} = 0.17 \text{ kN/m}^2$$
		This can be neglected.
		(If there is an imbalance, we find the average load and base our design load on this average load, or we may adjust the reaction pressure to balance the downward load. See Step 5, Example 11.3.)
	7	*Check for punching shear*
Chapter 2		$$\tau_p = 0.25\sqrt{f_{ck}} = 0.25\sqrt{20} = 1.12 \text{ N/mm}^2$$
		Max column load = 1.5 × 275 = 412.5 kN
		Assume D = 250 mm (as loading is light)
		d = 250 − 50 − 18 (2 layers of 12 mm steel) = 182 mm
		Punching shear resistance = V_p at $d/2$ from capital.
		= $(2a + 2b + 4d)d \times \tau_p$

Reference	Step	Calculation
		= 4(600 + 182)182 × 1.12
		= 637 > 412.5 kN. Hence safe.
		(Resistance without capital is as follows:
		= 4(300 + 182) × 182 × 1.12 = 393 < 412.5 kN
		Hence, an enlargement of column is needed.)
	8	*Find M_0 value*
		Column capital is square 600 × 600. (If it is circular, we convert it into an *equivalent square* of 0.85 times the diameter.)
Eq. (11.1)		$L_n = (3 - 0.6) = 2.4$ m, $L_2 = 3$ m, and $q = 42.5$ kN/m²
		$M_0 = WL_n/8 = \dfrac{(42.5 \times 3 \times 2.4) \times 2.4}{8} = 91.8$ kNm
	9	*Find longitudinal distribution of moments to support and middle spans and transverse distribution to C.S (col) and MS (mid)*
		M_0 = 91.8 kNm
		(a) *For interior span* (in kNm)
		–ve support = 65% = 60 kNm — Col 75% = 45 kNm; Mid 15% = 15 kNm
Table 11.1 and Table 11.2		+ve in span = 35% = 31.8 kNm — Col 60% = 19 kNm; Mid 40% = 13 kNm
		(b) *For exterior span:*
		Inner support = 70% = 64.3 kNm — 75% = 48 kNm; 25% = 16 kNm
		At mid span = 50% = 46 kNm — 60% = 28 kNm; 40% = 18 kNm
		At end support = 30% = 26 kNm — 75% = 19 kNm; 25% = 7 kNm
		[*Note:* Alternately, as the column loads and base reactions are known, we can draw the SF and BM diagram from single statics also. But the moments so obtained at supports and span are distributed to the middle strip and column strip according to Table 11.2.]
	10	*Check depth of slab for maximum BM*
		M_{max} = 48 kNm for the column strip at the inner support of the exterior span b = 1500 mm
Chapter 2		$d = \left(\dfrac{M}{0.14 f_{ck} b}\right)^{1/2} = \left(\dfrac{48 \times 10^6}{0.14 \times 20 \times 1500}\right)^{1/2} = 107$ mm
		Hence d = 182 mm is OK

Reference	Step	Calculation
	11	*Find steel for column strips and mid strips at interior and exterior spans*
		To illustrate the column strip. Let us take –ve steel at support of column strip. M_U = 45 kNm, width of strip is 1500 mm
Table B.1		$\frac{M}{bd^2} = \frac{45 \times 10^6}{1500 \times 182 \times 182} = 0.90$
		p = percentage of steel = 0.264%
		As per m width = $\frac{0.264 \times 1000 \times 182}{100}$
		= 480 mm^2 (12ϕ @ 200 gives 502 mm^2)
	12	*Analyse other sections for A_s*
Sec. 11.8	13	*Detail steel as prescribed in IS 456*
Figure 11.5	14	*Analyse edge beam*
		Depth of beam > 1.5 × thickness of steel
		D = 1.5 × 250 = 375 mm. Let B = 250 mm
		Load from the inner panel at 45° dispersion
		W_1 = (3/2) (1.5) × 42.5 = 95.6 kN as UDL
		Load from the apron edge 0.2 m wide
Figure 11.6		W_2 = 3 × 0.2 × 42.5 = 25.5 kN (UDL)
		Support moment for triangular load $\frac{W_1 l}{9.6}$
		$M = \frac{95.6 \times 3}{9.6} + \frac{25.5 \times 3}{24} = 33.0$ kNm
		Span moment = $(W_1 l/12) + (Wl/24)$
		$M = \frac{95.6 \times 3}{12} + \frac{25.5 \times 3}{24} = 27.0$ kNm
		Design the steel for these moments.
		Provide necessary shear steel also.
	15	*Design of span apron*
		See Example 12.4.
		[*Note:* We separately analyze the external frame *AB–B′A′* also in a similar manner].

(*Note:* As the span lengths are the same in both *X*- and *Y*-directions, the same results hold good for frames also. If the spans are different, we repeat the same method of analysis in *XX*-frames also. Note that the whole load on the span is taken once in the *Y*-direction and then in the *X*-direction also.)

EXAMPLE 11.3 (Approximate analysis of flat slab with eccentricity)

The layout of a flat slab with eccentric loading is shown in Figure E11.3. Indicate how to analyze this raft [4].

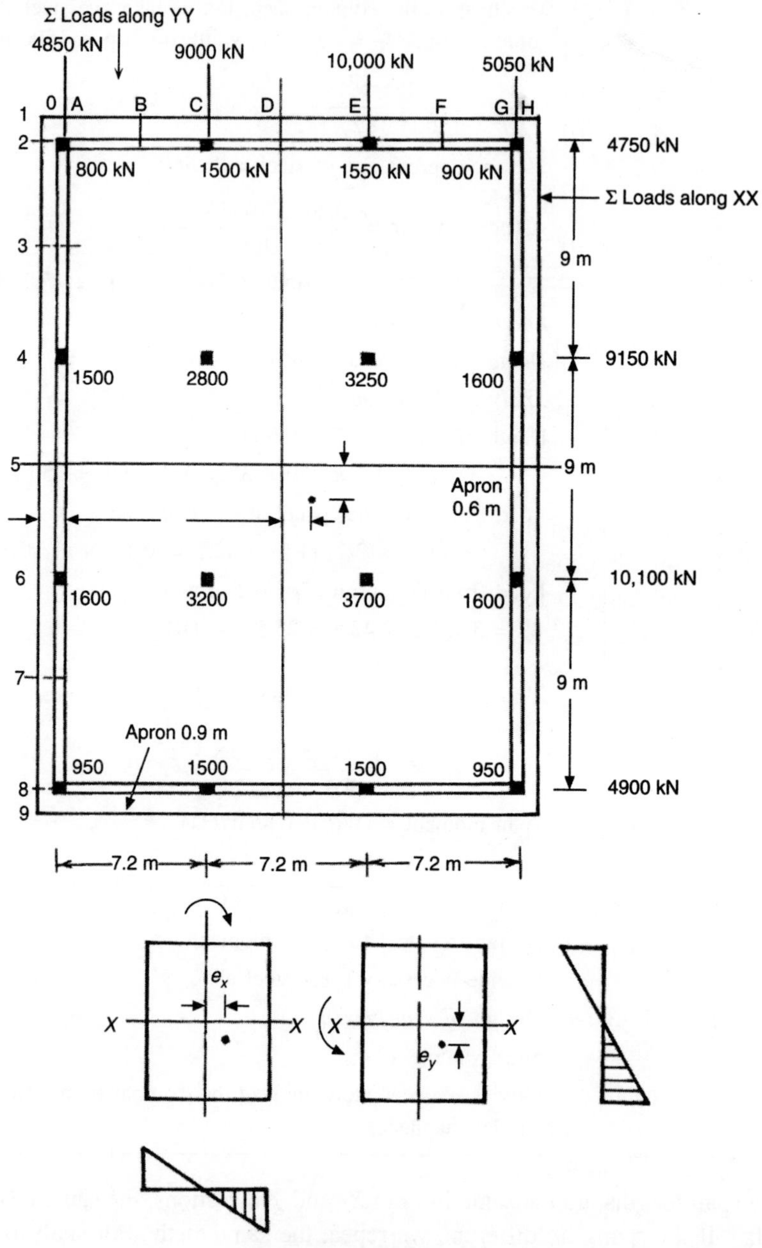

Figure E11.3 (Contd.)

(Note: Number beside columns indicate loads in kN)

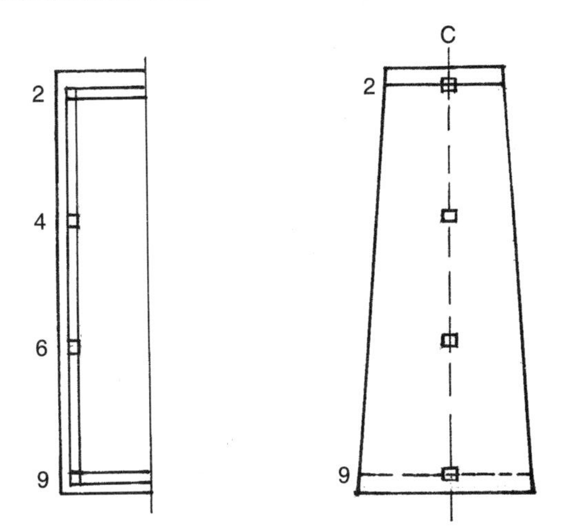

Figure E11.3 Rafts loaded with small eccentricity.

Note: In theory, the flat slab analysis is restricted to uniformly distributed load. However, if the eccentricity is small, we may proceed as follows. We adopt the following steps to analyze such a flat slab:

Step 1: Check eccentricity of the resultant load and find e_x and e_y with service loads. (As IS specifies the same load factor 1.5 for DL and LL, the eccentricity is the same for both characteristic and factored loadings.)

Step 2: Find M_X and M_Y with factored loads of all loads about *XX*- and *YY*-axes to calculate the varying base pressures.

Step 3: Find I_{XX} and I_{YY}.

Step 4: A flat slab has to be analyzed in frames formed by cutting the slab along midspans in the *XX*- and *YY*-directions. Let us first analyze the strips in the *YY*-direction. Find the ground pressures due to the eccentric loading at the end and mid points *O* to *H* of the slab at the top and along the bottom in Figure E11.3.

Step 5: Check whether the maximum pressure exceeds the safe bearing capacity or not.

Step 6: Take each strip along *YY*. As the pressure in the strip varies, find the average pressures over the top edge and also at the bottom edge. (This average pressure for the interior frame will be pressure along the column line.) Draw the load diagram with the column loads P_1, P_2, etc. and base reaction in the strip along its length.

Step 7: Check for balance of downward and upward forces. If they do not match, modify the ground pressures for equilibrium.

Step 8: Draw the SF and BM for the strip for design. As the loads from the top and the base pressure from below are known, we may use any of the following: (a) Simple statics or neglecting col. loads used, (b) DDM or (c) EFM (we may use the transverse distribution percentage in detailing of steel).

(As a quick and short procedure, we first find the pressures along the boundary lines of the spans take and design each span for the average pressure in the span. We will also *assume* for simplicity that the load from the apron below the peripheral column is transmitted directly to the beams.)

Step 9: Distribute the moments using transverse distribution (Table 11.2).

(*Note:* The following are the calculations according to the above steps. (By convention, we denote moment M_x as moment due to e_x *about YY*. Similarly, M_y is due to e_y *about XX axis.*)

Reference	*Step*	*Calculation*
		Let us provide an apron 0.6 m and 0.9 m as shown in Figure E11.3.
Figure E11.3	1	*Cut the slab into strips and along mid points of the slab. Mark points as in Figure E11.3. Find* e_x and e_y, *taking left hand corner load point as the origin*
		Taking moment of loads about *YY*-axis on the line of loads A9 to find e_x.
		$\bar{x} = 7.2(9000 + (2 \times 10000) + 3(5050))/24050 = 10.999$
		Hence, $e_x = 10.999 - 10.8 = 0.199$, say 0.2 m
		i.e. 0.2 m on right of the centre line.
		Similarly, take moments of loads about *XX*-axis about the bottom to find e_y.
		$(\bar{y}) = 9(1010 + (2 \times 915) + 3(475))/2890 = 13.282$
		Hence, $e_y = 13.282 - 13.50 = -0.218$ m below the centre line.
$1t = 10$ kN	2	*Find moments produced* M_X *and* M_Y *with factored loads*
		Factored load $P = 1.5 \times 28900 = 43350$ kN
		$M_{XX} = Pe_x = 43350 \times 0.2 = 8670$ kNm (I_{YY} controls)
		$M_{YY} = Pe_y = 43350 \times 0.218 = 9450$ kNm (I_{XX} controls)
	3	*Find* I_{XX} *and* I_{YY}
		$I_{XX} = BL^3/12 = 22.8(28.8)^3/12 = 45387$ m^4
		$I_{YY} = LB^3/12 = 28.8(22.8)^3/12 = 28446$ m^4
	4	*Find ground pressures in YY-direction*
	4a	*Find base pressures (p) at top and bottom lines of the strips for analysis of strips in YY-direction*
		Let p_x be due to M_X and p_y due to M_Y
		$p_0 = \frac{P}{A} \pm \frac{M_{XX}}{I_{YY}}(x) \pm \frac{M_{YY}}{I_{XX}}(y)$
		$= p_0 \pm p_x \pm p_y$
		$p_x = \frac{M_{XX}}{I_{YY}}(x) = \frac{8670}{28446}(x) = 0.305x$

Reference	*Step*	*Calculation*
		$p_y = \frac{M_{YY}}{I_{XX}}(y) = \frac{9450}{45387}(y) = 0.208y$ $p_0 = \frac{Q}{A} = 43350/(22.8 \times 28.8) = 66 \text{ kN/m}^2$ Tabulate values (Example point B1) x = 7.2 m; y = 14.4 m from centre line (4.5 + 9 + 0.9) $p_x = 0.305 \times 7.2 = 2.20 \text{ kN/m}^2$ (negative) $p_y = 0.208 \times 14.4 = 3.00 \text{ kN/m}^2$ (negative) (as shown in Table E11.3 below)

Table E11.3 gives the variation of pressures along *top* and *bottom* edges of the raft for the analysis of frames in the *YY*-direction.

[*Note:* For the analysis of strips in the *Y*-direction, we calculate the pressures on top and bottom to find the average pressure. Again, for the analysis of strips in the *X*-direction, we will find pressure on the LHS & RHS and take the mean for each strip.]

TABLE E11.3

Point	p_0	p_x *due to* M_x (x varies)	p_y *due to* M_y (y = const.)	p
Top edge				
O1 (corner)	66.01	–3.48	–3.00	59.53 (corner)
B1	66.01	–2.20	–3.00	60.81
C1	**66.01**	**–1.10**	**–3.00**	**61.91**
D1	66.01	0.00	–3.00	63.00
Note: B1 means point *B* along *X*-axis as marked and along *Y*-axis.				
E1	**66.01**	**+1.10**	**–3.00**	**64.11**
F1	66.01	+2.20	–3.00	65.21
H1 (corner)	66.01	+3.38	–3.00	66.49 (corner)
Bottom edge				
H9 (corner)	66.01	+3.38	_3.00	72.49 (corner)
F9	66.01	+2.20	+3.00	71.21
E9	**66.01**	**+1.10**	**+3.00**	**70.11**
D9	66.01	0.00	+3.00	69.01
C9	**66.01**	**–1.10**	**+3.00**	**67.91**
B9	66.01	–2.20	+3.00	66.81
O9 (corner)	66.01	–3.48	+3.00	66.52 (corner)

Note: Values p_0 and p_y remain constant as y remains constant for top and bottom; only p_x varies as varies from left to right.

Note: To make matters clear, we have tabulated only values for *YY*-strips)

Reference	*Step*	*Calculation*
	5	*Check bearing capacity*
		Max factored pressure under point H9 = 72.49 kN/m^2
		Service load = 72.49/1.5 = 48.32 kN/m^2 < 75 kN/m^2
		Min factored pressure = 59.5 kN/m^2 (no tension)
	6	*Find the base pressure at the two ends of the strip (upper and lower) and draw the loads on strip*
		Mean p of strip BCD at top = (60.8 + 61.9 + 63)/3 = 61.9 kN/m^2 (near col. C)
		Mean p of bottom of BCD = (69.0 + 67.9 + 66.8)/3 = 67.8 kN/m^2 (near col. C′)
		(The central value along the column line will obviously be the mean value.)
	7	*Check balance of loads along column line C*
		Total factored load P = 9000 × 1.5 = 13500 kN
		Upward reaction Q = 7.2(61.9 + 67.9)/2 × 28.8 = 13500 kN
		[The difference is very little. If there is a difference, adjust the reaction as UDL or adjust reaction and load as given in Step 9.]
	8	*We may find BM and SF using direct design method or the equivalent frame method [EFM] by moment distribution or simple statics*
Step 7		For all practical purposes, we may calculate the moments at the supports and spans of the strip with mean base pressure = 13500/(7.2 × 28.8) = 65.1 kN/m^2 as the uniform pressure as in Example 11.2.
	9	*Let us now consider the external frame bounded by A1B1 and A9B9*
		(a) *Find mean pressure on strip*
		Mean between A1 and B1 = (59.53 + 60.811)/2
		= 60.2 kN/m^2
Figure E11.3		Mean between A9 and B9 = (66.81 + 66.52)/2
		= 66.7 kN/m^2
		Average pressure along strip = (60.2 + 66.7)/2
		= 63.45 kN/m^2
		(b) *Check for balance of loads*
		1.5 (col. load) = 1.5(800 + 1500 + 1600 + 950) = 7275 kN
		Total base pressure = 63.45 × 3.8 × 28.8 = 6944 kN
		Upward pressure is lower.
		Unbalance = 7275 − 6944 = 331 kN
		Increase base pressure by 331/(3.8 × 28.8) = 3.0 kN/m^2

Reference	*Step*	*Calculation*
		Mean base pressure = 63.45 + 3.0 = 66.45 kN/m^2
		Design this external frame for a uniform base pressure of 66.45 kN/m^2 and the column loads.
	10	*Distribution of moments and detailing of steel*
		After finding the moments at the supports and the spans, we may laterally distribute the loads to the column and middle strips, as given in Table 11.2, for flat slabs and proceed with the detailing of steel.

REFERENCES

[1] IS 456, 2000, *Plain and Reinforced Concrete Code of Practice*, Bureau of Indian Standards, New Delhi, 2000.

[2] ACI 318 (1989), *Building Code Requirements for Reinforced Concrete*, American Concrete Institute, U.S.A., 1989.

[3] Varghese, P.C., *Advanced Reinforced Concrete Design*, 2nd Edition, Prentice-Hall of India, New Delhi, 2005.

[4] Das, B.M., *Principles of Foundation Engineering*, 5th Edition, Thomson Asia Pvt. Ltd., Singapore, 2003.

CHAPTER 12

Beam and Slab Rafts

12.1 INTRODUCTION

Beam and slab rafts are used as foundation under one or more of the following conditions:

(i) When there is an unsymmetrical load arrangement of columns,

(ii) When the column spacings are so large that the depth of the slab required for a flat slab raft becomes too large,

(iii) When the foundation soil is not very uniform and we expect soft spots in the foundation area.

The raft, as shown in Figure 12.1, consists of slab and beams supporting the column loads. This type of raft is more rigid than the flat slab rafts and more efficient in distributing loads. The beams can be upstanding (above the slab) or downstanding (below the slab in the ground). In practice, in many cases, they have to be built at least partly downstanding to allow the crossing of reinforcements in the beams where they intersect. In this chapter, we deal with the analysis of beam and slab rafts subjected to uniform pressure.

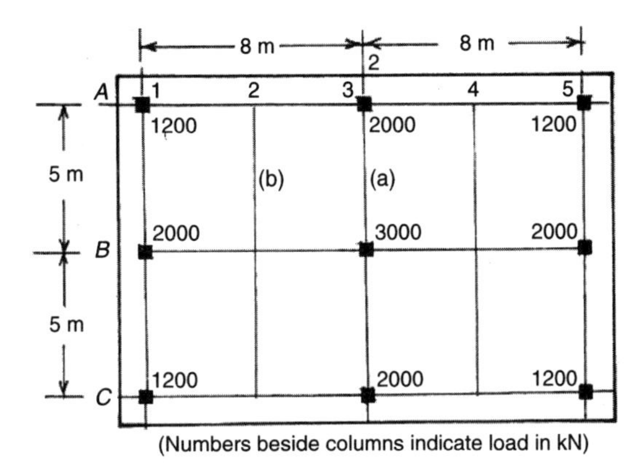

Figure 12.1 Layout of beam and slab rafts.

12.2 PLANNING OF THE RAFT

For uniform ground pressure to be exerted by the raft, it should be so planned that, as far as possible, the CG of loads and the raft should coincide. We assume uniform contact pressure under a *rigid foundation.* [If it cannot be avoided and there is eccentricity in loading we can proceed as indicated in Example 11.3.] It is also a good practice to extend the base slab beyond the outer column lines to form a peripheral *apron* of 0.5–1.5 m width beyond the peripheral beams. As shown in Figure 12.1, beams are laid out along the columns, both in the *X*- and *Y*-directions so that we will have *main longitudinal beams in the long direction and also transverse beams.* If the space between the transverse beams joining the columns is too large, we also divide the slab with *secondary beams.* (Transverse beams are also called and designated as secondary beams.) Generally, the longitudinal and the end transverse beams are taken for the full lengths and widths of the slab. Other secondary beams may or may not extend to the full width of the slab (see Figure 12.1).

12.3 ACTION OF THE RAFT

Depending on the L_y/L_x ratio (dimensions of the slab), the base slab will act as one-way or two-way slab. The conventional method of analysis is to assume the transfer of load upwards. The slab transfers its load onto the secondary beams and transverse beams, which, in turn, transfer the load to the main longitudinal beams. In our computations, such a simple concept of transfer of loads may not result in the upward reaction from the ground on the longitudinal beams equal to the downward load from the columns unless the loading is symmetrical. This inadequacy of equilibrium of forces, if small, is usually neglected by assuming that redistribution of loads will take place. However, some recommend that this discrepancy should be corrected and equilibrium of forces should be brought in the analysis of each element by introducing a uniformly distributed load (or proportionally estimated concentrated loads) on the longitudinal beams.

Another widely accepted method is to estimate the reactions by assuming the transfer of loads from the column to the longitudinal and then to the transverse and subsidiary beams by assuming that these reactions from transverse and secondary beams are in proportion to the slab areas supported by these beams. This assumption makes the analysis simple. We examine these two concepts in more detail in Sec. 12.4.2.

12.3.1 Approximate Dimensioning of the Raft

In civil engineering practice, especially for preliminary estimation, it is usual to dimension the members of the raft using rules of thumb such as *L*/*d* ratios and minimum sizes. However, we should remember, that whereas the total loadings of roof slabs of buildings will be of the order of 8–15 kN/m^2 only, rafts which support many floors can have very large loadings from all the floors, up to the safe bearing capacity of the foundation (which may be 100–200 kN/m^2). Hence the *L*/*d* ratios we apply for roofs cannot be applicable to rafts. The following values can be used for the first estimation to be finalized later by structural calculations:

1. **Main slabs.** For main slabs, we may use *L*/*d* (span/depth) ratios of 20–25 (that we use for beams for floors). As steel costs are high, to save steel (the percentage of steel in slabs need to be also around 0.25% only), it is wiser to use larger depths of slabs. However, the minimum depth of slabs with a cover of 50 mm (to mud mat) should be 250 mm for column loads 50–200 tons and about 600 mm for column loads above 400 tons. Generally, more than the theoretical value derived by calculations is given for providing adequate compression flange for the T beams and saving the cost of steel. Where the contact stresses vary due to eccentric loading, the slab thickness may also be varied depending on the pressures it has to resist.
2. **Aprons.** Usually, the slabs are extended beyond the peripheral beams to form an apron on all sides. This apron can be used to reduce the bearing pressure or adjust the centre of gravity of the raft. They also assist in resisting floatation in deep basement floors as they are loaded by earth from the above. As the aprons are cantilevers, the span/depth ratio of 7 (based on the *corner span* of the apron) is usually used in design (Figure 12.1).
3. **Transverse beams and secondary beams.** The simple difference between transverse beams and secondary beams is that the secondary beams have column loads at all junctions with the main beams (see Figure E12.4. S_1 and S_3 are secondary beams while S_2 is a transverse beam). These beams can be imagined as a means to collect the pressures from the slab and transmit it to the longitudinal beams. Their depth should be more than 1.5 times the thickness of the slab for beam action. Usually, a depth of over 400–700 mm is adopted. A breadth suitable for the depth, say, a minimum of 250–300 mm can be adopted for these members. A span/depth ratio of 10–16 is an approximate estimate of these beams.
4. **Longitudinal beams.** The transverse beams transmit the reactions to the main beams which are assumed to carry the column loads. Usually, a depth of over 800–1000 mm is adopted for the central beams and also for the side beams when they are used for supporting the load from the apron. They should naturally be deeper and stiffer than the transverse and secondary beams. A span/depth ratio of 5–8 may be needed for these beams.

5. **Estimate of steel required.** Generally, the percentage of steel in beam and slab works out to 60–120 kg/m^3 of concrete. The minimum steel in slabs should be 0.23% and in secondary and main beams not less than one per cent (average 1.75 per cent of the section as steel reinforcement).

12.4 DESIGN OF THE BEAM AND SLAB RAFT UNDER UNIFORM PRESSURE

The first condition for the approximate design of these rafts is that the CG of the raft and the loads should coincide. At least, the eccentricity should be small. The conventional design of the basic elements, the slab, transverse and secondary beams, as well as the main beams, is carried out by the conventional structural analysis of cutting the whole structure into base elements. A brief description of their analysis is given here.

12.4.1 Structural Analysis for the Main Slab

In general, we find three types of slabs in the design of raft foundation.

- Main slab
- Apron slab
- Corner edge slab which needs special consideration.

1. **Design of main internal slab.** This slab is bounded by the transverse or secondary beams and the longitudinal beams. It can act as one-way or two-way slabs, depending on the ratio of spans in the *XX*- and *YY*-directions. One-way slabs can be analysed by the basic principles and two-way slabs by using the table of coefficients given in Table 26 of IS 456 [1].

   Two-way slabs are divided into central and edge strips. The edge strips need to be reinforced only with nominal steel. [ACI uses the flat slab method for the design of two-way slab also.] Using the table of coefficients, we calculate

$$M_x = \alpha_x w (L_x)^2 \quad \text{for span } L_x$$

$$M_y = \alpha_y w (L_x)^2 \quad \text{for span } L_y$$

   (*Note:* Both expressions use L_x, *the short span* for moments α_x and α_y are the coefficient and w is the uniformly distributed load.)

2. **Design of apron or edge slab.** These are cantilever slabs and their depth is governed by the depth of the corner edge slab. A span/depth ratio of 7 used for cantilever beams in roofs, based on the corner distance, can be tried for initial value. The aprons are designed as cantilevers, and the corner is specially designed as described here.

3. **Design of corner portion of apron slab.** If the beam does not extend to the end of the apron, we have a corner slab. For design of this corner slab, the corner span length is normally used

$$M_c = q(L_c)^2/2 \quad \text{per metre width}$$

The necessary steel is provided by the extensions of steel from the two sides.

A_s (required) is obtained from $\sqrt{2}[A_{s1} + A_{s2}]$, i.e. the steel from the slabs from the X- and Y-directions.

12.4.2 Design of Secondary and Main Beams

There are usually secondary beams, transverse beams, (similar to secondary beams) and the main or longitudinal beams. Generally, transverse beams are also treated as secondary beams.

1. **Design of transverse and secondary beams.** The loads from the ground slab are assumed to be transferred to the transverse and secondary beams. If the slab is a one-way slab, all loads from the slab are carried directly to these beams. If the slabs are two-way slabs, the loads are assumed to be dispersed at 45° to the beams as shown in Figure E12.4. This will result in triangular or trapezoidal loading as shown in Table 12.1. Two different assumptions can be made as to how the load is transferred by the transverse beam to longitudinal beams, one from column downwards to secondary beams and the other from secondary beams upwards to columns. Depending on these assumptions we can use the following methods of analysis.

   Method 1 By simple statics. (*Analysis based on the area of the slab attached to the secondary beam. Area of support*): In this method, we consider the *flow of loads from column downwards*. We assume that all the loads from the columns are transmitted through the main beams to the secondary beams and that the *reactions from the secondary beams are in proportion to the area of the slab supported by the secondary beams*[2]. This method is simple and is illustrated in Example 12.2.

   Method 2 (*Approximate analysis using table of coefficients*): In this method, we assume load transfer *from the slab to the secondary beam and then from the secondary to the longitudinal beam.* We analyze the secondary beams for the ground reaction transmitted *from the slab to these beams.* These secondary beams will be subjected to UDL (in one-way slabs) or trapezoidal loads in the case of two-way slabs. These beams may be also continuous. We use the table of coefficients given in Table 12.1 for BM and SF calculation instead of theoretical methods such as moment distribution to determine the bending moments and reactions. This method is shown by Example 12.3.

   Method 3 (*Use of moment distribution method or other classical analysis*): When the fixing moments calculated from the two sides at the same support are different, we use this method of analysis as for example there is a cantilever at the end of the beams. In this method, we assume the same load transfer as in method 2, but we use moment distribution method to determine the exact moments. This principle is illustrated in Example 12.5.

2. **Analysis of longitudinal (main) beams.** There are two types of longitudinal beams, viz. the side or external longitudinal beams and the internal longitudinal beams. Both

are analysed using the same principles as explained above for secondary beams according to the layout of these beams. Thus we may use any of the following three methods as in secondary beams. Usually when using method 2 of coefficients we first analyse the internal main beams.

TABLE 12.1 Table of Bending Moment Coefficients for Analysis of Raft Loadings[3]

Case No.	Loading	B.M. diagram	Bending moment: Support	Bending moment: Centre
1.	W; L		$\frac{WL}{8}$	$\frac{WL}{8}$
2.	$wL = W$		$\frac{WL}{12}$	$\frac{WL}{24}$
3.	W		$\frac{WL}{9.6}$	$\frac{WL}{6}$
4.	$W/2$, $W/2$		$\frac{WL}{12}$	$\frac{WL}{24}$
5.	$\bar{w}$		$\frac{\bar{w}L^2}{3}$	–
6.	w; A, B, C, D, E; L, L		$M_B = 0.07\ wL^2$ $M_C = 0.125\ wL^2$ $R_A = 0.375\ wL$ $R_C = 1.25\ wL$	
7.	αL, W, αL; A, B; L		$M_A = \frac{WL}{12}(1 + \alpha - \alpha^2)$ W = Total load	

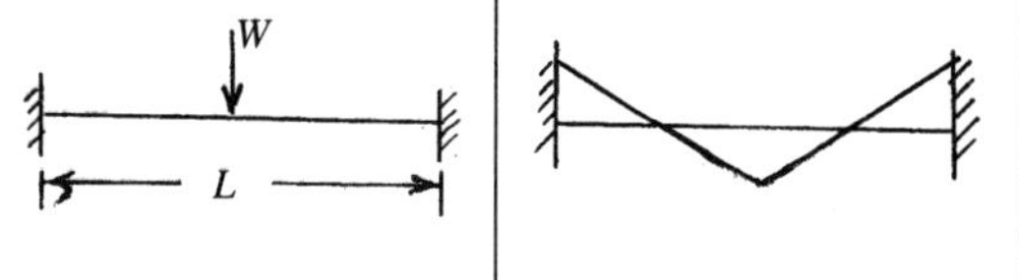

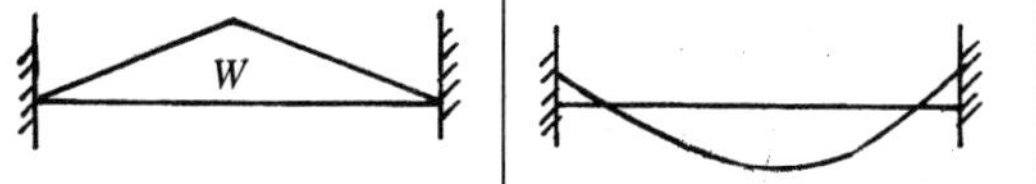

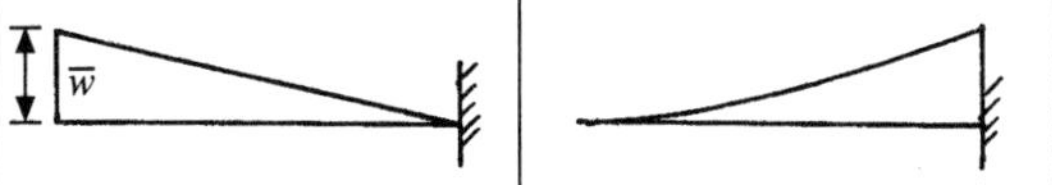

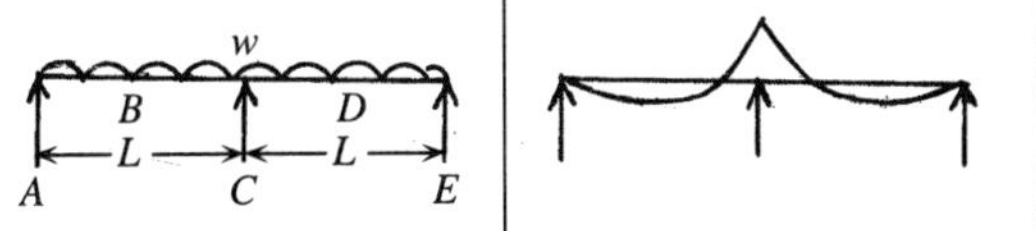

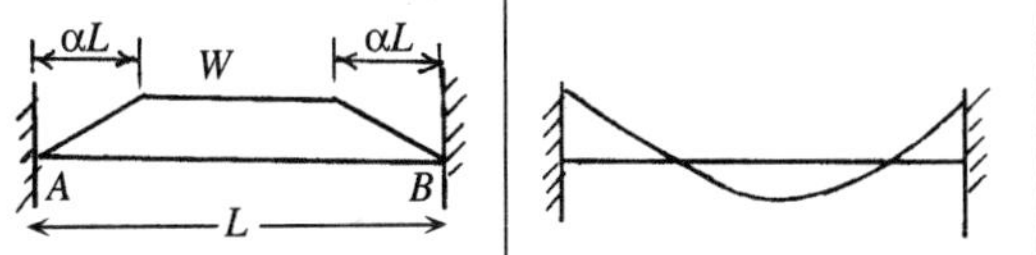

Method 1: *By simple statics* For an approximate and conservative analysis, we proceed as follows. After finding the reactions from the secondary beams, we may assume that each longitudinal beam is independent and statically determinate as we know the load from the columns acting downwards and also the upward reactions.

Theoretically, the upward reaction should be equal to the load on the beam from the columns. If there is any difference, an additional load is introduced for equilibrium. This load may be a UDL from below acting upwards or additional concentrated load along the column line acting downwards in proportion to the column load. In this case, the beam becomes statically determinate and the BM and SF can be easily calculated.

***Method 2:** By table of coefficients* Depending on the support conditions, we may use the table of moment coefficients (Table 12.1) to calculate the fixing moments as indicated in Example 12.4, item (E).

***Method 3:** By moment distribution* As explained in Sec. 12.4.2 (Method 3) the classical method is to use the moment distribution method. We analyse the longitudinal beams loaded from the reactions of the secondary beams and other condition loads such as apron load for a side longitudinal beam. No checking of the downward load of columns and upward reactions is made as it is assumed that the load will be redistributed in the system as a whole.

12.5 ANALYSIS BY WINKLER MODEL

Exact soil structure interaction methods can give us better results than the rigid method approach described above. But the difficulty is in finding the exact soil constants to be used in the analysis. An approximate method of using the Winkler model for raft analysis is to cut the raft (even the beam and slab type), as in all two-way slabs in ACI method, along the mid-points between beams both in the *X*- and *Y*-directions and design them as beams on elastic foundation. We should be aware that the two-way slab analysis using the ACI method is different from IS and BS methods for two-way slabs. In the ACI method, we assign the beam moments according to the rigidity of the beam.

12.6 MODERN METHODS BY USE OF COMPUTERS

The most exact method is the use of computer methods, such as finite element method after modelling the raft and foundation for soil-structure interaction. However, they are needed in only difficult cases as the approximate rigid method has been found to give satisfactory results in the field. Dimensioning the thickness of slab, depth of beam etc., providing the required minimum steel, and detailing the steel are more important than the exact analysis for a successful performance of the raft.

12.7 DETAILING OF STEEL

As we cannot know the exact distribution of the ground pressures and we assume they are uniformly distributed, we must be very conservative in providing the necessary rigidity to the foundation. The minimum span/depth ratios and steel should be taken care of. *A certain minimum amount of reinforcement should be provided both ways on top and bottom of slabs and beams*. Many recommend steel of 0.25% in slabs and 1% in beams at top and bottom, which works out to about 1.75% steel on an average in the raft.

The design should also be in conformity with the soil conditions such as the nature of the soil, the likely presence of soft pockets, and the amount of shrinkage of concrete that can take place in dry weather. We should remember that any defect that can happen in foundation structures cannot be rectified as easily as in the superstructure. Hence, we should be very conservative in our design of foundations.

12.8 SUMMARY

The initial layout of a raft foundation is as important as the design. The CG of loads and rafts should coincide. The slab is designed assuming uniform ground pressure. The concept of downward transfer of column loads through the main beams to the secondary beams in proportion to the area supported by the secondary beams is a simple concept, leading to an easy analysis (Example 12.2). However, in the classical method, we imagine that the loads transfer is from slab to the secondary beam and from the secondary beam to the main beam to the column. We make elaborate calculations by moment distribution or other exact analysis to find the reactions from secondary beams to the main beams and also for the analysis of the longitudinal beams. This method is tedious and time-consuming. Simple methods of estimation of load transfer with a judicious choice of minimum steel for each member and correct detailing of reinforcement will always give satisfactory results. In all our designs, the basic dimensions, minimum steel ratios and detailing of steel are more important than sophisticated methods of analysis.

In the following examples, we will examine only the methods of analysis of beam and slab rafts. For brevity of presentation, only the analysis will be explained. The design of these reinforced concrete beams and slabs after finding the BM and SF diagrams is carried out using the basic principles of the limit state design explained in Chapter 2. These details of design are not presented.

EXAMPLE 12.1 (Example on preliminary dimensioning of a beam and slab raft)

The layout of the columns of a multi-storied building is shown in Figure 12.1. The column loads are 3000 kN at the centre, 1200 kN at the corners, and 2000 kN at the sides. Give an approximate dimensioning of beam and slab raft foundation for the buildings. Assume the bearing capacity is 100 kN/m^2.

Reference	*Step*	*Calculation*
	1	*Give a layout of the slab and check bearing capacity* Total characteristic load = 3000 + 4(2000) + 4(1200) = 15800 kN Add weight of raft @ 10% = 17380 kN We will also provide 1 m apron all around so that the base slab is 12 m × 18 m. Contact pressure = $\frac{17380}{12 \times 18} = 80\ \text{kN/m}^2 < 100\ \text{kN/m}^2$ (SBC) The layout of the slab is OK.

Reference	Step	Calculation
	2	*Division of slab with beams*
		We first connect the column by *longitudinal and transverse* beams. As the distance between the columns is 8 m, we will divide them by *secondary* beams as shown in Figure 12.1. As the resultant slabs will be 5 m × 4 m they will act as two-way slabs.
	3	*Depth of slab*
		Two-way slabs of (5 m × 4 m). Adopt 1/25 spans (as for beams in roof system).
		Thickness of slab, $d = \frac{4 \times 1000}{25} = 160 \text{ mm}$
		D = 160 + 6 + 12 + 50 = 228 mm; Adopt 230 or 250 mm slab.
	4	*Layout of apron slab (cantilevers)*
		For the apron slab, base depth on the corner distance of the slab.
		Cantilever corner length $L_c = \sqrt{2} \text{ m} = 1.4 \text{ m}$
		Adop $L/d = 7; d = \frac{1400}{7} = 200 \text{ mm}$
		D = 200 + 6 + 50 = 256 mm; adopt 260 mm.
	5	*Transverse beams*
		Assumeand L/d = 10 and minimum D = 500–700 mm
		$d = \frac{4000}{10} = 400 \text{ mm}$. Using 16 mm rod
		D = 400 + 8 + 16 + 50 = 474 mm, say 500 mm
IS 456 (2000) Cl. 22.2		Choose a compatible width of the beam so that the widths do not become wide supports for the slabs. b < span/12 (as per IS). The beam width should suit the column size also, say $300 \text{ mm} < \frac{4000}{12}$.
		Adopt 500 × 300 beams.
	6	*Secondary beams*
		For secondary beams, adopt the same size as transverse beams.
	7	*Central longitudinal beam*
		Try L/d = 8 or minimum D = 1000–1800 mm
		$d = \frac{8000}{8} = 1000 \text{ mm}$
		$D = 1000 + 74 \approx 1100$ mm; b = 400 mm
		(b should be < 5000/12 = 416)
		Adopt 1100 × 400 beams.

Reference	*Step*	*Calculation*
	8	*Side longitudinal beams* $d = L/10$; D = 800–1000 mm $d = \dfrac{8000}{10}$ = 800 mm; D = 900 mm; b = 400 mm *Notes:* (a) If the apron is wide and the loads are heavy, we may have to adopt the side longitudinal beam as the same size as the central longitudinal beam. (b) All these dimensions have to be checked by actual calculations as described in the text.

EXAMPLE 12.2 (Design of beam and slab raft by Approximate method 1. Reactions based on areas of support. Flow of load from columns downwards)

A building is built on 12 columns, whose layout and loading is as shown in Figure E12.2. Indicate how the bending moment and shear forces for the design of the various elements are to be calculated. Assume SBC is 75 kN/m^2 [2].

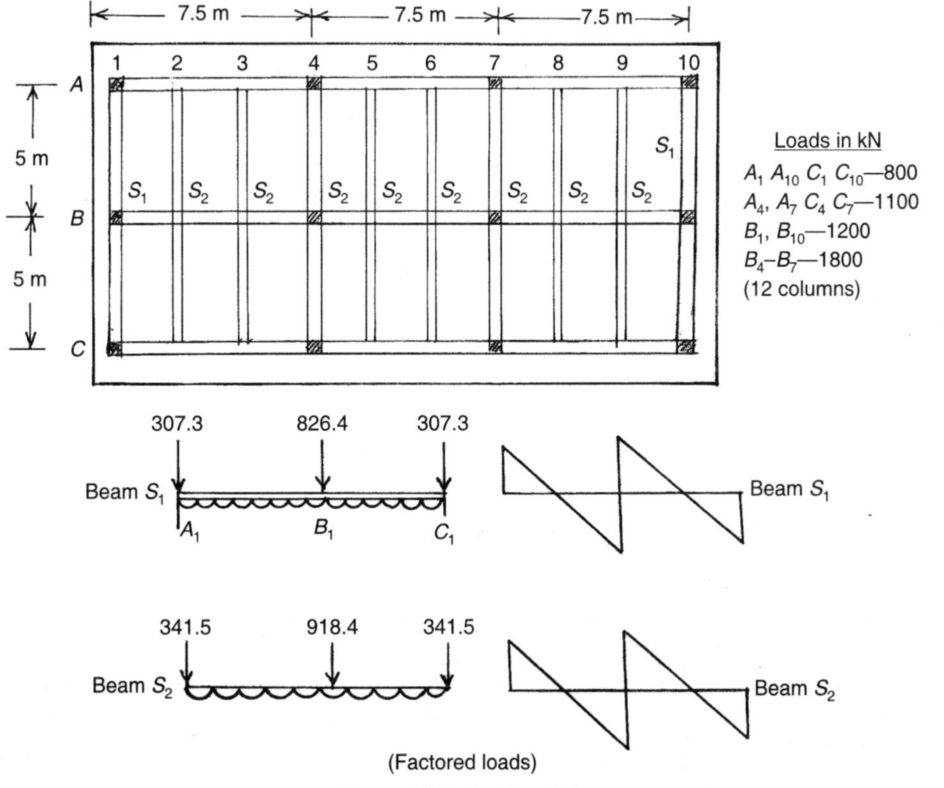

(Factored loads)

Figure E12.2 (Contd.)

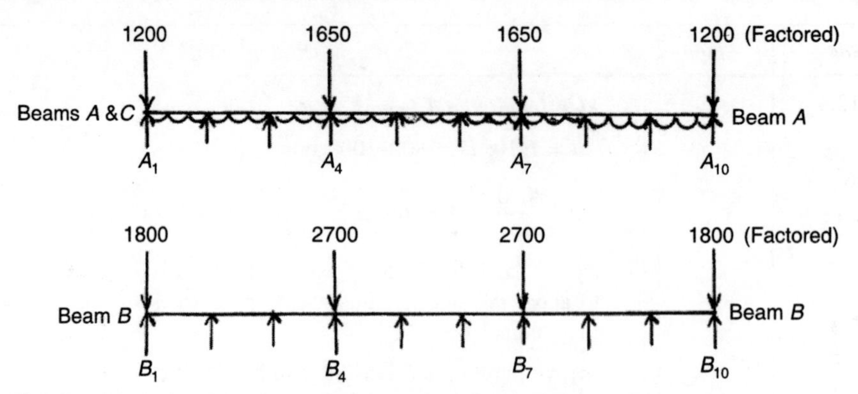

Figure E12.2 Analysis of beam and slab path assuming reactions on secondary beams are proportional to area they support.

Reference	*Step*	*Calculation*
	1	*Determine the layout of the raft*
		We will provide aprons by extending the raft to 1.5 m on each side of the breadth and 1 m on each side along the length (larger in the breadth).
		$L = (3 \times 7.5) + 2 = 24.5$ m
		$B = (2 \times 5.0) + 3 = 13$ m
		Total area = 24.5 × 13 = 318.5 m^2
		As the loads are symmetrical, the CGs of loads and raft coincide.
	2	*Check bearing pressure under characteristic load in* kN
		Total load = 2(1800) + 2(1200) + 4(1100) + 4(800) = 13600 kN
		Load + 15% self wt. = 13600 + 2040 = 15640 kN
		Bearing pressure = 15640/318.5 = 49.10 kN/m^2
		This is less than 75 kN/m^2 (based on working load)
	3	*Find factored design load for limit state design of raft*
		Factored load to be distributed = 1.5 × 13600 kN
		(Self-weight need not be distributed by raft.)
		Factored contact pressure $= \dfrac{1.5 \times 13600}{318.5} = 64.05 \text{ kN/m}^2$
	4	*Arrive at layout of transverse and secondary beams*
		We will connect columns by *longitudinal beams* and *transverse beams*. The greater span of 7.5 m between columns and transverse beams will be divided into three spans by *secondary beams* at 2.5 m, resulting in 2.5 × 5.0 m two-way slab spanning between the secondary beams.

Reference	Step	Calculation
Sec. 12.4.1	5	*Design of slabs* (2.5 × 5.0 m) using standard design method
		These are *one-way slabs* continuous over a series of beams.
IS 456		We can use the BM coefficients as given below.
Table 12		+ve end span = 1/10; +ve interior = 1/12
		–ve end span = –1/9; –ve interior = –1/10
		Let us use the coefficient 1/10 both for positive and negative moments.
		BM = $wl^2/10$ = 64.05(2.5)^2/10 = 40 kNm width
		Minimum depth, $d = \sqrt{M/0.14 f_{ck} b}$
		$= \left(\dfrac{40 \times 10^6}{0.14 \times 20 \times 1000}\right)^{1/2} = 119.5 \text{ mm}$
		D = 161 + 6 + 50 = 217 mm; Adopt 220 mm
	6	*Design the cantilever slab*
		BM for 1.0 m cantilever = 64.05(1)^2/2 = 32 kN/m
		Adopt the same depth, vary steel.
		These slabs can be tapered to the end to 125 mm at the outer ends.
	7	*Design of corner of apron slab*
		We design this part of the slab for the corner length = $\sqrt{1 + (1.5)^2}$
		= 1.8 m (apron widths are 1 and 1.5 m)
		As already explained in Sec. 12.4 the reinforcements from both sides are extended in this corner.
	8	*Design of transverse beams by simple statics*
		Note: The design of the complex raft structure by approximate methods gives approximate results, which give guidance for its detailing. In this example, we will use method 1 of reactions based *on the areas of support.*
	9	**(a) Central long beam**
		Analysis for reactions from transverse and secondary beams to central longitudinal beam B
		We work from the column load downwards. We assume that the reactions from transverse beams on to the *central longitudinal beam* are directly proportional to the soil contact area of the slab area it supports.
		Area covered by end transverse beam S_1
		S_1 = (2.5/2 + 1) = 2.25 units
		Area covered by S_2 (interior beams = 2.5 units)

Reference	Step	Calculation
		Total units $2S_1 + 8S_2 = 2 \times 2.25 + 8 \times 2.5 = 24.5$
		Total downward load of *central longitudinal beams* only with load factor (loads from B_1, B_3, B_7 and B_{10}).
		$$1.5(2)(1200 + 1800) = 9000 \text{ kN}$$
Figure E12.2		Reaction of $S_1 = \dfrac{9000 \times 2.25}{24.5} = 826.5 \text{ kN}$ (i)
		Reaction of $S_2 = \dfrac{9000 \times 2.5}{24.5} = 918.4 \text{ kN}$ (ii)
		[*Note:* The design loads on the central longitudinal beam will be as shown in Figure E12.2. The total reaction from the transverse beams below the longitudinal beam has been made equal to the downward loads from the column on it.]
	10	**(b) Analysis of side longitudinal beams**
		Next consider reactions from S_1, S_2, and others *on side longitudinal beams*
		Consider beam S_1 along its length (neglecting apron)
Step 9		UDL on S_1 from slab = beam spacing × factored slab load
		= 2.25 × 64.05 = 144.11 kN/m (length = 10 m)
		Total load on S_1 = 144.11 × 10 = 1441.1 kN (span of S_1 = 10 m)
		Hence, the reactions on far ends of S_1 are obtained as
		$$\frac{\text{Total load} - \text{Central reaction on central beam}}{2}$$
		$= \dfrac{1441.1 - 826.53}{2} = 307.3 \text{ kN}$ (iii)
See Step 8	11	*Next consider beam* S_2 *in a similar way*
		UDL on S_2 = 2.5 × 64.05 = 160.13 kN/m
		Total load on S_2 = 160.13 × 10 = 1601.3 kN
		Hence, the reactions on the far ends of S_2 are equal to
		$\dfrac{1601.3 - 918.36}{2} = 341.5 \text{ kN}$ (iv)
	12	*Draw the BM and SF diagrams*
		The BM and shears can be calculated as shown in Figure E12.2.
	13	*UDL on the side longitudinal beams (load from apron + correction load)*
		For the side longitudinal beam, the upward and downward loads may not equal. For finding correction to be applied, we first find the total reaction from ends of transverse and secondary beams, i.e.

Reference	Step	Calculation
		Two end beams S_1 + 8 internal beams S_2 (from Step 9)
		= 2(307.28) + 8(341.47) = 3346.2 kN (v)
		Ground pressure of adjacent apron = 1.5 × 64.05 = 96.07 kN/m^2
		Total load from adjacent apron = 96.07 × (3 × 7.5) = 2161 kN (vi)
		Total upward loads = 3346.2 + 2161 = 5507.2 kN col. load
		Total col. loads = 1.5 × 2(800 + 1100) = 5700 kN
		Difference = 5700 – 5507 = 193 kN upwards
		This is small and can be neglected or added as a UDL along the beam.
		We will add with the UDL from the apron.
		UDL on side beams = UDL of apron + 193/21.5
		(unbalanced load)
		= 96.05 + 8.98 = 105 kN/m (vii)
	14	*Drawing of SF and BM diagrams*
		The loading will be as shown in Figure E12.2. As the loads are all known, the values for BM and SF diagrams can be easily calculated and the diagrams drawn.

EXAMPLE 12.3 (Analyse the beams of the beam and slab raft in Example 12.2 by Method 2 assuming load transfer from slab to secondary beams and from secondary to main beams analysis by using BM and SF coefficients)

The beam and slab raft shown in Example 12.2 is to be analysed by alternate method.

Reference	Step	Calculation
	1–7	As in Example 12.2
		Factored contact pressure = 64 kN/m^2
	8	*Analysis of transverse and secondary beams (2 spans)*
		In this procedure, we work upwards from the slab to the secondary beam to the main beams to the columns. The transverse beam is assumed as a continuous beam of equal span over three supports with UDL from two-way slab. The following coefficients can be used for moments and shears.
		Taking l as the span of each beam and w as UDL
		[Coefficients to be used – span = l
		BM at ends = 0
Table 12.1		*max –ve BM at the centre* = $0.125wl^2$
		max +ve BM in the span = $0.07wl^2$

Reference	Step	Calculation
		R_1 and R_3 at ends = 0.375*wl*
		R_2 at the middle = 1.25*wl*]
Step 9		*(a) Analysis of end transverse beams S_1;*
Example 12.2		w = 144 kN/m and l = 5 m
		$M_B = 0.125wl^2 = 450$ kNm
		BM in spans = $0.07wl^2$ = 252 kNm
		$R_A = R_C = 0.375wl$ = 270 kN [307 in Example 12.2, Eq. (iii)]
		$R_B = 1.25wl$ = 900 kN [826 in Example 12.2, Eq. (i)]
		(b) Analysis of middle beams S_2[w = 160.13 kN/m, l = 5 m]
		$M_B = 0.125wl^2 = 500.4$ kNm
		Span $M_{AB} = M_{BC} = 0.07wl^2 = 280.2$ kNm
		$R_A = R_C = 0.375wl$ = 300.2 kN [341 in Example 12.2]
		$R_B = 1.25wl$ = 1000.8 kN [918 in Example 12.2]
		BM and SF can be easily drawn.
	9	*Analysis of side longitudinal beams*
		The reactions from the transverse and secondary beams to the supports on to the central and side longitudinal beams are known. For the side beams, we check whether there is equilibrium of the upward reaction and the downward column load. If there is no equilibrium of forces, we add a UDL *or* proportionate concentrated load to bring equilibrium of forces. By this procedure, all the upward and downward forces acting on the beams are known, so we can easily draw the BM and SF diagrams.
		We proceed as follows for the side longitudinal beam:
		Reaction from S_1 = 270 kN; reaction from S_2 = 300 kN
		Reaction from below = $2S_1 + 8S_2$ = 2940 kN
		Load from 1.5 m apron = 1.5 × 64.05 = 96.07 kN/m
		Total load from apron 96.07(3 × 7.5) = 2161 kN
		Total upward load = 2940 + 2161 = 5101 kN
		Sum of col. loads = 1.5(3800) = 5700 kN
		Unbalanced load 5700 – 5101 = 599 kN
		Hence,
		Upwards UDL on length (3 × 7.5) m for equilibrium
		= 599/(3 × 7.5) = 26.62 kN/m
		Total UDL = 96.07 + 26.62 = 122.7 kN/m [105 in Example 12.2]

Reference	Step	Calculation
		As all the column loads and base reactions are known, we can draw the SF and BM diagrams. The column loads act downwards, and the reactions from the transverse and secondary beams and a UDL of 122.7 kN/m act upwards.
	10	*Analysis of central longitudinal beams*
		The same procedure can be used for the longitudinal beams also.
		Mid-reaction from S_1 (@ B = 900 kN
		Mid-reaction from S_2 = 1000 kN
		Total upward reaction = $2S_1 + 8S_2$ = 9800 kN
		Total downward col. load = 1.5 × 2(1200 + 1800) = 9000 kN
		Unbalanced = 9800 – 9000 = 800 kN upwards
		Apply a downward UDL $= \frac{800}{3 \times 7.5} = 35.5$ kN/m
		The BM and SF of the central longitudinal beams can be easily calculated as all the loads and reactions are known.

EXAMPLE 12.4 (Conventional Analysis of beam and slab rafts). Indication of Procedure to be used.

Figure E12.4 shows the layout of a beam and slab raft with 13 columns and with a 1.5 m apron. The raft is to be built on a stratum of medium sand. Indicate the procedure to analyse the transverse, secondary and main beams of the raft foundation by the conventional method assuming that *factored* base pressure on the raft is 66 kN/m^2 [4].

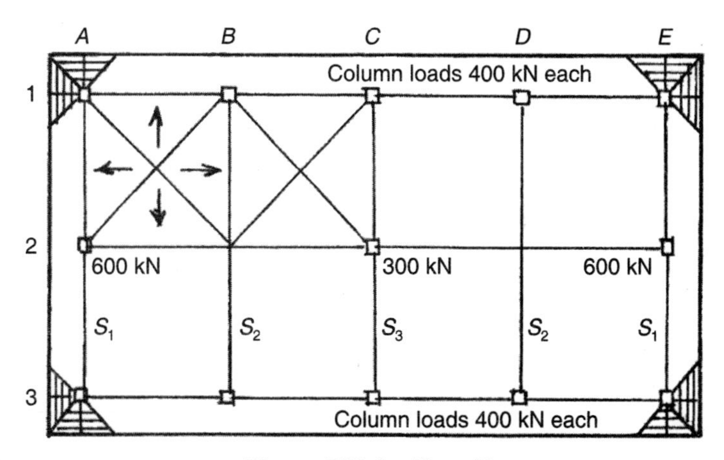

Figure E12.4 (Contd.)

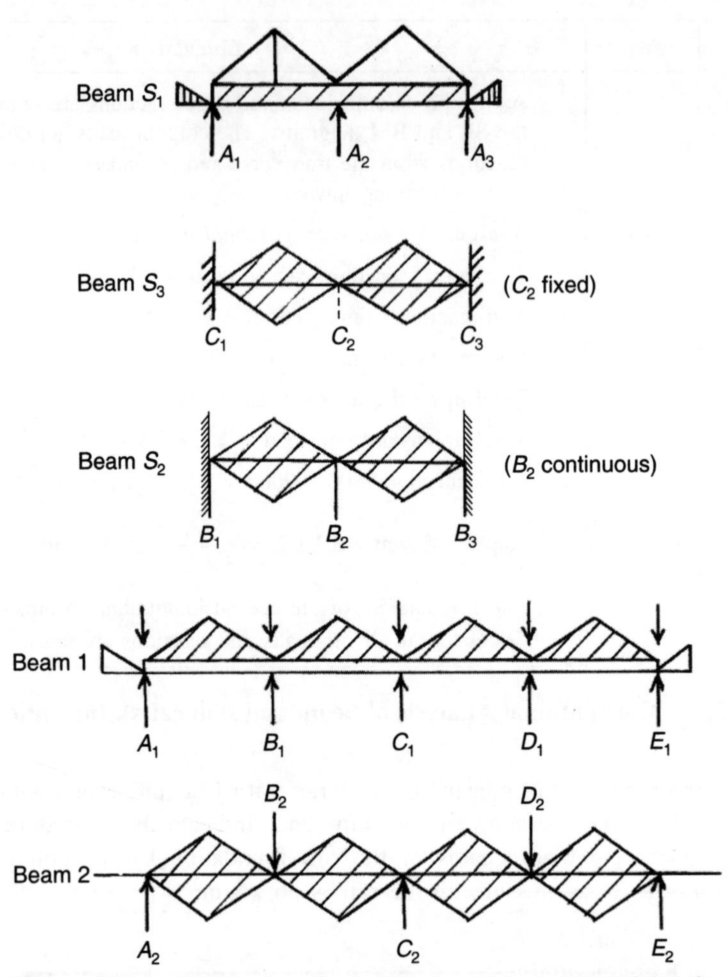

Figure E12.4 Classical method of design of beam and slab rafts.

Reference	*Step*	*Calculation*
	1	*Design the base slab, apron, etc.* (q = 66 kN/m^2) The slabs are two-way slabs designed in the conventional way. However, the minimum steel must be 0.25%.
	2	*Analysis of secondary and transverse beams* (S_1, S_2 and S_3) There are three types of junctions: (a) We consider a junction where columns, longitudinal beams and *transverse beams* meet as rigid supports (S_3). (b) We consider junction of *secondary beams* and longitudinal beams without column as simply supported if they are not continuous beyond the junction. They should be also detailed as such. If the beam is continuous, we consider it as continuous.

Reference	Step	Calculation
		(c) We consider junctions which are continuous over the longitudinal beam and continue over the apron as cantilevers (S_1). The loading diagrams of the end transverse beam and the other secondary beams are shown in Figure E12.4. Beam S_1 (A_1 A_2 A_3) can be analysed by moment distribution method. Beam S_3 (C_1 C_2 C_3) can be analysed by coefficients method for a two-span continuous beam and fixed at ends. Beam S_2 (B_1 B_2 B_3 and D_1 D_2 D_3) can be analysed by coefficient method as continuous beams, with simply supported ends.
	3	*Design of longitudinal beams* The loadings on the central and side longitudinal beams are as shown in Figure E12.4. As the loads are symmetrical, for a simple analysis of these beams, we can check for equilibrium of the upward reactions and the downward column loads. If they are not in equilibrium, we add either a UDL or concentrated loads proportionate to the reactions at the junctions of secondary beams. As all the downward and upward loads are known, we can draw the BM and SF diagrams from statics. For unsymmetrically loaded rafts, after analysing the slab and secondary beams and their reactions, we have to use theoretical methods such as moment distribution to analyse these main beams with the reactions we obtained by assuming that the load is transmitted from below upwards. In such an analysis, the need for equilibrium of forces on the beam from calculated reactions and the column loads is usually ignored; however, if necessary, it can also be taken into account.

EXAMPLE 12.5 (Analysis of the end transverse beam and longitudinal beams of raft shown in Figure E12.4 assuming transfer of loads from slab to beams and then to columns.)

[*Note:* For analysis of the slabs we carry out normal calculations with the ground pressures acting up. Proper care should be taken in the interpretation of the signs of the moments. In general, we use the following two methods depending on the conditions of support and Method 2 use of the table of coefficients. Method 3 use of moment distribution method.]

Reference	Step	Calculation
	A	**Analysis of end transverse beam S_1 supported at A_1 A_2 A_3.** *Note:* As the end transverse beam cantilevers into the apron slab, we will take it as a cantilever and continuous over other supports. The cantilever takes the load from an area bounded by 45° lines. The cantilever moment is known. As explained in Sec. 12.4.2 under method 3 we will use the moment distribution to find the moments. We analyse the beam as supported on column points and loaded as shown in Figure E12.4.

Reference	Step	Calculation
	1	*Determine loads on beam A1 A2 A3*
		a) Cantilever w = 66 kN/m
		End load $\bar{w} = 3 \times 66 = 198$ kN/m
		$M_A = wl^2/3 = 198(1.5)^2/3 = 148$ kNm (cantilever M)
		(b) Bending moments due to the triangular load from the raft slab
		$W = 66 \times 1.5 \times 1.5 = 148.5$ kN
Table 12.1		$M_{\text{Fixed}} = WL/9.6 = \dfrac{(148.5 \times 3)}{9.6} = 46.5$ kNm
		$M_0(\text{SS}) = WL/6 = 74.3$ kNm
		(c) Moments due to UDL from the apron outside
		$= 1.5 \times 66 = 99$ kN
		$M_F = wL^2/12 = 99 \times 9/12 = 74.5$ kNm
		$M_0 = wL^2/8 = 99 \times 9/8 = 111.5$ kNm
		(d) Total fixing moment = 46.5 + 74.5 ≈ 121 kNm
		Total free moment = 74.3 + 111.5 ≈ 186 kNm
		[*Note:* The rough value of span +ve moment (with fixing moment at both ends and M_0 = 186) = 186 – 121 = 65 kNm]
		Note: We may proceed to analyse the beam more exactly since at A, the cantilever BM = 148 and the inside total fixing moment is 121 kN. This can be done by moment distribution as follows.
	2	*Find distribution factors for end transverse beam (ABC)*
		At A_1 = 0 : 0
		At A_2 = 0.5 : 0.5 [each 3/4 stiffness]
	3	*Carry out moment distribution for beam A_1 A_2 A_3*

	A_1		A_2	
Junction Span		A_1A_2	A_2A_1	A_2A_3
DF	0	0	0.5	0.5
	+148	–121	+121	–121
		–27	–13.5	+13.5
	+148	–148	+107.5	–107.5

Step					
4	*Shear calculations*				
	Simply supported shear	+297	223 kN	223	223
	Shear change for fixing moment		+ 13.5*	–13.5	
	Final	297	236.5	209.5	209.5

* Change in shear = (148 – 107.5)/3 = 13.5

Step	Calculation
5	*Find value of mid-point moment (+ve moment)*
	Mean of BM at mid-point = (148 + 107.5)/2 = 127.75

Reference	Step	Calculation
		From Step 1, total M_0 = 186.5 kNm (SS moment)
		+ ve moment = 186.5 – 127.75 = 58.75 kNm (approx.)
	6	*Find design shear value*
		Determine shear as shown in the tabular representation above.
	B	**Analysis of transverse beam** S_3 supported at C_1 C_2 C_3 (with columns)
		We have columns at C_1, C_2 and C_3. Hence, beams can be assumed as fixed at these points and hence can be *analysed by Method 2 coefficients.*
	1	*Calculate the loads and moments*
		Triangular load at both sides $W = 2 \times 148.5 = 297.0$ kN
		We use coefficients for fixing moments
		$M_F = (297 \times 3)/9.6 = 92.8$ kNm
		$M_0 = 297 \times 3/6 = 148.5$ kNm
		+ve moment = 148.5 – 92.8 = 55.7 kNm
		Max shear = 148.5 kN
		Design the beam for these moments and shear.
	C	**Analysis of secondary beam** S_2 supported at B_1 B_2 B_3
		No column in the middle
		These beams have columns at B_1 and B_3 at the two ends (which can be considered as fixed) and are continuous with no column in the middle. Use Method 2 coefficients for beams fixed at ends, continuous in middle support.
		[If the ends had no columns at the ends, they can be considered as simply supported and solved by using the corresponding coefficients.]
	D	**Analysis of side longitudinal beams No. 1 supported on** A_1 B_1 C_1 D_1 E_1 with cantilever on both ends.
		We analyse this beam by moment distribution as in the case of the transverse beams A_1 A_2 A_3.
	1	Find distribution factors (EI = const.)

Joint	*Member*	*Relative stiffness*	*Sum*	*Distribution factor*
B_1	B_1A_1	$3/4I$	7/12	3/7
	B_1C_1	I		4/7
C_1	C_1B_1	I	2	1/2
	C_1D_1	I		1/2
D_1	D_1C_1	I	7/12	4/7
	D_1E_1	$3/4I$		3/7

Reference	*Step*	*Calculation*
	2	*Calculate moments and distribute*

Joint	A_1		B_1		C_1	
	X_1A_1	A_1B_1	B_1A_1	B_1C_1	C_1B_1	C_1D_1
DF	0	0	3/7	4/7	1/2	1/2
FM	148	–121	+121	–121	+121	–121
		–27	–14			
	148	–148	+107			
			+6	+8	+4	–4
	+148	–148		–113	+125	–125

Proceed as in item A to find positive moment and shear.

E — **Analysis of central longitudinal beams supported on A_2 B_2 C_2 D_2 E_2 with no cantilever**

There are only two spans symmetrically loaded. There are columns at all the three supports.

We assume fixity at column points and determine the fixing moments by Method 2 by coefficients

Step 1 — *Determine fixing moment for triangular load*

$$W/2 = (2 \times 66 \times 1.5) \times (3/2) = 297 \text{ kN}$$

$$W = 2 \times 297 = 594 \text{ (from 2 triangular loads)}$$

Reference: Table 12.1

Fixing moment from Table 12.1, item (4)

$$M_F = \frac{WL}{12} = \frac{594 \times 6}{12} = 297 \text{ kNm}$$

$$M_S = \text{Span moment } = \frac{WL}{24} = 148.5 \text{ kNm}$$

$$M_0 = M_F + M_S = 297 + 148.5 = 445.5 \text{ kNm}$$

Step 2 — *Determine fixing moment due to concentrated load from the secondary beam S_2 = 297 kN*

Reference: Table 12.1

$$M_F = \text{Fixing moment } = \frac{WL}{8} = \frac{297 \times 6}{8} = 222.75 \text{ kNm (Case 1)}$$

$$M_S = \text{Span moment } = \frac{WL}{8} = 222.75 \text{ kNm}$$

$$M_0 = \frac{WL}{4} = 445.5 \text{ kNm}$$

Step 3 — *Combined effect of triangular and concentrated loads*

Total fixing moment = 297.0 + 222.75 = 519.75 kNm

Reference	Step	Calculation
		Total M_0 = 445.5 + 445.5 = 891 kNm
		+ve moment = 891.0 – (519.75)/2 = 631.0 kNm
	4	*Find design moment*
		–ve moment = 520 kNm
		+ve moment = 631 kNm
	5	*Determine shear and check for shear*
		As shown in item (A), Step 4.

REFERENCES

[1] Varghese, P.C., *Advanced Reinforced Concrete Design*, Prentice-Hall of India, New Delhi.

[2] *Examples of Design of Buildings to CP110 and allied codes*, Cement and Concrete Association, London, 1978.

[3] Reynolds, C.E. and J.C. Studman, Reinforced Concrete Designer's Handbook, A Viewpoint Publication, 1988.

[4] Brian, J. Bell and M.J. Smith, *Reinforced Concrete Foundations*, George Godwin Ltd., London, 1981.

CHAPTER 13

Compensated Foundations, Cellular Rafts and Basement Floors

13.1 INTRODUCTION

Basement rafts and buoyancy rafts are compensated foundations. They are adopted when (a) the safe bearing capacity of the soil in the foundation of a building is very low to support the building, or (b) when the calculated settlement of the building is excessive. In the latter case, it is possible to have a partly compensated foundation by supporting the building on an excavation, where the weight of the soil excavated is equal to only a part of the weight of the building. The raft foundation for such cases is a *basement raft*. If the weight of the soil excavated is fully equal to that of the building, *it is called a fully compensated or floating foundation or buoyancy raft*.

Where a foundation is to be constructed in very deep soft clay, the buoyancy raft (or floating foundations) is the only possible type of foundations. Other shallow or deep foundations (e.g. pile foundations) are not possible in such deep soft soils. A number of such compensated foundations have been successfully built all over the world, starting with the one that was planned by Prof. Casagrande for the Albany building in the USA in about 1940.

The difference between *basement rafts* and *buoyancy rafts* should be clearly understood. *True buoyancy rafts* are made in weak clayey soils. In such cases, compensation of weight is made by construction of a basement solely for compensation by buoyancy of the displaced soil without any regard for utilizing the space for any other purpose. In practice, it will not be possible to fully balance a building because of reasons such as the variation of loading from

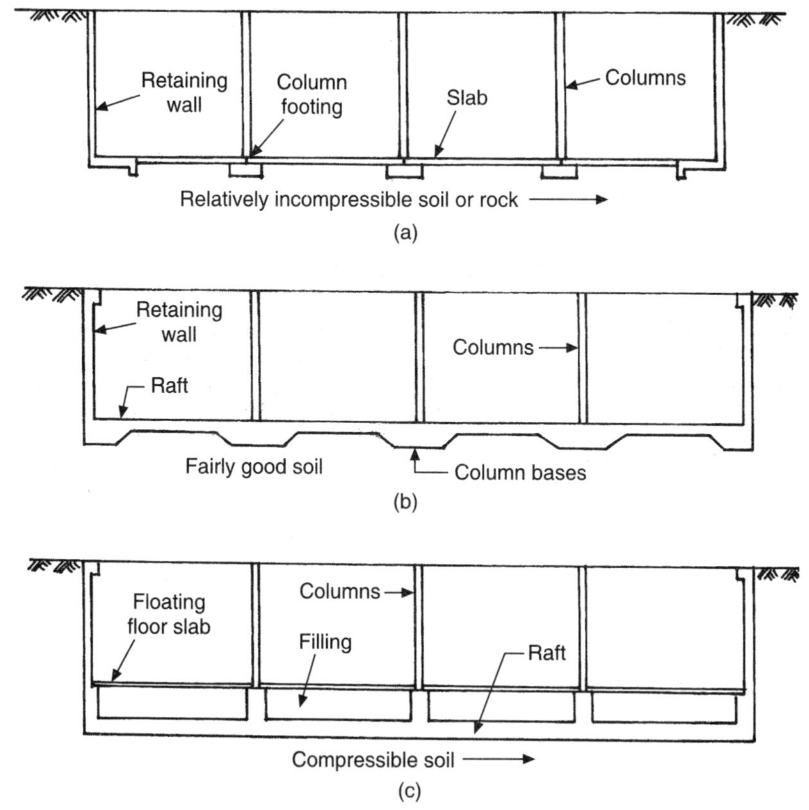

Figure 13.1 Footing and raft foundations for basement floors: (a) Type 1. Basement floor with independent footings for columns and cantilever walls, (b) Type 2. Basement floor with flat slabs, (c) Type 3. Basement floor with beam and slab floors. (*Note:* In the first type there is no increase in bearing capacity at foundation level due to depth factor which occurs in bearing capacity formulae for sands and clays.

the superstructure on the foundation, and variation of the ground water level. Hence, floating rafts, especially in soft clays, will always be subjected to some settlement and consequent differential settlement. It is very important that such a foundation should be a stiff raft foundation. These rafts should be *light and rigid to avoid large differential settlements and to evenly distribute the load from all the floors above*. Thus, they are invariably made of cellular construction and are known as *cellular rafts*.

On the other hand, *basement rafts* are those rafts constructed below the ground level as shown in Figure 13.1. The basement is also put to use for other purposes and forms part of the building proper. They are used in stronger soils. Their foundations may be ordinary rafts, or in extreme cases they may be made of rigid Virendeel girders, which not only act as rigid frames but also allow passage through them so that the basement can be put to use. The structural design of basement rafts follows the same principles of structural design of ordinary slab and beam raft foundations explained in Chapter 12. The basement floor may also have piles or piled raft foundation as shown in Figure 13.2. The general nature of construction of cellular rafts is shown in Figure 10.2 given in Chapter 10.

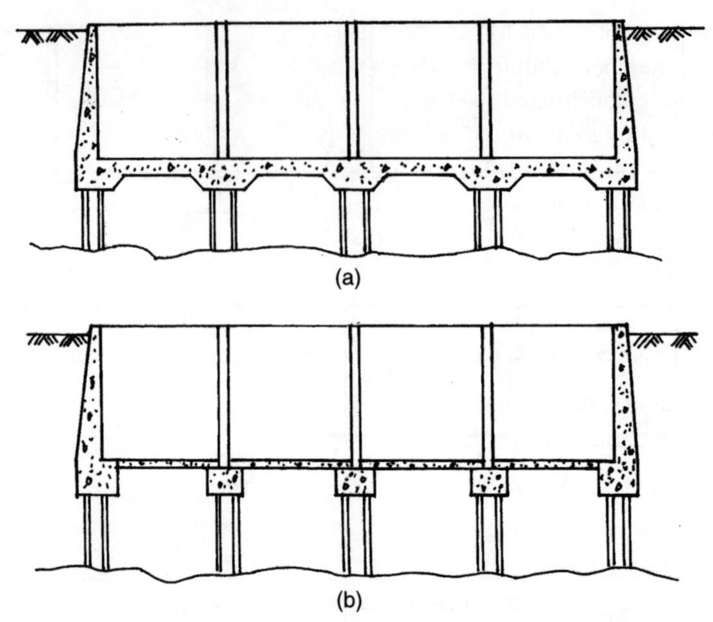

Figure 13.2 Deep pile foundations for basement floors: (a) Piled raft foundation where bearing capacity is satisfied but settlements are excessive, (b) Pure pile foundation for column loads and retaining walls.

13.2 TYPES OF COMPENSATED FOUNDATIONS

As already indicated, we may thus divide the *compensated foundation* into two types, depending on the soil conditions.

In Type I are the basement rafts. In this type of compensated foundations, the bearing capacity is not too bad and the excavation of the soil is not difficult. But, the settlement due to soil condition is large. In such cases, we have to reduce the settlement, particularly the differential settlement. The large settlement, in some cases, may be due to the presence of a lower soft layer of the soil. In this case, the foundation usually consists of a mat kept rigid, using Virendeel type of construction for the frames. The basement space can be put to use. The net effect of this type of basement raft is to reduce the load on the foundation, thus the safe bearing capacity is not exceeded and the differential settlement is also reduced.

In Type II are the fully compensated buoyancy rafts. In compensated foundations, the shear strength of the foundation soil is low and the *only way to build on the soil is by means of floating the foundation*. In such cases, *a box section type cellular raft* is sunk and made into a rigid raft to reduce the differential settlement. In this case, only the raft is *a truly cellular raft*. These are sometimes referred as buoyancy rafts. [These principles are well explained in Ref. 1.]

13.3 CONSTRUCTION OF CELLULAR RAFTS

We have seen that true cellular rafts are cellular in construction so that the foundation is as rigid and as light as possible. This will assist in reducing differential settlement to a minimum. The hollow space inside is not meant for use. A special arrangement for drainage of any water that may leak into the chambers should be also provided in these basements [1].

Cellular rafts are constructed in the form of caissons in soft clays with high water levels. The soils are removed by grabs as the raft-walls sink down under their own weight. [See Ref. 1 for details.] Construction in the open is possible in sites where the ground water level can be depressed by methods of ground water lowering.

The necessary precautions to be taken in excavations in clays, the heaves to be expected, etc. are explained in books on Soil Mechanics and Foundation Engineering.

13.4 COMPONENTS OF CELLULAR RAFTS

We know the magnitudes of the superstructure (column) loads on the ground floor. These loads are transmitted to the soil through the cellular raft, which is composed of the following:

- The ground floor slab *forming the top slab* of the cellular raft
- The peripheral wall which has to transmit vertical loads as well as the earth pressure and the water pressure
- The cross walls of the cellular raft
- The *bottom slab* resting on the soil at the foundation level

All these elements acting together give the necessary strength and rigidity. It should be clearly noted that even though the *net pressure on the foundation* at the foundation level is small, the raft has to be designed for the full (gross) vertical and other loads than act on the structure. We must remember that the full uniformly distributed dead load and the appropriative live loads (with allowable reduction for the number of storeys) are to be taken by the cellular system as a whole.

The slab in contact with the ground is to take the full upward pressure. The slenderness ratio of the peripheral and cross walls should be such that they are short braced walls with slenderness ratio less than 12. These walls should have at least 0.5% reinforcements, one half of the steel placed on each side of the wall. In conventional design, they are also to be considered as deep beams spanning in the horizontal direction.

13.5 ANALYSIS

The overall analysis can be carried out in two ways. The first simple elemental design method will be found to be very safe and is an overestimate of the real conditions. But, the design is simple to work out.

Method 1: The conventional method for analysis is to cut the structure to its elements as indicated (in Sec. 12.4) and to design the slabs, wall, etc. in the conventional way as in the design of a beam and slab raft.

Method 2: A less conservative design is to cut the cellular raft itself into a number of I sections through the middle of the slab both in the *XX*- and *YY*-directions as in the case of a flat slab and determine the requirements for bending moment and shear.

13.6 PRINCIPLES OF DESIGN OF CONCRETE WALLS

IS 456 (2000), Cl. 32 deals with the design of concrete walls. Walls with reinforcement less than 0.4% are classified as plain walls. In cellular rafts, the walls should be provided with a minimum of 0.5% steel (0.25% on each side) [2].

13.7 PLANNING AND DESIGN OF BASEMENT FLOORS

When basement floors are planned on rock or good soil, the differential settlement will be small and it will be necessary to design the floors only to resist the upward water pressure. If no water pressure can develop, the columns and walls can be built on independent foundations and the floor slab need to be only of nominal thickness, as shown in Figure 13.1(a).

Where appreciable differential settlement is expected, the bottom slab should be flat slab raft or beam and slab raft, as shown in Figure 13.1(b) and Figure 13.1(c).

Where the basement has to be supported on piles, we design the floor as piled raft or as pile foundation shown in Figure 13.2. Design of piled rafts is dealt with in Chapter 14.

13.8 SUMMARY

This chapter gives the description of the foundations of floors below the ground as compensated foundation, cellular raft or as basement floors.

EXAMPLE 13.1 (Analysis of a buoyancy raft)

A ground plus three-storeyed flat 12 m × 9 m in plan (as shown in Figure E13.1) is to be constructed on a site, where the safe bearing capacity of the soil is 30 kN/m^2. Indicate how to plan a buoyancy foundation for the building. [For a reinforced concrete framed block of flats we may assume a total load of 14 to 16 kN/m^2 per storey and the unit weight of the soil is 17 kN/m^3.]

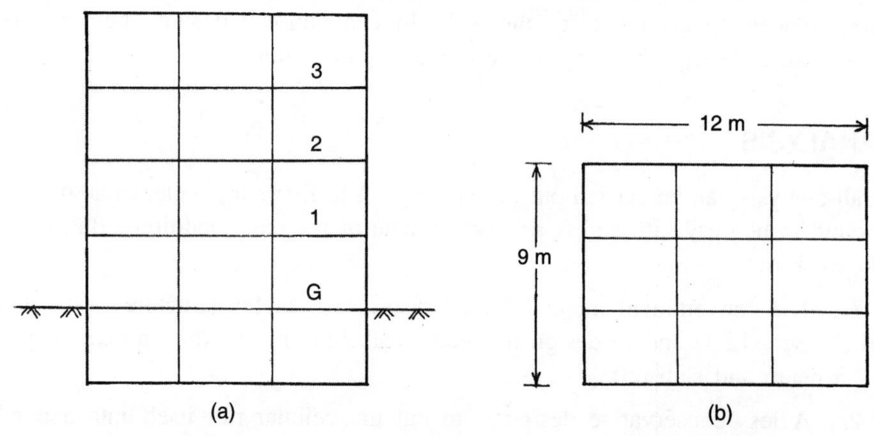

Figure E13.1 Design of cellular rafts: (a) Elevation, (b) Plan.

Reference	*Step*	*Calculation*
	1	*Estimate foundation pressure* As the load from the building is high and the SBC is very low, we will choose a buoyancy raft for the foundation. Assume the foundation pressure for the basement + (ground + three floors + roof), i.e. 4 floors and 1 roof, with no reduction for basement = 10 + (14 × 5) = 80 kN/m^2 (approx.) is very much higher than SBC of 30 kN/m^2. We have to adopt a buoyancy raft 12 m × 9 m in plan. Total load = 80 × 12 × 9 = 8640 kN
	2	*Estimate depth of foundation for compression* Ground pressure from structure = 80 kN/m^2 Assume soil wt. = 17 kN/m^2 Let depth of basement be z $(30 + 17z) = 80$ gives $z = 2.94$ m Adopt 3 m basement. Allowable pressure = 30 + 51 = 81 kN/m^2 > 80 kN/m^2 (applied)
	3	*Calculate factored load* Assume 500 mm apron all around the foundation. Dimension of base = (12 + 1) × (9 + 1) = 13 × 10 m Design load on soil = Total load – Basement slab weight. (The weight from the basement slab can be assumed to be 6 kN/m^2 opposed by direct reaction from the ground.) Factored design load $= (1.5) \times \dfrac{(80-6)\times 12 \times 9}{13 \times 10} \approx 92.2\ \text{kN/m}^2$ [*Note:* In practical problems, we know the column loads to ground floor level with reduction applied for a number of storeys above the floor. To this, we add the loads of the ground floor and the walls for the basement to find the *design pressure*. As the weight of the slab in contact with the ground will be directly taken by ground reaction, it need not be included in the design load.] *Note:* The following is the description of the conventional method of analysis. As a detailed design of the structure will be lengthy, only indications of the methods are given below.
	4	*Design of basement slab* [*Note:* It should be emphasized that even though the additional ground pressure is only 30 kN/m^2 on the foundation, we have to design the raft for the full design load we got in Step 3 (92.2 kN/m^2) as it is this load that it has to transmit to the ground.] We design the slab 3 × 5 m as a two-way slab supported on the rigid walls above them.

Reference	Step	Calculation
		(a) *Design the slab for BM* By use of coefficients. (b) *Design for bending shear* The slab should be safe in bending shear without shear reinforcements. (c) *Design for punching shear* As the bottom cell is rigid, the slabs are supported throughout by rigid wall (or beams in the case of a Virendeel girder). Hence, the slab should be checked for punching shear around the wall beams. If necessary, splays are added at the bottom of the wall.
	5	*Design of cross walls* Cross walls act as beams between columns. The cross walls can be considered as fully braced concrete walls with allowable slenderness ratio. We provide at least 0.25% nominal vertical steel on each face totalling 0.5% in the wall [2]. Cross walls are also designed as beams between columns with pressure from the slab below minus the downward load due to the self-weight of the wall beam and the ground slab. A simple coefficient of $\pm \frac{wl^2}{16}$ or moment distribution can be used for its design as a beam.
	6	*Design of side walls* Side walls have to be designed for the superstructure load and, in addition, for the earth pressure and water pressure assuming a ground water level.
	7	*Design of slab at the ground level (ground floor slab)* The top ground floor slab can be designed as slabs continuous over the walls below. *Note:* This type of elemental design with proper detailing of reinforcement is reported to give satisfactory results.

REFERENCES

[1] Thomlinson, M.J., *Foundation Design and Construction ELBS*, Longman, Singapore, 1995.

[2] Varghese, P.C., *Advanced Reinforced Concrete Design*, 2nd Edition, Prentice-Hall of India, New Delhi, 2005.

CHAPTER 14

Combined Piled Raft Foundation (CPRF)

14.1 INTRODUCTION

The number of high rise buildings being built in cities all over the world is on the rise. While conventional framed construction can be used for RC buildings only up to 20 storeys in height, with modified structural frames (e.g. tubes in a tube system), RC buildings can be built up to 80 storeys or more in height. With specially designed steel frames, buildings higher than a hundred storeys are being built in large cities. In India, the definition of a high rise building varies with the Metro Development Authorities. Buildings over 15.25 m are called high rise buildings in Chennai, but the height to such buildings is over 24 m in Bangalore, and 18 m in Hyderabad. Most of the RC buildings built in Indian cities at present are, however, only less than 60 m in height or 20 storeys.

Pile raft foundation, as shown in Figure 14.1, is a modern concept developed in Europe, especially in Germany, for foundations of tall buildings in places like Frankfurt where the subsoil is hard clay extending to a depth of over 40 m. The 166 m high Dresdon Bank Tower (Germany) rests on such a foundation. In CPRF, the load from the superstructure is shared by both the raft and the piles. The pile load can be as much as 60% of the total load. An exact analysis of this three-phase system including raft, pile and soil can be made only by means of computer methods. Specialized literature and publications given at the end of this chapter or other chapters should be consulted when using this procedure. In this chapter, we examine only the basic concepts and the conceptual method of analysis of simple piled raft system for moderately high buildings.

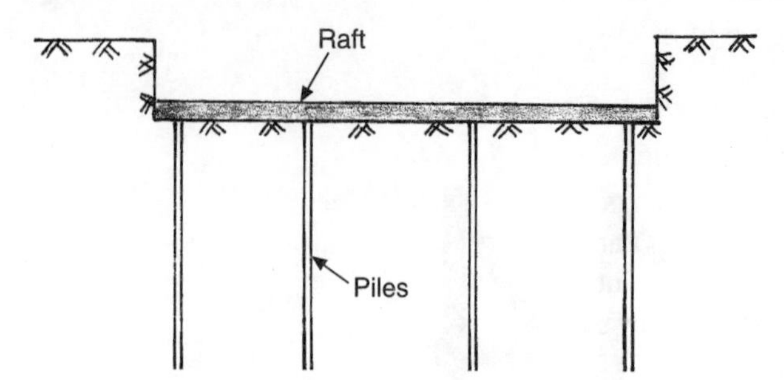

Figure 14.1 Piled-raft foundation.

14.2 TYPES AND USES OF PILED RAFTS

We may adopt CPRF for the following situations.

Case 1: *Piled rafts adopted to reduce settlement:* This is a classical case where a raft foundation (with basement floors, if needed) is found to have enough strength, but the calculated settlement is more than the allowable settlement as it usually happens in hard clay soils as in Frankfurt. In such cases, we introduce piles to take off up to 60% of the load from the superstructure so that the settlement will be within allowable limits. The distances between the piles can be made large (> 6 times diameter of piles) to avoid group effects. We should also calculate the settlement needed to fully mobilize the *ultimate capacity of piles*, which will be rather small. Hence, the pile can be assumed to carry load up to its ultimate capacity in soils. This is one of the great advantages of CPRF over conventional pile foundation as in conventional pile foundations, the working load, which is taken as much lesser than its ultimate capacity, is taken as the carrying capacity of the piles. (It is important to note that in piled rafts, the piles are loaded to ultimate capacity.)

The length of these piles need not go to the rock level. It can be also varied according to the requirements to develop the required strength. This is a great advantage in situations where the building has higher loads at certain points (such as due to the occurrence of a higher portion like a tower), where longer piles can be used to take higher loads.

Case 2: *Piled raft adopted to satisfy bearing capacity consideration:* In those cases in which the bearing capacity of the raft is not enough, piles can be added to satisfy the bearing capacity. The excess load can be carried by piles. In this case, as there should be enough safety factors against failure, we cannot assume the full, ultimate bearing capacity of the piles but only the safe bearing capacity which is much less than the ultimate capacity.

Case 3: *Piled raft adopted to increase lateral resistance in high rise buildings:* When we have a high rise building on raft foundation, the lateral forces due to wind or earthquake will have to be resisted by the ground. In pure raft foundation, we may add basement floors to fully take advantage of the passive resistance of the soil. Alternatively, we may use piles along with the raft to overcome these lateral forces. The piles can be assumed to take part of the horizontal loads.

14.2.1 Beneficial Effects of CPRF

The beneficial effects of CPRF are the following:

1. A reduction in settlement of the system.
2. In the case of buildings with varying loads on the foundations (as with a tower in one part of the buildings) and consequent eccentricities, the provision of piles in the areas of high loads can reduce the eccentricity of the load on the foundation. Under such heavy loads, suitable piles may be used to carry higher loads.
3. An increase in the overall stability as well as stability against lateral loads (as due to wind). A raft foundation with a sufficient number of piles can increase the lateral stability of high rise buildings.
4. A reduction of bending moment for the raft.
5. General economy of foundation.

14.3 INTERACTION OF PILE AND RAFT

A full theoretical consideration of the system has to recognize the following effects of a CPRF:

1. Interaction between the raft and the soil (raft–soil interaction).
2. Interaction between the pile and the soil (pile–soil interaction).
3. Pile–raft interaction. (The super load of the raft on the soil produces confinement of the soil and it affects the reaction of soil on the pile.)
4. Pile–pile interaction. (The spacing of the pile affects its behaviour depending on whether the pile will fail individually or by block failure.)

The first two effects are the most important. Without complicated computational models, the third and fourth effects are difficult to estimate. But we must be aware of their effects. For example, the confinement of the soil due to load on the soil from raft may increase the bearing capacity of the piles. Similarly, the effect of the pile on pile may affect the bearing capacity of the raft. However, in most of the current analyses, these secondary effects can be neglected, bearing in mind these effects in a qualitative way. When we cannot ascertain soil properties in any exactness, such approximation is fully warranted.

14.4 ULTIMATE CAPACITY AND SETTLEMENT OF PILES

We know that piled rafts can work only if the piles do move very slightly under the load, so that there is load transfer from piles to rafts. Piles ending in rock cannot transfer much load easily to a raft as it does not yield. The piles in CPRF are usually spaced apart sufficiently so that the group action of the pile (by which the strength of n piles in a piled foundation is less than n times the strength of one pile) does not take place. The deformation of the raft and piles should match and, in such deformation the loads are shared.

The first requirement is, therefore, to estimate the ultimate carrying capacity of the piles in given soil strata and also the settlement of the pile at the ultimate load. The ultimate load carrying capacity can be determined by the Static Formulae given in IS 2911. These are given in all books on Soil Mechanics [1]. The settlement of the pile under the ultimate load is difficult given to estimate with precision, except by field tests. A very rough estimate can be made as given in the following section.

14.4.1 Estimation of Settlement of Piles

There are many methods to estimate the settlement of single piles under working loads and under ultimate loads. In this section, we describe the principle of rough estimation of the settlement of single piles under load [1]. Even though the following expressions are for the calculation of elastic settlement, we may extend their use to ultimate load. This is also illustrated by Example 14.1. We should also remember that the settlement of a group of piles will be different from that of a single pile.

Estimation of settlement of piles: An empirical method for the calculation of settlement *up to ultimate load* is given now. The total elastic settlement of the pile can be expressed as follows.

$$S_t = S_f + S_b + S_p$$

where S_t = Total settlement
S_f = Settlement due to friction by pile shaft
S_b = Settlement due to bearing by the end of the pile
S_p = Elastic settlement of piles due to load

The following empirical methods for the calculation of S_f, S_b and S_p have been proposed. Calculate the ultimate frictional and bearing capacity of the pile from the IS recommended formulae for their calculation:

Q_f = Load carried by shaft by friction
Q_b = Load carried by base by bearing
$Q_t = Q_f + Q_b$

The following values are recommended for the deformation for these loads.

1. **Calculation of friction settlement**

$$S_f = \frac{Q_f}{\text{(Surface area)}}\left(\frac{\text{Diameter of the pile}}{E_s \text{ of soil}}\right)(1-\mu_s^2)(I_f) \tag{14.1}$$

The value of influence factor I_f is given as

$$I_f = 2 + 0.35\sqrt{L/D}$$

where Q_f = Frictional resistance of the pile
L = Length of the pile
D = Diameter of the pile
πDL = Surface area of pile

Usually, this value will be small (about 0.2–0.6% diameter of the piles as shown in Figure 18.2). Piles will mobilize full friction with very small movements in driven piles in sand and with a little more movements in clays and in bored piles.

2. **Calculation of settlement in bearing** (S_b) Assuming elastic settlement pro portional to diameter of pile

$$S_b = \left(\frac{Q_b}{A}\right)\left(\frac{D}{E_s}\right)(1-\mu^2)I_b \tag{14.2}$$

where Q_b = Load in bearing
A = Area at tip of pile
E_s = Modulus of elasticity of soil
μ = Poisson's ratio = 0.35
I_b = Empirical constant = 0.85

(This value can be as much as 4 to 10% of the diameter of the pile depending on the type of soil.)

3. **Elastic compression of the pile** (S_p)

The elastic compression will depend on the distribution of the friction along the length of the piles, as shown in Figure 14.2.

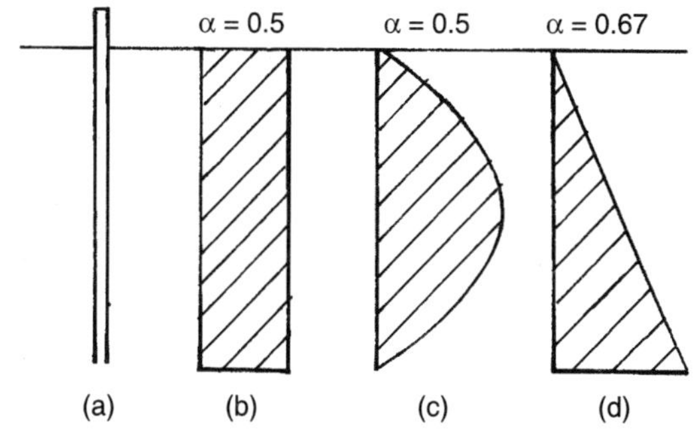

Figure 14.2 Distribution of friction along pile depending on soil strata: (a) Pile, (b) Uniform distribution of friction, (c) Parabolic distribution of friction, (d) Triangular distribution of friction.

S_p = Elastic compression of the pile

$$= \frac{(Q_p + \alpha Q_f)L}{\text{Area } E_c} \tag{14.3}$$

where

E_c = Modulus of elasticity of concrete = 20×10^6 kN/m^2
α = A factor depending on frictional resistance along the pile = 0.5 for uniform and parabolic distribution and 0.67 for triangular distribution (as in sands)

4. **Total settlement.** The total settlement at the ultimate load can be assumed to be the sum of the above three settlements, that is, due to ultimate friction transfer, ultimate base load, and the pile compression.

14.5 ESTIMATION OF SETTLEMENT OF RAFT IN SOILS

The immediate or elastic settlement of an area resting on soil can be computed as:

$$s_i = qB \frac{1 - \mu^2}{E_s} I_w \tag{14.4}$$

where q = Intensity of loading
B = Width of the area
E_s = Modulus of elasticity
μ = Poisson's ratio ≈ 0.35
I_w = Influence factor

Table 14.1 gives the influence factors recommended. Table 14.2 gives the E_s and μ values.

TABLE 14.1 Influence Factor I_w (IS 8009, Part I, 1976, Table 2)

	Flexible foundation			
Shape	*Centre*	*Corner*	*Average*	*Rigid foundation*
Circle	1.00	0.64	0.85	0.86
Square	1.12	0.56	0.95	0.82
Rectangle				
L/B = 1.5	1.36	0.68	1.20	1.06
L/B = 2.0	1.52	0.76	1.30	1.20
L/B = 5.0	2.10	1.05	1.83	1.70
L/B = 10.0	2.52	1.26	2.25	2.10

TABLE 14.2 Approximate Modulus of Elasticity (E_S) and Poisson's Ratio (μ) for Various Soils

Type of soil		E_s (10^4 kN/m^2)	μ
1. Clay			
	Very soft	0.20–1.50	0.30–0.5
	Soft	0.50–2.50	
	Medium	2.50–4.50	
	Stiff	4.50–9.50	
	Sandy	2.50–20.00	
2. Sand			
	Loose	1.05–2.50	0.2–0.40
	Medium dense	1.73–2.76	0.25–0.40
	Dense	4.80–8.10	0.30–0.45
	Silty	1.05–1.75	0.20–0.45
3. Others			
	Gravel	10–20	
	Broken stone	15–30	

14.6 ALLOWABLE MAXIMUM AND DIFFERENTIAL SETTLEMENT IN BUILDINGS

IS 1904, Table 1 gives us the allowable and differential settlement in buildings. This is shown in Table 14.3.

TABLE 14.3 Allowable Maximum and Differential Settlement of Buildings (mm) **(IS 1904, Table 1)**

No.	Type of structure	Steel frames	Reinforced concrete frames	Plain brick walls multi-storeyed		Water towers/ silos
				$L/H \leq 3$	$L/H > 3$	
1	**Isolated foundations**					
	(a) *Sand and hard clay*					
	Maximum (mm)	50	50	60	60	40
	Differential (mm)	$0.0033L$	$0.0015L$	$0.00025L$	$0.0033L$	$0.0015L$
	Angular distortion	1/300	1/666	1/4000	1/3000	1/666
	(b) *Plastic clay*					
	Maximum (mm)	50	75	80	80	75
	Differential (mm)	$0.0033L$	$0.0015L$	$0.00025L$	$0.0033L$	$0.0015L$
	Angular distortion	1/300	1/666	1/4000	1/3000	1/666
2	**Raft foundations**					
	(a) *Sand and hard clay*					
	Maximum (mm)	75	75	–	–	100
	Differential (mm)	$0.0033L$	$0.002L$	–	–	$0.0025L$
	Angular distortion	1/300	1/500	–	–	1/400
	(b) *Plastic clay*					
	Maximum (mm)	100	100	–	–	125
	Differential (mm)	$0.0033L$	$0.002L$	–	–	$0.0025L$
	Angular distortion	1/300	1/500	–	–	1/400

3 Masonry wall buildings

Degree of damage	Width of wall crack	Description
0. Negligible	< 0.1 mm	Will disappear when painting
1. Very slight	$\not>$ 1 mm	Can be treated during maintenance
2. Slight	$\not>$ 5 mm	Needs repair
3. Moderate	5 to 15 mm	Requires opening up and patching
4. Severe	15 to 25 mm (also depends on number of cracks)	Requires extensive repairs, especially near door and window openings
5. Very severe	> 25 mm (also depends on number of cracks)	Requires very major repair amounting to reconstruction

Note: Wall cracking is only one of the many items to be considered in reporting damage in buildings.

To conclude, we may allow a total *maximum settlement* of not more than the following for individual footings: 65 mm on clay and 40 mm on sand. For rafts, the values can be 70 to 100 mm for clays, and 45 to 65 mm for sand. The permissible values for *differential settlement* can be about 40 mm for clay and 25 mm for sand. It is the angular distortion that is important,

especially in clays. The permissible angular distortion in framed structures should be only 1/500 to 1/1000, depending on their use and importance. Angular distortion more than 1/150 can produce cracking in panel walls.
(Note: For classification of damages in masonry buildings see Section A.16)

14.7 DESIGN OF CPRF SYSTEM

No commonly accepted direct design method is available at present for the design of CPRF. The commonly used analytical methods can be classified as follows. [2][3]

- Empirical or computational method
- Refined computational methods
- Computer-based numerical analysis

Here, we deal with the principles of the conceptual methods only.

14.7.1 Conceptual Method of Design

The four quantities involved in our computations are:

- Total load on the foundation
- Proportion of load shared by piles
- Settlement of raft without piles
- Reduction of settlement by piles.

We define the ratio of the load carried by the pile to the total load on the foundation for *a given settlement* as α_{CPRF} or α_{PR}. The reduction of the settlement of the CPRF compared to the raft without piles is called *settlement reduction ratio* S_R.[4]

$$S_R = \frac{\text{Settlement of piled raft}}{\text{Settlement of raft without piles}}$$

Naturally, the *settlement reduction will depend on the type of the pile used, the pile capacity*, and also the settlement of piles at ultimate load. If the settlement of the piles with a large ultimate load capacity of the pile is small, there will be a larger reduction of the total settlement of the foundation. The relation between α_{PR} and S_R settlement can be represented by Figure 14.3. It shows the reduction with two types of piles (representing the range) with different ultimate settlement values.

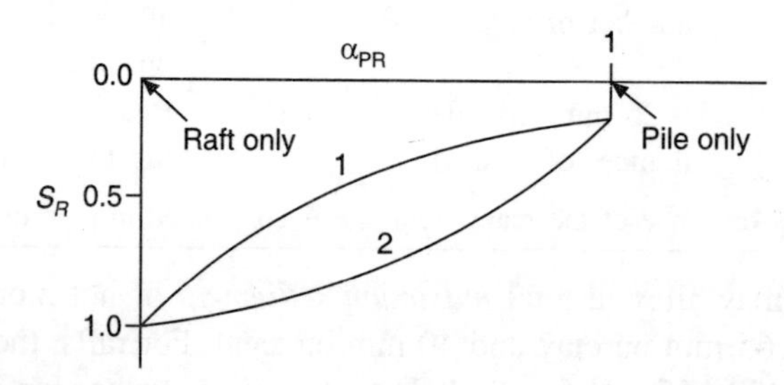

Figure 14.3 Relation between α_{PR} (ratio of load carried by pile to total load on the foundation) and S_R (settlement reduction ratio in piled rafts).

The aim of our calculation is to arrive at the value of α_{PR} for the allowable settlement of the structure.

14.8 CONCEPTUAL METHOD OF ANALYSIS

When only a few piles which are spaced sufficiently apart are used (so that group action will not take place), we may assume that they act as single piles. It has been reported that in such piles, the major part of the settlement occurs as immediate settlement, and consolidation settlement in clays by group action is relatively unimportant [1]. Consolidation settlement is not present in sand. With these assumptions, we can easily select a pile whose ultimate load will match the required settlement and also find the number of such piles required to reduce the required settlement of the raft as shown in Example 14.3.

14.9 DISTRIBUTION OF PILES IN THE RAFTS

We generally assume that the spacing of the piles is large enough ($s > 6D$) so that the pile will act independently. If the number of piles is such that they can be accommodated near the column load, the relief of load to mat can be easily worked out. Otherwise, we have to distribute them in other places in the slab to prevent any group action. In such cases, the raft slab around the pile should be designed to transfer the load to the pile. Example 14.1 illustrates the principles involved.

14.10 THEORETICAL METHODS OF ANALYSIS

Computer methods by representing raft as an elastic plate, soil as a bed of equivalent springs at nodal points and piles as stiff springs are available and can be used for design [6].

14.11 SUMMARY

Piled raft systems are becoming popular in high rise buildings in stiff clays to reduce the total settlement of the building. They are far more economical than the pure pile foundation in very deep deposits of hard clays. The diameter and length of the pile must be selected by detailed calculations. In clays when the piles are near each other, consolidation settlements should be also taken into account. Advanced methods of analysis will be required to solve these problems in important cases. In this chapter, we have given a brief account of the principles involved. This is a subject on which large amount of research is being carried out [5][6].

EXAMPLE 14.1 (Rough estimation of settlement of single piles at ultimate load)

A bored pile of total length 13.2 m is with enlarged base has a shaft diameter of 600 mm and in the last 1.2 m, it is enlarged to 1200 mm diameter. If the SPT (N) value of clay in which the shaft is installed is 13 and that of the enlarged portion is 15, estimate the settlement we may expect at the ultimate load of the pile.

Reference	*Step*	*Calculation*
	1	*Estimate the soil characteristics from N values*
		Clay. Unconfined strength q_c = 10N kN/m^2 (assumed)
		Cohesions of clay along the shaft = (10 × 13/2) = 65 kN/m^2
		Cohesion of clay at enlarged end = (10 × 15/2) = 75 kN/m^2
		E_s of soil, medium clay = 1 × 10^4 kN/m^2
		E_c of concrete = 20 × 10^6 kN/m^2
	2	*Areas and perimeters*
		Diameter of the top part = d = 0.6 m, Area = 0.2826 m^2
		Diameter of the bottom = D = 1.2 m, Area = 1.130 m^2
	3	*Calculate friction capacity (neglecting bottom* 1.2 m*)*
		Capacity in friction = (πdL) (0.33 cohesion)
		= (3.14 × 0.6 × 12) × (0.33 × 65) = 485 kN
		Capacity in bearing = 9c × Area
		= 9 × 75 × 1.130 = 763 kN
		Total capacity = 485 + 763 = 1248 kN
	4	*Estimate settlement due to friction*
Eq. (14.1)		$S_f = \frac{Q_f}{(\pi dL)}\left(\frac{D}{E}\right)(1-\mu^2)I_f$ (Let $\mu = 0.35$)
		$I_f = 2 + 0.35\sqrt{L/d} = 2 + (0.35)\sqrt{12/0.6} = 3.56$
		$S_f = \left(\frac{485}{\pi \times 0.6 \times 12}\right)\left(\frac{0.6}{1 \times 10^4}\right) \times 3.56 \times 10^3 = 4$ mm
		(This is about 400/600 = 0.66% of the diameter)
	5	*Estimate settlement due to bearing*
Eq. (14.2)		$S_f = \left(\frac{Q_f}{\pi D^2}\right)\left(\frac{D}{ES}\right)(1-\mu^2)I_b$, and $I_b = 0.85$
		$= \left(\frac{763}{\pi \times 1.2 \times 1.2}\right)\left(\frac{1.2}{1 \times 10^4}\right)(0.877)(0.85) \times 10^3 = 60$ mm
		(This is about 5% diameter of the base)
	6	*Estimate settlement due to bearing*
		d = 0.6 m and L = 12 m
		Total load = (0.5 × 485) + 763 = 1005.5 kN
Eq. (14.3)		$S_p = \left(\frac{1005.5 \times 4}{\pi \times 0.6 \times 0.6}\right)\left(\frac{12}{20 \times 10^6}\right) \times 10^3 = 6.7$ mm
	7	*Total settlement*
		$S_t = S_f + S_b + S_p = 4 + 60 + 6.7 = 71$ mm

Reference	Step	Calculation
	8	*Draw load settlement diagram* Total capacity = 485 + 763 = 1248 kN Draw a parabolic curve connecting the origin, settlement due to friction and the ultimate and settlement points.

EXAMPLE 14.2 (Planning of a pile raft foundation)

Estimate the ultimate load bearing capacity of a raft 15 m × 15 m founded on a clay soil with $c_u = 70$ kN/m^2. Estimate its settlement with $E_s = 1 \times 10^4$ kN/m^2 and $\mu = 0.35$. If it is loaded with 35000 kN, what percentage of load should be transferred to piles in a CPRF (Combined Pile Raft Foundation) to satisfy the settlement requirements?

Reference	Step	Calculation
	1	*Estimate ultimate bearing capacity* $BC = q_u = 5.69c_u = 5.69 \times 70 = 398.3$ kN/m^2 Ultimate load capacity = 15 × 15 × 398 = 89550 kN Safe bearing capacity FS = 2 = 44775 kN Hence, it is safe under 35000 kN.
	2	*Estimate settlement as a rigid raft* $s = \frac{qB}{E_s}(1-\mu^2)I_f$ Assume $I_f = 0.85$ $q = \frac{35000}{15 \times 15} = 156 \text{ kN/m}^2$ $s = (0.85 \times 156) \times \frac{15 \times (1-0.35^2)}{1 \times 10^4} \times 10^3 \text{ mm}$ = 175 mm (Allowable settlement is only of the order of 75 to 100 mm in raft foundations.)
Table 14.3	3	*Recommended foundation* In this clay soil, as the safe bearing capacity is satisfied but the settlement is not satisfied, a piled raft will be a suitable foundation. Referring to Table 14.3, let us assume the allowable settlement is 75 mm. Then, Load allowable for this allowable settlement = p $75 = (0.85 \times p \times 15 \times 0.877)/10$ $p = 67.1$ kN/m^2 out of 156 kN/m^2, i.e. $\frac{156-67.1}{156} = 0.57$ or 57%

Reference	Step	Calculation
		Load to be taken by raft = 43%, piles take 57% (say, 60%) of the total load.
	4	*Find the number of piles required*
		Assume the ultimate capacity of the pile = 1250 kN (calculated)
		Settlement at the ultimate capacity = 71 mm (calculated)
		Hence, number of piles to carry 60% of load $= \dfrac{0.6 \times 35000}{1250} = 17$
		only (allowing piles to be loaded at the ultimate load)
	5	*Find the number of piles for an ordinary piled foundation*
		Assume working load (in contrast to the ultimate load allowed in piled raft) = 1250/2 = 625 kN
		Number of piles required $= \dfrac{35000}{625} = 56$ Nos.
		(56 is much larger than the 17 required for BPRF. However, the settlement of a conventional pile foundation will be much smaller than this).

EXAMPLE 14.3 (Rough design of piled raft)

A square raft of size 15 m is to rest on a deep clay deposit with SPT value of 14. If the total load on the raft is 35000 kN, find the preliminary layout of a foundation so that the settlement does not exceed 100 mm. Assume E_s = 6.3 N/mm^2.

Reference	Step	Calculation
	1	*Estimate cohesion from SPT values*
		Unconfined strength = SPT/100 = 14/100 = 0.14 N/mm^2
		Cohesion = 0.14/2 = 0.07 N/mm^2 = 70 kN/m^2
	2	*Estimate safe bearing capacity*
		Assume ultimate capacity (N_c = 5.14)
		$= cN_c = 70 \times 5.14 = 360$ kN/m^2
		Total ultimate bearing capacity
		= 360 × 15 × 15 = 81000 kN
		$\text{FS} = \dfrac{81000}{35000} = 2.31$ (safe)
	3	*Estimate settlements*
Eq. (14.4) and Table 14.1		Settlements (Δ) $= \dfrac{P}{B^2}(B)\left(\dfrac{1-\mu^2}{E_s}\right)I_p = \dfrac{P}{15}\left[\dfrac{1-(0.35)^2}{6300}\right] \times 0.95$
		$\Delta = 35000 \times 8.82 \times 10^{-6} = 0.31$ m = 310 mm
		Allowable Δ for the raft = 100 mm (clay) much less than Δ.
		Adopt piled raft to reduce settlement.

Reference	*Step*	*Calculation*
		Raft load for Δ 100 = $\frac{35000 \times 100}{310} = 11290$ kN
		Balance load on piles ≐ 23710 kN, i.e. 67% of the load.
	4	*Estimate size of pile and its capacity*
		We should have the load settlement curve of a pile. In clay, we assume the load remains constant after the ultimate load with increase in settlement. We may assume that the *ultimate load is usually reached at a settlement of about 6% of the diameter of the pile.* Let us adopt 600 mm diameter pile 30 m in length.
		Δ at ultimate load $= \left(\frac{6}{100}\right) 600 = 36 \text{ mm} < 100 \text{ mm}$
		Pile capacity = $2/3\, c(\pi DL) + cN_c(\pi D^2/4)$
		$= 2/3 \times 70 \times \pi(0.6) \times 30 + (9 \times 70 \times \pi \times 0.6 \times 0.6)/4$
		$= 2637 + 178 = 2815$ kN
	5	*Estimate the number required*
Step 3		Number of piles required $= \frac{23710}{2815} = 9$ piles
	6	*Estimate the number of piles for a full pile foundation*
		Pile capacity with FS = 3 = $\frac{\text{ULT}}{\text{FS}} = \frac{2815}{3} = 938$ kN
		Number of piles required to take total load
		$= \frac{35000}{938} = 38$ piles
		For a raft on pure pile foundation, we need about 40 piles.
		[*Note:* The above calculations neglect consolidation settlement and are only approximate; they are meant only to illustrate the principle involved.]

REFERENCES

[1] Poulos, H.G. and E.G. Davis, *Pile Foundation Analysis and Design* (Chapter 10), John Wiley and Sons, New York, 1980.

[2] Randolph, M.F., *Design Methods for Pile Groups and Pile Rafts*, 13th International Conference on Soil Mechanics and Foundation Engineering (Vol. 4), New Delhi, 1994.

[3] Katzenbach, R. and O. Reul. *Design and Performance of Piled Rafts*, Proceedings of 14th International Conference on Soil Mechanics and Foundation Engineering, Hamburg, Geneva, 1997.

[4] Rolf, Katzenbach, *Soil Interaction by High Rise Buildings*, Vol. I, Proceedings of Indian Geotechnical Conference, 2006, IIT Madras, Chennai.

[5] Balakumar, V. and K. Illamparuthi, *Performance of Model Pile Raft on Sand*, Vol. I, Proceedings of Indian Geotechnical Conference, 2006, IIT Madras, Chennai.

[6] Ramasamy, G. and Veera Narayana Pandia, *Settlement Analysis of Piled Raft Foundations*, Proceedings of (Vol. 2) Indian Geotechnical Conference, 2006, IIT Madras, Chennai.

CHAPTER 15

Circular and Annular Rafts

15.1 INTRODUCTION

Structures such as chimneys, which are in the form of a cylindrical shaft rising from the ground, are generally founded on an annular or circular raft (Figure 15.1). Similarly, in the case of an elevated circular water tank supported by a number of columns positioned along a circle, the columns rest on a ring beam with a slab below the beam. The slab below acts as a circular raft. For the conventional design of this water tank foundation, the design of the ring beam is taken separately and assumed to transmit a uniform line load on to the annular raft. The annular raft is designed to carry this line load. In other places where the dimension of the annular raft is much greater than that of the vertical shaft as may be the case in a chimney, a frustum of a cone can be introduced in between as shown in Figure 15.1(c).

In this chapter, we first examine the formulae for the design of the annular raft subjected to a "circular line load." We will also examine the formulae for the design of the ring beam supported on columns. A detailed analysis of the subject is given in Refs. [1] and [2]. Computer-based elastic design is given in Ref. [3].

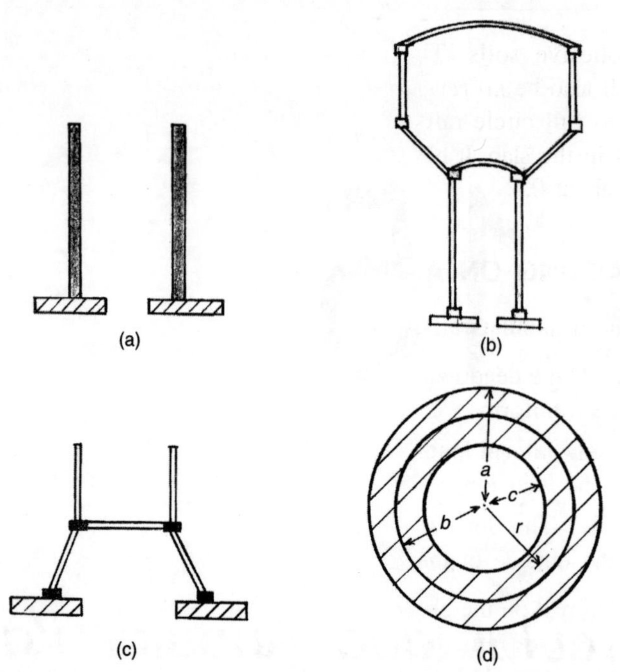

Figure 15.1 Annular rafts in foundations: (a) Chimney foundation, (b) Water tank with ring beam on annular raft foundation, (c) Enlargement by conical structure on annular foundation, (d) Definition of outer radius a, loading radius $b = \beta a$ and inner radius $c = \alpha a$. Any general radius $r = \rho a$.

15.2 POSITIONING OF CHIMNEY LOAD ON ANNULAR RAFT

When designing the annular raft of outer radius a and inner radius c it is considered a good practice to locate the load on the raft along the centre of gravity of the uniform ground pressure at radius b.

In order to arrive at dimensionless quantities, we use the notations used in Ref. [2] Figure 15.2.

$$\frac{\text{Inner radius}}{\text{Outer radius}} = \frac{c}{a} = \alpha \quad \text{or} \quad c = \alpha a$$

$$\frac{\text{Loading radius}}{\text{Outer radius}} = \frac{b}{a} = \beta \quad \text{or} \quad b = \beta a$$

Taking q_0 as uniform ground pressure, the value of b is given by the following formula:

$$b = \beta a = \frac{\int_c^a q_0 r^2\, dr}{\int_c^a q_0 r\, dr} = \frac{2}{3}\left(\frac{a^3 - c^3}{a^2 - c^2}\right) = \frac{2}{3} a \frac{(1-\alpha^3)}{(1-\alpha^2)} \tag{15.1}$$

This positioning will give uniform pressure under dead load. We can also prove that an annular raft with the above positioning of load is much superior to a full circular raft without a central hole. It will not only save material but also lead to a more uniform distribution of the base pressure in cohesive soils. This prevents tilling due to non-uniform pressure and settlement. There will also be *no reversal* of bending moment (in such annular rafts), along its radius as compared to full circle rafts. The reversal of bending moment makes it difficult to place reinforcements in the slab. It has been also found that it is best to keep the size of annular ratio (inner to outer) about 0.4.

15.3 FORCES ACTING ON ANNULAR RAFTS

The forces acting on an annular raft such as those for a chimney are the following:

1. **Dead load.** If the dead load is applied at the CG of the raft, the ground pressure can be assumed as uniform.
2. **Moment due to lateral loads.** This load is transmitted to the raft at the radius where the raft and the wall join.
3. **Backfill load.** This load is due to the difference in the backfill level between the inside and the outside of the foundation as in the case of a chimney.

The raft itself can rest on the ground or in the case of weak soils, on piles. In this chapter, we deal with only the case of rafts resting on soil. [A detailed treatment of annular rafts on soil and on piles is given in Ref. [1].]

15.4 PRESSURES UNDER DEAD LOAD AND MOMENT

Let the uniform pressure under dead load with proper positioning, as explained in Sec. 15.2, be q_0.

$$q_0 = \frac{\text{DL}}{A} \tag{15.2}$$

Let the maximum value of varying pressure be q_1 under moment M.

$$q_1 = \frac{M}{I} y = \frac{M \times a}{\pi / 4(a^4 - c^4)} \tag{15.3}$$

For full circle

$$c = 0 \quad \text{and} \quad q_1 = \frac{4M}{\pi a^3} \tag{15.4}$$

15.5 METHODS OF ANALYSIS

For the design of annular rafts, the stress resultants to be determined are:

- Radial moments = M_r along the radius
- Tangential moment = M_t along the tangent
- Radial shear = Q

The following methods of analysis are commonly used:

- Conventional analysis for annular rafts [1], [2].
- Results of Chu and Afandi for circular and annular rafts [1], [2].

15.6 CONVENTIONAL ANALYSIS OF ANNULAR RAFTS

Approximate or conventional methods can be used for annular rafts, but they cannot be applied to the full circular raft. In annular rafts, the approximate values of the *radial shears* and *radial moments* can be computed by assuming that any radial element of the slab acts as a cantilever beam fixed at the chimney wall. [However, the tangential moments cannot be determined by this method and have to be determined by more refined methods.]

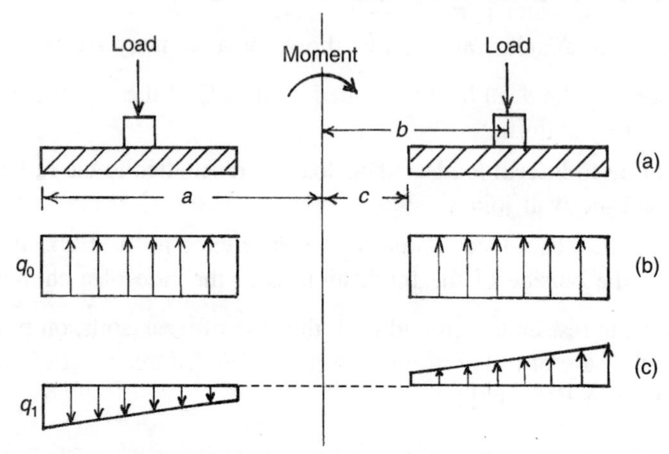

Figure 15.2 Ground reaction pressure due to vertical load and moment: (a) Load and moment on annular raft, (b) Uniform ground pressure due to vertical load, (c) Nonuniform ground pressure of opposing signs due to moment.

For a uniform pressure under the dead load and the varying pressure under moments, the formulae for the calculation of shear and moments by approximate method will be as follows.

Let q_0 be the uniform pressure due to the dead load and $\pm q_1$ the maximum pressure due to wind moment as shown in Figure 15.2.

The following are the values for shears and moments per unit length at any radius r. Let $r = \rho a$ or $\rho = r/a$. The pressure is from the bottom of the slab upwards and we take positive moment as that which produces tension at the bottom.

Case 1: Q and M_r due to *dead load producing uniform pressure* q_0 [2]

$$(r < b)\ Q \text{ inner part } = q_0 a \frac{\rho^2 - \alpha^2}{2\rho} \tag{15.4a}$$

$$(r > b)\ Q \text{ outer part } = -q_0 a \frac{1-\rho^2}{2\rho} \tag{15.4b}$$

$$(r < b)\ M_r \text{ inner part } = q_0 a^2 \left(\frac{\rho^3 - (3\rho - 2\alpha)\alpha^2}{6\rho}\right) \tag{15.4c}$$

$$(r > b)\ M_r \text{ outer part } = q_0 a^2 \left(\frac{\rho^3 - (3\rho - 2)}{6\rho}\right) \tag{15.4d}$$

Case 2: Approximate shears and moments (Q and M_r) *due to external moment* producing maximum pressure q_1

$$(r < b)\ Q \text{ inner part } = \frac{q_1 a}{3\rho}(\rho^3 - \alpha^3)\cos\theta \quad (15.5a)$$

$$(r > b)\ Q \text{ outer part } = \frac{-q_1 a}{3\rho}(1 - \rho^3)\cos\theta \quad (15.5b)$$

$$(r < b)\ M_r \text{ inner part } = q_1 a^2 \left(\frac{\rho^4 - (4\rho - 3\alpha)\alpha^3}{12\rho}\right)\cos\theta \quad (15.5c)$$

$$(r > b)\ M_r \text{ outer part } = q_1 a^2 \left(\frac{\rho^4 (4\rho - 3\alpha)}{12\rho}\right)\cos\theta \quad (15.5d)$$

15.7 CHU AND AFANDI'S FORMULAE FOR ANNULAR SLABS [2]

Chu and Afandi [2] have derived more general formulae for M_r, M_t and Q for full circular chimney foundation by assuming the slab as viewed upside down supported along a circle at the centre line of the chimney and subjected to uniform dead load or linearly varying load produced by wind load moment. They assumed two types of rafts as follows: (a) If the supporting wall section is not designed to resist the unbalanced moments between the inside and outside portions of the slab, it is considered as *simply supported*; (b) if the supporting wall section is capable of resisting the unbalanced moment, the slab is considered as fixed. [For example, a relatively thin shell meeting a thick raft can be considered as hinged.]

In the paper by Chu and Afandi referred above, they have given the formulae for the moments and shears caused by vertical loads and moments acting on the raft. The quantities to be determined are listed in Table 15.1.

TABLE 15.1 Moments and Shears to be Determined (Zone 1 is where $r < b$ and Zone 2 is where $r > b$)

Quantity	*Due to DL*	*Due to moment*
(1) Radial moment M_r	M_{r1}, M_{t1}, Q_{r1} — zone 1, $r < b$	M'_{r1}, M'_{t1}, Q'_{r1} — zone 1, $r < b$
(2) Tangential moment M_t and (3) Shear Q_r	M_{r2}, M_{t2}, Q_{r2} — zone 2, $r > b$	M'_{r2}, M'_{t2}, Q'_{r2} — zone 2, $r > b$

Note: The formulae have been derived in terms of radius 'a' and the following non-dimensional parameters (Figure 15.1(d)):

$\alpha = c/a$ = Ratio of inner to outer radius
$\beta = b/a$ = Ratio of load circle to outer radius
$\rho = r/a$ = Ratio of radius chosen to outer radius

The value of μ can be taken as 0.15.

This non-dimensional representation makes the expression useful for computer computation for various values of outer radius 'a'. For a circular raft diameter, $c = 0$ or $\alpha = 0$.

Different formulae for the two cases—the slab is assumed as simply supported by the chimney wall, and as fixed at the chimney wall—have been worked out in the above paper. The results for the simply supported case are given below. Formulae for other cases can be obtained from Ref. [1] or [2].

15.7.1 Chu and Afandi's Formulae for Analysis of Circular Rafts Subjected to Vertical Loads (Chimney simply supported by slab; for circular raft, put $b = 0$) [2]

$$\alpha = \frac{c}{a};\ \beta = \frac{b}{a};\ \rho = \frac{r}{a};\ \mu = 0.15 \quad \text{(Only } \rho \text{ is variable)}$$

q_0 = uniform foundation pressure.

Zone 1 ($r < b$) **(Inner part)** The values of radial moment, tangential moment and shear for zone 1 can be calculated by the following formulae in terms of α and ρ.

$$M_{r1} = \frac{q_0 a^2}{64}[4(3 + \mu)\rho^2 - 16\alpha^2(1 + \mu) \log_e \rho - 8\alpha^2(3 + \mu) + 2(1 + \mu)K_1 - (1 - \mu)K_2\rho^{-2}]$$

(Radial moment) (15.6a)

$$M_{t1} = \frac{q_0 a^2}{64}[4(1 + 3\mu)\rho^2 - 16\alpha^2(1 + \mu) \log_e \rho - 8\alpha^2(1 + 3\mu) + 2(1 + \mu)K_1 + (1 - \mu)K_2\rho^{-2}]$$

(Tangential moment) (15.6b)

$$Q_{r1} = \frac{q_0 a}{2}(\rho - \alpha^2 \rho^{-1}) \qquad \text{(Shear) (15.6c)}$$

where

$$K_1 = 2(1 + \alpha^2)\left(\frac{3 + \mu}{1 + \mu}\right) - \frac{8\alpha^4 \log_e \alpha}{1 - \alpha^2} - 4\beta^2\left(\frac{1 - \mu}{1 + \mu}\right) - 8(1 + \log_e \beta) \qquad (15.6d)$$

$$K_2 = 4\alpha^2\left(\frac{3 + \mu}{1 - \mu}\right) - \frac{16\alpha^4 \log_e \alpha}{1 - \alpha^2}\left(\frac{1 + \mu}{1 - \mu}\right) - 16\alpha^2\left(\frac{1 + \mu}{1 - \mu}\right)(1 + \log_e \beta) - 8\alpha^2\beta^2 \qquad (15.6e)$$

(*Note:* Different constants not given here are to be used for chimney fixed to the slab.)

Zone 2 ($r > b$) Radial moments, tangential moments and shear for zone 2 can be calculated from the following formulae in terms of α and ρ:

$$M_{r2} = \frac{q_0 a^2}{64}[4(3+\mu)\rho^2 - 16(1+\mu)\log_e \rho - 8(3+\mu) + 2(1+\mu)K_1' - (1-\mu)K_2'\rho^{-2}]$$

(Radial moment) (15.7a)

$$M_{t2} = \frac{q_0 a^2}{64}[4(1+3\mu)\rho^2 - 16(1+\mu)\log_e \rho - 8(1+3\mu) + 2(1+\mu)K_1' + (1-\mu)K_2'\rho^{-2}]$$

(Tangential moment) (15.7b)

$$Qr_2 = \frac{q_0 a}{2}(\rho - \rho^{-1}) \quad (15.7c)$$

where

$$K_1' = K_1 + 8(1 + \log_e \beta)(1 - \alpha^2) \quad (15.7d)$$

$$K_2' = K_2 - 8\beta^2(1 - \alpha^2) \quad (15.7e)$$

15.7.2 Chu and Afandi's Formulae for Analysis of Circular Rafts Subjected to Moment (Chimney simply supported to slab; for circular raft, put $b = 0$)

Zone 1 ($r < b$) (Let q' = maximum pressure due to moment when $\cos\theta = 1$ and $\theta = 0$ with respect to the direction of the moment, say, due to the wind, θ is the angle with xx axis.)

The wind can blow from any direction. Hence wind from right to left and left to right must be considered along with the effect of q_0.

Values of radial moment, tangential moment and shear can be calculated from the following formulae for zone 1: (These varies with $\cos\theta$.)

$$M_{r1}' = \frac{q'a^2}{192}[4(5+\mu)\rho^3 + 2(3+\mu)K_3\rho + 2(1-\mu)K_4\rho^{-3} + (1+\mu)K_5\rho^{-1}] \quad (15.8a)$$

$$M_{t1}' = \frac{q'a^2}{192}[4(1+5\mu)\rho^3 + 2(1+3\mu)K_3\rho - 2(1-\mu)K_4\rho^{-3} + (1+\mu)K_5\rho^{-1}] \quad (15.8b)$$

$$Q_{r1}' = \frac{q'a}{192}(72\rho^2 + 8K_3 - 2K_5\rho^{-2}) \quad (15.8c)$$

where

$$K_3 = \frac{3}{\beta^2} - 3\beta^2\left(\frac{1-\mu}{3+\mu}\right) - 8\left(1 + \frac{\alpha^4}{1+\alpha^2}\right)\left(\frac{2+\mu}{3+\mu}\right) \quad (15.8d)$$

$$K_4 = 3\beta^2\alpha^4 - \frac{3\alpha^4}{\beta^2}\left(\frac{3+\mu}{1-\mu}\right) + 8\left(\frac{\alpha^4}{1+\alpha^2}\right)\left(\frac{2+\mu}{1-\mu}\right) \quad (15.8e)$$

$$K_5 = 12\alpha^4 \quad (15.8f)$$

Zone 2 ($r > b$) Values of M_{t2}, M_{r2} and Q_{r2} can be obtained from the *above equations* also by using the *following arbitrary constants* instead of K_3, K_4 and K_5

$$K_3' = K_3 - \frac{3}{\beta^2}(1 - \alpha^4) \tag{15.9a}$$

$$K_4' = K_4 + 3\beta^2(1 - \alpha^4) \tag{15.9b}$$

$$K_5' = 12.0 \tag{15.9c}$$

15.7.3 Nature of Moments and Shear

The following remarks are important.

1. Regarding the sign of moments for positive moments of M_r and M_t, we place the steel at the bottom of the raft. For negative moments, we place the steel at the top of the raft. Hence, by convention, when plotting the moment diagram with respect to the radius, we plot the positive moment downwards and the negative moment values upwards from the centre line of the slab.
2. It is worthwhile to remember that the radial moment M_r and the shear values for annular rafts for the points inside the loading circle obtained by the conventional method (Sec. 15.6) for simply supported chimney are very close to the values obtained by the exact method. The values got for all points by conventional method lie between the theoretical fixed and simply supported chimney cases. For the outer part of the loading circle M_r, values got by approximate method are close to the fixed chimney condition.
3. Values of the tangential moment M_t cannot be calculated by conventional method. We have to calculate them by using the formulae given above.
4. For the full circular raft, the approximate method is not applicable and for these rafts the value of M_t is also rather high.
5. *In annular raft*, M_r values are mostly +ve, but M_t values can be positive or negative depending on the load position. *In full circle raft*, M_r values can also be +ve or –ve along the radius.
6. As the wind can blow in any direction, M_r, M_t and Q at the various points should be taken as that produced by the worst sum of the uniform positive pressure due to the dead load and the corresponding pressures caused by the wind load. If there is any reversal of moments, steel should be placed for such reversal on each side separately.

15.8 ANALYSIS OF RING BEAMS UNDER CIRCULAR LAYOUT OF COLUMNS

Beams curved in plan such as balcony beams, and ring beams are subjected to torsion in addition to bending and shear. Their analysis is dealt with in books on structural analysis, and design of R.C. members for torsion is dealt in detail in textbooks on reinforced concrete [3], [4]. A brief account of the analysis is given below.

15.8.1 Analysis of Ring Beams Transmitting Column Loads to Annular Rafts

It is common for circular water tanks to be supported by columns along their circumference, and these columns receive their loads at the top and also transmit their loads at the bottom to the annular raft through a ring beam. In such cases the distribution of bending moment and torsion in these circular beams depends on the number of column supports.

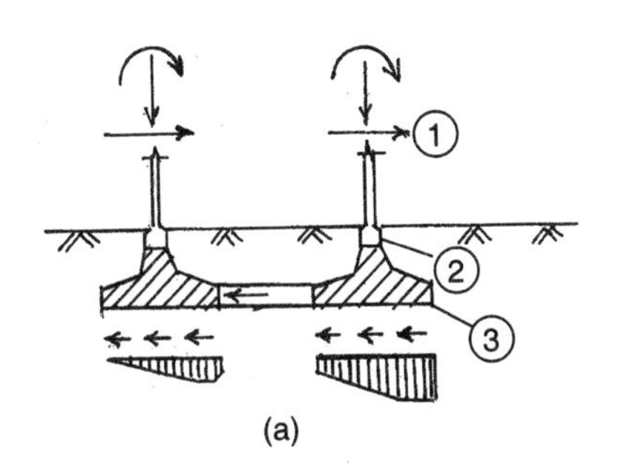

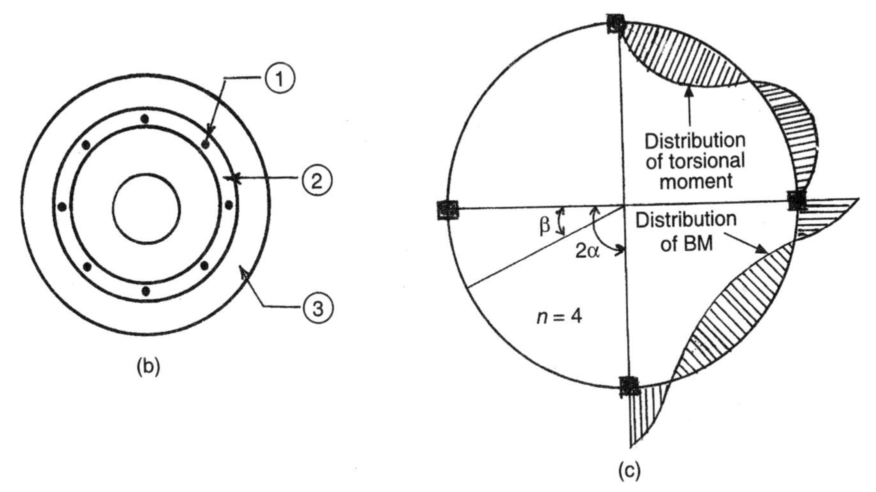

Figure 15.3 Annular raft foundation: (a) General arrangement for a circular water tank. [① Column, ② Ring beam, ③ Annular raft], (b) Ring foundation, (c) Bending moment and torsional moment in a ring beam on four supports (n = 4).

Let the radius of the loading circle = r
The angle between columns = 2α
The UDL on the beam = w

The formulae for bending moments and torsion for a point situated at "*angle* β" from a column support are as follows:

$$M_\beta = wr^2 (\alpha \sin \beta - 1 + \alpha \cot \alpha \cos \beta) \quad (15.10a)$$

$$T_\beta = wr^2 (\beta - \alpha + \alpha \cos \beta - \alpha \sin \beta \cos \alpha) \quad (15.10b)$$

$$T_\beta \text{ is maximum when } M_\beta = 0 \quad (15.10c)$$

As the beam is fixed at the column, the maximum positive (sagging) moment will be at the centre of the beam, the maximum negative (logging) moment at the support. The torsional moment is maximum between the columns. The angle at which the torsional moment is maximum measured from the centre line of supports (columns), is designated as θ_o. The distribution of bending moments and torsion is shown in Figure 15.3 for four supports. These maximum values can be obtained from the following expressions:

Sagging (+ve) moment at mid-span $= 2wr^2\alpha\lambda_1$ (15.11a)

Hogging (–ve) moment at support $= 2wr^2\alpha\lambda_2$ (15.11b)

Maximum torsional moment $= 2wr^2a\lambda_3$ (15.11c)

where λ_1, λ_2 and λ_3 are coefficients, given in Table 15.2.

The values of λ_1, λ_2, λ_3 and θ_0 for various conditions of supports can be tabulated as in Table 15.2.

TABLE 15.2 Coefficients for Bending and Twisting Moments in Circular Beams

No. of supports (*n*)	2α (degrees)	λ_1	λ_2	λ_3	θ_0 (degrees)
4	90	0.070	0.137	0.021	19.25
5	72	0.054	0.108	0.014	15.25
6	60	0.045	0.089	0.009	12.75
7	51.43	0.037	0.077	0.007	10.75
8	45	0.033	0.066	0.005	9.5
9	40	0.030	0.060	0.004	8.5
10	36	0.027	0.054	0.003	7.5
12	30	0.023	0.045	0.002	6.25

15.9 DETAILING OF ANNULAR RAFT UNDER COLUMNS OF A CIRCULAR WATER TANK

The details of reinforcement of the beam under the column and the annular raft for a circular water tank are shown in Figure 15.4.

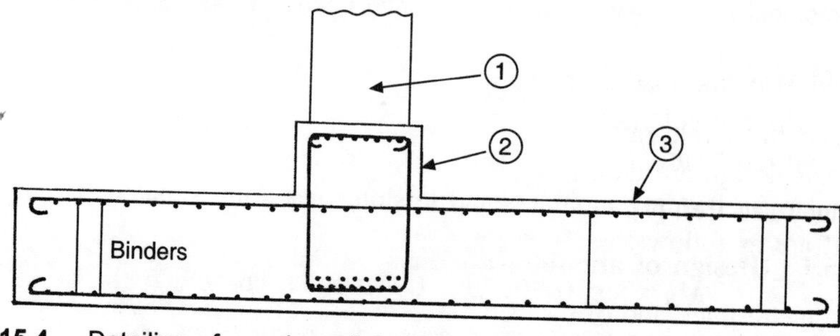

Figure 15.4 Detailing of annular raft foundation for a circular water tank: (1) Column, (2) Ring beam, (3) Annular raft.

15.10 CIRCULAR RAFT ON PILES [1]

If the foundation of the soil is weak, a pile foundation can be given for the raft which will behave as a pile cap. Piles are located in concentric circles as shown in Sec. 17.8.1. Such an arrangement will give the maximum moment of inertia. The pile cap may be a full raft or annular raft. [Reference [1] gives the design procedure to be adopted.]

$$\text{Pile load} = \frac{\text{Vertical load}}{\text{Number of piles}} \pm \frac{M}{I}(r\cos\theta) \tag{15.12}$$

where

$I = \sum (r\cos\theta)^2$ in m^2

θ = angle of each pile from the reference

We can use the theory of circular plates for solution of the problem. There can be two cases of such foundations, namely, full raft and annular raft.

Case 1 *Full raft:* As in Afandi's analysis referred in Sec. 15.7, here also, the raft is considered as a plate placed upside down, simply supported along the loading circle. We then analyse each ring of piles separately for ring load condition due to the reaction from the piles. The results are added to get the final value.

Case 2 *Annular raft:* For an annular raft of outer radius a and inner radius b, we first consider it as a full circular raft. From this analysis, for the position of the inner edge of the annular raft, we will have a moment $M_{r\alpha}$. As, in reality, this is a free edge, for an annular raft, we apply an equal and opposite moment along the inner circle. The sum of the two effects will give as the final value. (Ref. [1] can be consulted for fuller details. Circular pile caps on pile is shown in Figure 17.9 and described in Sec. 17.8.)

15.11 ENLARGEMENT OF CHIMNEY SHAFTS FOR ANNULAR RAFTS

As shown in Figure 15.1, the chimney shafts may have to be enlarged to fit the annular raft foundation. A reinforced concrete conical structure with ring beams at top and bottom is used for this purpose. The design of this enlargement is a structural design problem and is fully described in Ref. [1].

15.12 SUMMARY

Annular rafts and full circle rafts are used in many situations such as foundations. The basic principles of their design are described in this chapter. For an example for design of circular raft for a chimney see Ref. [1].

EXAMPLE 15.1 (Design of annular raft for a chimney)

A chimney one metre thick and of mean radius 8 m has a weight of 30000 kN. The weight of the soil above the foundation is estimated as 1000 kN. The moment due to wind is 8600 kNm. *Design an annular raft* for the foundation assuming a safe bearing capacity of 85 kN/m^2 for the dead load and an increase of 20% for the wind load. Grade 20 concrete and Fe 415 steel are to be used.

Reference	*Step*	*Calculation*
	1	*Find dimension of the annular raft*
		Let a = outer radius
		c = inner radius
		b = radius of loading of chimney = 8 m (fixed)
		We must position the chimney at the CG of the raft
		Total foundation area required $= \dfrac{30000+1000}{85} = 365\ \text{m}^2$
		Fixed loading circle of chimney weight = 8 m = radius
Sec. 15.2		Thus, CG $b = 8\ \text{m} = \dfrac{2(a^3-c^3)}{3(a^2-c^2)}$
		But the area required = $\pi(a^2 - c^2) = 365$
		$a^3 - c^3 = 1414$
	2	Assume the outer radius a = 11.3 m, then the inner radius c = 3.1 m
		Find area and moment of inertia of raft
		$A = \pi\left\lfloor (11.3)^2 - (8.1)^2 \right\rfloor = 370\ \text{m}^2$
		$I = \dfrac{\pi}{4}(a^4 - c^4) = 12.726\ \text{m}^2$
	3	*Calculate maximum and minimum bearing pressure (allow 20% increase in wind load)*
		$q = \dfrac{31000}{370} \pm \dfrac{8600 \times 11.3}{12.726}$
		$= 83.78 \pm 7.64$
		Max pressure = 91.42 kN/m^2 < 1.20 × 85 (i.e. < 102 kN/m^2)
		Min pressure = 75.14 kN/m^2 is +ve.
	4	*Calculate pressure with factored loads for design*
		$q_0 = \dfrac{1.5 \times 30000}{370}$ (Earth load omitted) = 121.6 kN/m^2
		q_1 = pressure by moment = 1.5 × 7.64 = 11.46 kN/m^2
	5	*Data for annular raft design*
		a = 11.3 m; c = 3.1 m; b = 8 m
		$\beta = 8/11.3 = 0.708$
		$\alpha = 3.1/11.3 = 0.274$; $\rho = r/a = r/11.3$

Reference	Step	Calculation
		[*Note:* The types of tabular forms Tables E15.1 to E15.4 are necessary for the analysis of a full circular raft as there can be reversals of M_r and M_t.] However, for an annular raft, we need to calculate the following quantities only:
		(1) Radial bending moment M_r at the junctions of the chimney with the raft (inner and outer junctions). $M_{r\beta}$ and also at M_{r1} (outer edge)
		(2) Tangential bending moment M_{t1} (outer end); $M_{t\alpha}$ at inner edge and $M_{t\beta}$ at loading points
		(3) Q at Q_β
8 m assumed in Step 1		For the given annular raft with the inner radius 3.1 m, we start our calculation with radius 3.5 m. We calculate for radii 3.5 m, 4 m, 6 m, 7.5 m, 8.5 m, 9 m, 10 m and 11.3 m. For r = 3.5 m, r = 3.5/a = 3.5/11.3 = 0.31. Let the chimney have a thickness of 1 m with centre line at 8 m. Thus, we find the moments, shears at 7.5 and 8.5 m the meeting points of the chimney with the raft.
	6	*(Let chimney be loaded at* 7.5–8.5 m *radius)*
		Tabulate values of M_r, M_t and Q_r using formulae and also maximum values as in Tables E15.1 to E15.4.
		[*Note:* As the wind can blow from any direction, we should consider (a) wind from left to right and (b) wind from right to left to find whether there is any reversal of moments in the raft. If there is reversal, designed steel should be provided on top and bottom of the raft. The following tabular form is needed for the design of full rafts only].

TABLE E15.1 Calculation of Radial Moment M_r by Eqs. (15.6) and (15.7) for Factored Load

Load ↓ / r →	Inner				Outer			
	3.5	4.0	6.0	7.5	8.5	9.0	10	11.3
DL								
WL								
Total								

TABLE E15.2 Calculation of Tangential Moment M_t by Eqs. (15.6) and (15.7)

Load ↓ / r →	Inner				Outer			
	3.5	4.0	6.0	7.5	8.5	9.0	10	11.3
DL								
WL to left								
WL to right								
Total								

TABLE E15.3 Calculation Shear for Q_r kN by Eqs. (15.6) and (15.7)

Load ↓ / r →	Inner				Outer			
	3.5	4.0	6.0	7.5	8.5	9.0	10	11.3
DL								
WL to left								
WL to right								
Total								

TABLE E15.4 Summary of Max +ve and –ve Moments and Shear Along Radius (Assumed Values)

Radius r (m)	$r = r/a$	M_r = (kNm/m)	M_t = (kNm/m)	Q (kN)
3.5	0.310	30	–487	200
4.0	0.354	–	–	–
6.0	0.531	–	–	–
7.5	0.664	1025.0	+19.3	480
8.5	0.752	580.0	+18.0	430
10.0	0.885	103.0	–	–
11.3	1.000	–0.3	105	0

(*Note:* For positive moment, we provide steel at the bottom and for negative steel, we provide steel at the top.)

Reference	Step	Calculations
	7	**Evaluate M_r by approximate method at r = 7.5 m** (M_t can be evaluated by approximate method)
		$r = r/a = 0.664;\ \alpha = 0.274$
Eq. (15.4) Eq. (15.5)		$M_r = q_0 a^2 \dfrac{[\rho^3 - (3\rho - 2\alpha)\alpha^2]}{6\rho} + q_1 a^2 \dfrac{[\rho^4 - (4\rho - 3\alpha)\alpha^3]}{12\rho}$
		= 721 + 460 = 1181 kNm
		(It will be very nearly equal to the value obtained by the formula. Similarly, Q can also be estimated. However, M_t values cannot be obtained by the approximate method.)
	8	*Design of section*
		Let max radial moment = 1025 kNm
		$d = \left(\dfrac{M_u}{0.14 \times f_{ck} \times b}\right)^{1/2} = \left(\dfrac{1025 \times 10^6}{0.14 \times 20 \times 1000}\right)^{1/2}$
		= 605 mm. Adopt 615 mm.
		D = 615 + 10 + 50 = 675 mm
Table B.1		$\dfrac{M}{bd^2} = \dfrac{1025 \times 10^6}{1000 \times 615 \times 615} = 2.70$

Reference	Step	Calculation
		Percentage of steel = 0.928
		$$A_s = \frac{0.928 \times 1000 \times 615}{100} = 5707 \text{ mm}^2$$
		32 mm @ 125 mm spacing gives 6434 mm^2
	9	*Design for shear*
Table B.4		Allowable shear @ 0.90% steel = 0.61 N/mm^2
		$$d = \frac{\text{Max shear}}{1000 \times 0.61} = \frac{480 \times 10^3}{1000 \times 0.61} = 786 \text{ mm}$$
		(Adopt depth d = 800 mm and let us provide 1% steel so that we do not have to provide shear reinforcement as in slabs)
		$$A_s = \frac{1 \times 1000 \times 800}{100} = 8000 \text{ mm}^2$$
	10	*Design for* M_t
		M_t is –ve (steel at top near the interior) and M_t is +ve in the interior.
		Let M_t @ inner edge = 480 kNm/m
Table B.1		$$\frac{M}{bd^2} = \frac{480 \times 10^6}{1000 \times 800 \times 800} = 0.75$$
		Percentage of steel required = 0.218
		$$A_s = \frac{0.218 \times 1000 \times 800}{100} = 1744 \text{ mm}^2$$
		Provide 20 mm @ 175 mm giving 1795 mm^2
Table C_2	11	*Find or estimate tangential steel at other places and detail steel*
		Nominal steel @ 0.12%
		$$= \frac{0.12 \times 1000 \times 800}{100} = 960 \text{ mm}^2$$
		Provide distribution steel as 16 mm @ 200 mm giving 1005 mm^2.

EXAMPLE 15.2 (Design of ring beam over an annular raft for a circular water tank)

The load from a circular water tank supported by six columns rests on a ring beam, which, in turn, rests on an annular raft. Assuming the mean radius of the centres of column line is 8 m and the total load from the tank is 30,000 kN, design the ring beam.

Reference	Step	Calculation
	1	*Find data for analysis*
		Radius of ring beam = r = 8 m
		Number of columns 6; $2\alpha = 60°$

Reference	*Step*	*Calculation*
		$\alpha = 30° = \dfrac{3.14 \times 30}{180} = 0.5233$ radians
		$w = \dfrac{30{,}000}{\pi \times 16} = 597$ kN/m
	2	*Find max +ve moment at centre of beam*
		(a) M +ve = $2wr^2\alpha\lambda_1$ = $(2 \times 597 \times 64 \times 0.5233)\lambda_1$
		= 39991 × 0.045 = 1799 kNm
Sec. 15.8		(b) M –ve = 39991 × λ_2 (0.089) = 3559 kNm
		(c) Max torsion at β_0 = 12.75°
		T = 39991 × λ_3 (0.009) = 360 kNm
		(d) Shear $= \dfrac{\text{Load}}{2}\left(\dfrac{2 \times \pi \times 8 \times 597}{6 \times 2}\right)$ = 2500 kN at support
	3	*Design of beam*
		The beam is above the annular foundation. It can be designed as a T beam with the tension reinforcement placed just above the bottom reinforcement of the annular raft. The raft itself can be of sloping section as shown in Figure 15.3(a).
		Assume width more than the size of the column and design the beam for the above bending moments, shear and torsion (as per normal cases).

REFERENCES

[1] Manohar, S.N., *Tall Chimneys*, Tata McGraw Hill (Torsteel Research Foundation of India), New Delhi, 1985.

[2] Chu, R.H. and O.F. Afandi, Analysis of Circular and Annular Slabs for Chimney Foundations, *Journal American Concrete Institute*, Vol. 63, December 1966.

[3] Hemsley, J.A., *Elastic Analysis of Raft Foundations*, Thomas Telford, London, 1998.

[4] IS 11089 (1984), *Code of Practice for Design and Construction of Ring Foundation*, BIS, New Delhi.

CHAPTER 16

Under-reamed Pile Foundations

16.1 INTRODUCTION

The theory and design of conventional types of piles are dealt with in detail in all books on soil mechanics and foundation engineering [1]. In this chapter, we examine exclusively the use of under-reamed piles. IS 2911, Part III deals with the design of under-reamed piles [2].

Under-reamed piles (which do not go deep unlike conventional piles) are used extensively for the foundations of walls and columns of one or two-storeyed residential buildings. They are also used extensively for foundations of compound walls. In places where excavation of foundations for ordinary buildings or extension of existing buildings is difficult, under-reamed piles are quite popular. The most important and special use of these piles is for foundations for ordinary buildings and compound walls in *expansive soils*. This has led to its popularity all over India where expansive soils are present.

Under-reamed piles can be single or double under-reamed as shown in Figure 16.1. The diameter of the bulb (D_v) is usually 2.5 times the diameter of the shaft and the centres of the bulbs are separated by a distance equal to 1.2–1.5D_v. In non-expansive soils, the first bulb is usually made at a minimum depth of 1.75 m from the ground level. In expansive soils, the moisture content of soils at the level of the last bulb and if possible at the level of the first bulb should not vary with seasons. The bulbs are installed by manual labour by special boring tools. For carrying light column loads, these piles are detailed in the same way as conventional piles. As it is not possible to produce bulbs in sandy soil or under-water level, under-reaming of piles

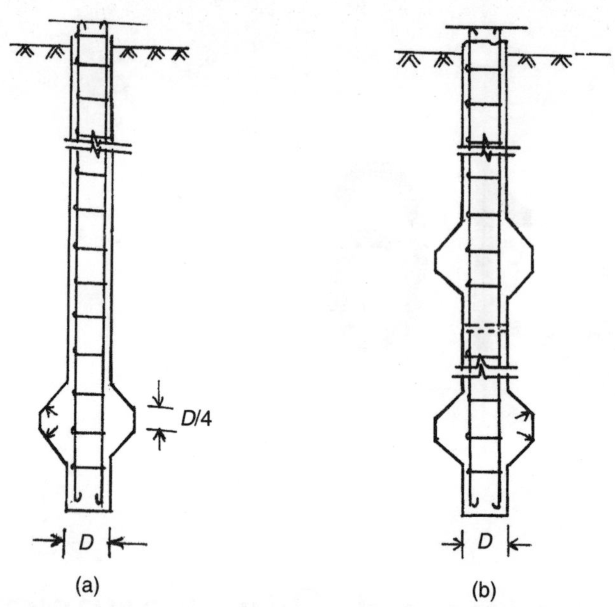

Figure 16.1 Bored cast in-situ under-reamed pile (The bulb should be formed below the level where moisture does not change with season).

should not be used in such situations. The type of layout of these piles for carrying wall loads of residential houses is as shown in Figure 16.2 and their design, is as follows. The above-mentioned considerations regarding the positioning of piles should be very clearly borne in mind while planning these foundations.

16.2 SAFE LOADS ON UNDER-REAMED PILES

Under-reamed piles are used under walls with grade beams (Figure 16.2) or under columns with pile caps (Figure 16.3). Safe load on pile can be determined by:

1. Calculation of safe load from soil properties as calculated from the theory given in books on Soil Mechanics with suitable factor of safety.
2. Actual load test on piles installed at site.
3. From safe load tables in IS 2911, Part 3 (Table 16.1).

16.3 DESIGN OF UNDER-REAMED PILE FOUNDATION FOR LOAD BEARING WALLS OF BUILDINGS

Under-reamed *foundations for walls* in ordinary soils consist of a series of under-reamed piles spaced at suitable intervals, and a continuous grade beam is laid over these piles; the walls are built on these grade beams on piles. Special care should be taken to provide these piles at junctions of walls, as shown in Figure 16.2. For the design of such a foundation, the following factors should be taken into account:[3]

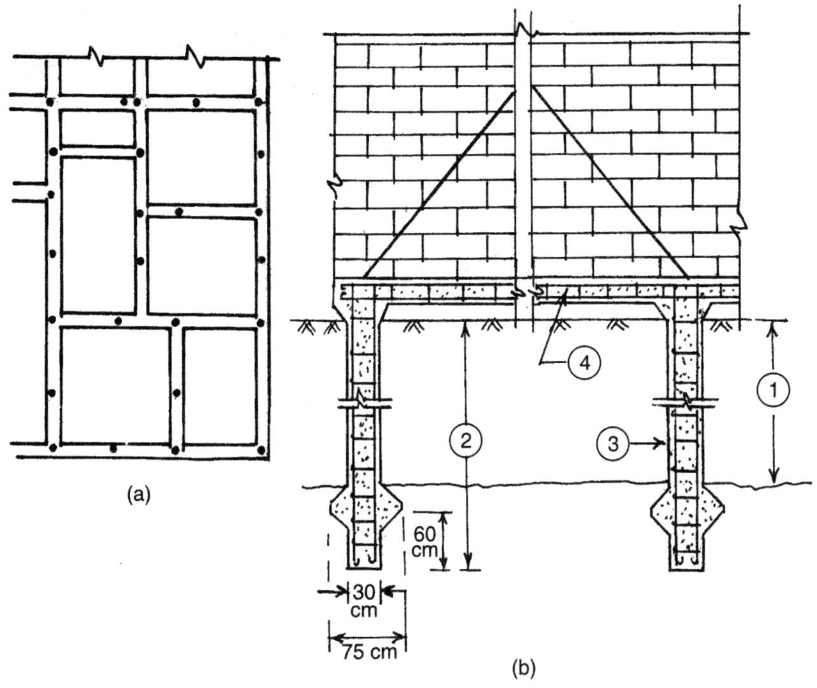

Figure 16.2 Layout of under-reamed pile for load bearing walks: (a) Layout of piles, (b) Details, ① Bulb to be formed below zone of seasonal volume change, ② Length of pile, ③ Pile shaft, ④ Grade beam placed above ground in expansive soils (see Figure 16.3 for details).

1. The spacing of the piles, usually between 1 m and 3 m, should depend on its capacity and the load it has to carry. Generally, it should not exceed 3 m to limit the sizes of the grade beams. The minimum spacing allowed for the under-reamed piles is $2D_v$. In critical cases, it can be reduced to $1.5D_v$, in which case, the capacity of the pile should be reduced by 10%. For a building, these piles are so planned that a pile is always placed at the junctions or very close to the junctions of the walls, as shown in Figure 16.2.
2. The carrying capacity of these piles, as given in IS 2911, is based on the field tests conducted by the Central Building Research Station, Roorkee. They have standardized the pile diameters and the reinforcements to be provided (not less than 0.4% for high yield deformed steel). The safe load capacity uplift and lateral loads for these under-reamed piles in medium compact sandy soils (N values, 10–30), and clayey soils of medium consistency (N value, 4–8) and with bulb diameters equal to 2.5 times the shaft diameter and 3.5 m–4.5 m deep are given in Table of IS 2911 (Part III). The safe vertical load capacities for single under-reamed piles are reproduced in Table 16.1. *It can be seen that this table assumes a nominal safe bearing capacity of about* 40 T/m^2 *and about* 2 m *depth below the ground level calculated on the nominal area of the bulbs*. For softer soils, the capacity of the pile should be reduced and, for harder soils,

it may be increased correspondingly. The safe load on double under-reamed piles is taken as 1.5 times that assumed for single under-reamed piles. Their capacities can also be calculated by using the standard static formula for piles.

16.3.1 Design of Grade Beams

1. Grade beams are meant to transfer the wall load to the ground. The width of the beam is made slightly larger than the pile diameter, and not less than the width of the wall it has to carry. In expansive soils, these grade beams (especially the grade beams of the external walls) are isolated (not supported) on the soil below (Figure 16.3). They should also be designed as not bearing on the soil. In non-expansive soils it is better that the grade beams are placed on an 8 cm sand filling and a levelling course of 1:5:10 concrete 8 cm thick so that it can also transfer part of the load (up to 20%) to the ground below. A span/depth ratio of 10 to 15 between piles, with a minimum depth of 15 cm, is used for practical construction for these grade beams.
2. For the design of these beams, according to IS 2911 (Part III), Cl. 5.3, the maximum bending moment in the case of beams *supported on the ground* during construction is taken as $wL^2/50$. However, for concentrated loads on the beam, the full bending moment should be considered. It should be noted that these reduced moments are due to arching or deep beam action of the supported masonry which takes place only if the height of the wall is at least 0.6 times the length of the beam.

   Alternatively, even though not suggested in the code, the grade beams may be designed as in the case of lintels so that only the wall load inside a 60° triangle is transferred on to the grade beam. As the height of the triangle of load will be $\sqrt{3}(L/2)$ (i.e. 0.87 times length), the weight of the *triangle of masonry* (assuming the unit weight of the wall as 19 kN/m^3) becomes

$$W = \frac{\sqrt{3}L^2}{4}(t \times 19) \text{ kN}$$

   For this action, the height of the wall should be at least 0.86 times the span. The bending moment due to this load will work out to be

$$M = \frac{WL}{6} \text{ kNm}$$

   In either case, the depth as also the reinforcement calculated, based on the above theory, will be small. A minimum depth of 15 cm should be always adopted for the beams as shown in Table 16.2. This table may be used for empirical design.
3. For detailing of steel in these beams, *IS* 2911 (Part III) *recommends that an equal amount of steel should be provided at the top and bottom of the grade beam.* It should also be not less than three bars of 8 mm high yield steel at top and bottom. The stirrups should consist of 6 mm MS bars at 300 mm spacing, which may be reduced to 100 mm at the door openings near the wall up to a distance of three times the depth of the beam.

A better method of detailing would be to bend two of the four bars from the under-reamed piles to each side of the pile into the grade beams as top steel and extend it to a length up to a quarter span of the beam on both sides of the pile. In addition, two top bars are provided continuous at top of the grade beam as hanger bars. The reinforcements required at the bottom (minimum of 3 Nos. of 8 mm bars) are provided as continuous bottom reinforcement. The recommended sizes of grade beams are given in Table 16.2. Thus, an empirical design of under-reamed piles and the grade beams can be made using Tables 16.1 and 16.2.

TABLE 16.1 Safe Loads for Vertical Under-reamed Piles [(L = 3.5 m) in Sandy Soils and Clayey Soils Including Black Cotton Soils] (Refer IS 2911, Part III (1980))

Diameter of pile (mm)	*Longitudinal steel-HYDS*	6 mm *stirrups; spacing*	*Safe load single under-reamed* (tons)	*Strength increase in* 300 mm *length* (tons)	*Strength decrease in* 300 mm *length* (tons)
200	3–10 mm	180 mm	8	0.9	0.7
250	4–10 mm	220 mm	12	1.15	0.9
300	4–12 mm	250 mm	16	1.4	1.1
375	5–12 mm	300 mm	24	1.8	1.4
400	6–12 mm	300 mm	28	1.9	1.9
450	7–12 mm	300 mm	35	2.15	1.7
500	9–12 mm	300 mm	42	2.4	1.9

Notes:

(i) Applicable for soils with N value (10–30) for sand and (4–8) for clay.

(ii) Diameter of the bulb is 2.5 times the diameter of the pile.

(iii) Capacity of double under-reamed is 1.5 times that of single under-reamed pile.

(iv) Spacing of piles should be such that there is undisturbed soil for a distance of not less than 0.5 times the diameter of the bulb between the bulbs.

(v) Uplift resistance is usually taken as one half the bearing value.

(vi) Lateral resistance can be taken as 1/8 the bearing value.

TABLE 16.2 Recommended Sizes of Grade Beams over Under-reamed Piles
(The following specified steel is to be provided both at the top and the bottom of the grade beam)

Load on beam (kN/m)	*Effective span of beams* 1.5 m		1.8 m		2.1 m		2.4 m	
	D (mm)	A_{st} (Nos. dia)	D (mm)	A_{st} (Nos. dia)	D (mm)	A_{st} (Nos. dia)	D (mm)	A_{st} (Nos. dia)
1. *Beam width* 230 mm								
15	150	4 T 8	150	4 T 8	150	4 T 8	150	4 T 8
30	150	4 T 8	150	4 T 8	150	4 T 8	180	4 T 8

(Contd.)

TABLE 16.2 Recommended Sizes of Grade Beams over Under-reamed Piles (The following specified steel is to be provided both at the top and the bottom of the grade beam) (*contd.*)

Load on beam (kN/m)	1.5 m D (mm)	1.5 m A_{st} (Nos. dia)	1.8 m D (mm)	1.8 m A_{st} (Nos. dia)	2.1 m D (mm)	2.1 m A_{st} (Nos. dia)	2.4 m D (mm)	2.4 m A_{st} (Nos. dia)
45	150	4 T 8	150	4 T 8	180	4 T 8	200	4 T 8
60	150	4 T 8	180	4 T 8	200	4 T 8	200	4 T 8
75	180	4 T 8	200	4 T 8	200	4 T 8	200	4 T 8
2. *Beam width* 345 mm								
15	150	4 T 10	150	4 T 10	150	4 T 10	150	4 T 10
30	150	4 T 10	150	4 T 10	180	4 T 10	180	4 T 10
45	150	4 T 10	150	4 T 10	180	4 T 10	180	4 T 10
60	150	4 T 10	150	4 T 10	180	4 T 10	200	4 T 10
75	150	4 T 10	150	4 T 10	200	4 T 12	200	4 T 12
90	180	4 T 10	150	4 T 12	200	4 T 12	200	5 T 12
105	180	4 T 10	150	4 T 12	200	4 T 12	200	6 T 12
3. *Beam width* 460 mm								
60	150	4 T 12	150	4 T 12	150	4 T 12	180	4 T 12
75	150	4 T 12	150	4 T 12	180	4 T 12	200	4 T 12
90	150	4 T 12	180	4 T 12	200	4 T 12	200	4 T 16
105	150	4 T 12	180	4 T 12	200	4 T 16	200	4 T 16
120	180	4 T 12	200	4 T 12	200	4 T 16	200	4 T 16
135	180	4 T 12	200	4 T 12	200	4 T 16	200	4 T 16
150	180	4 T 12	200	4 T 16	200	4 T 16	200	4 T 16

Note: 1. T denotes high yield steel (Fe 415). 4T12 means 4 nos 12 mm T bars at top and bottom.
2. The areas of steel specified above works out to 0.5 to 1% of section depending on the span.

16.4 DESIGN OF UNDER-REAMED PILES UNDER COLUMNS OF BUILDINGS

Under-reamed piles as a group are also used under lightly loaded columns. These piles are provided with conventional pile caps. Details of these arrangements are shown in Figure 16.3.

16.5 USE OF UNDER-REAMED PILES FOR EXPANSIVE SOILS

The best solution for foundations of ordinary buildings in highly expansive soil, where the strength of the *soil does not decrease with depth*, is the under-reamed pile and grade beam

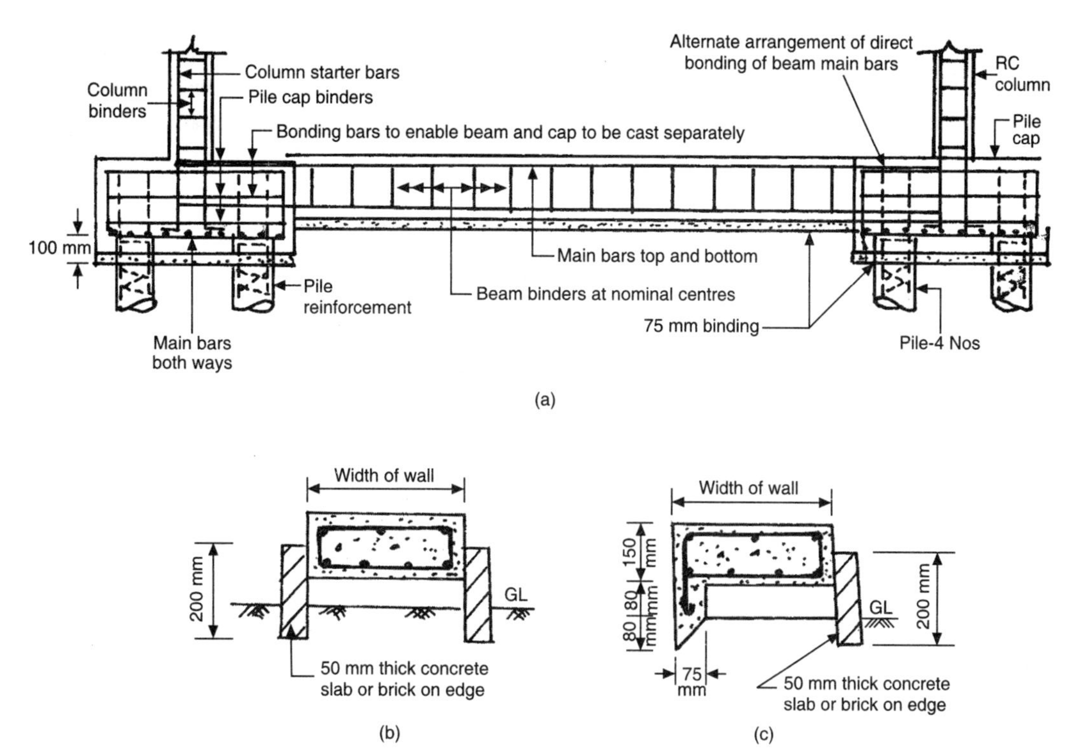

Figure 16.3 Details of under-reamed piles, pipe cap and grade beam: (a) General arrangements; (b) interior grade beams; (c) exterior grade beams (IS 2911 Part III).

foundation. It is considered as one of the most efficient environmental and structural solutions for dealing with light to moderate loads in expansive soils. The foundation acts independent of the ground movements. These piles can be used (a) with grade beams directly under walls, as shown in Figure 16.2, or (b) under columns with the pile cap, as shown in Figure 16.3. The following principles are important for their successful performance.

1. The depth of the pile should be at least equal to the depth where no loss in moisture occurs due to seasonal changes in that site. This depth is taken as approximately 3–3.5 m below the ground level in most places in India. Where large trees are present or expected to be planted, it should be taken approximately 5 m deep. If the piles are founded in regions where moisture changes take place, the pile will also move along with the soil in that zone. This should be strictly avoided. The piles should also be anchored by enlarged bulbs formed below the neutral point (zone of no moisture change with season) with sufficient pullout strength to offset the uplift pressure on the pile.
2. These under-reamed piles are generally not isolated from the surrounding soil. Therefore, when the surrounding soil heaves up, the pile also tends to lift up. The under-reaming and steel provided is meant to anchor the pile at the base so that the pile can take up this upward movement in tension. Thus, it is very important that the under-reaming is made in the zone of no climate change and also the pile has enough steel to develop the tension capacity.

3. The grade and plinth beams on the top of the piles should be tied properly on these piles, and the pile reinforcements should be properly anchored into these beams at the pile supports *with enough negative steel continuous over the piles.* This is very important in expansive soils to resist the uplift of the foundation. Many foundation failures have been reported due to lack of attention to this detailing.
4. In expansive soils, as already pointed out, the grade beams themselves should be preferably isolated from the expansive soils by suitable methods. IS 2911, Part III on *Under-reamed Piles* gives us the details of the design and construction of this structural system.
5. It is recommended that in South India, the construction of these piles should commence immediately at the end of the principal (NE) monsoon (i.e. January–February) and the full load should be allowed to act before the onset of the next monsoon (July–August). Such piles have been found to give good results.

16.6 SUMMARY

Under-reamed piles of diameter 200–500 mm are extensively used in India for building construction. The design of these beams has been briefly explained in this chapter.

EXAMPLE 16.1 (Design of under-reamed pile foundation for a load bearing wall)

The main brickwall of a room of a residential building is 225 mm thick and has a loading of 40 kN/m at the foundation level. Another crosswall of the same thickness joins it and transmits a concentrated load of 35 kN. Design a layout of under-reamed piles and grade beam for the foundation of the main wall.

Reference	*Step*	*Calculation*
	1	*Layout of foundation*
		Place one pile P_1 at the junction of the walls and pile P_2 at 2 m spacing along the walls.
	2	*Loads on piles and grade beams*
		Load on P_1 = (40 × 2) + 35 = 115 kN
		Load on P_2 = (40 × 2) = 80 kN
		Load on grade beam = 40 kN/m
	3	*Design of piles*
Text Table 16.1		Piles of 200 mm diameter, single under-reamed have a capacity of 80 kN. Adopt for P_2.
		Piles of 200 mm diameter, double under-reamed have a capacity of 120 kN. Adopt for P_1 (pile at the junction).
	4	*Design of grade beam*
		Load on grade beam = 40 kN/m
		Effective span = 2000 – 200 = 1800 mm

Reference	*Step*	*Calculation*
Text Table 16.2		Adopt 150 mm deep beam with 4 Nos. of 10 mm at the bottom and the same at the top.
	5	*Always check bearing capacity of soil*
		Diameter of under-ream = 2.5d = 2.5 × 200 = 500 mm
		Bearing area of pile $= \dfrac{\pi(0.5)^2}{4} = 0.196\ \text{m}^2$ (area of pile)
		Expected safe bearing capacity at 3 m depth is equal to
Sec. 16.3 (2)		$\dfrac{P}{A} = \dfrac{80}{0.196} = 408\ \text{kN/m}^2$

[*Note:* Pile caps for piles under columns are designed as described in Chapter 17.]

REFERENCES

[1] Varghese, P.C., *Foundation Engineering*, Prentice-Hall of India, New Delhi, 2005.

[2] IS 2911 (Part III) 1980, *Code of Practice for Design and Construction of Pile Foundations–Under-reamed Piles.*

[3] Varghese, P.C., *Limit State Design of Reinforced Concrete*, Prentice-Hall of India, New Delhi, 2002.

17

CHAPTER

Design of Pile Caps

17.1 INTRODUCTION

Piles, wells and caissons are the three basic forms of deep foundations. Piles can be displacement piles (driven into the ground) or replacement piles (installed by making a bore and then filling it with concrete). The main difference between piles and wells or caissons is the cross-sectional size. Wells and caissons can be looked upon as large rigid piles installed by excavation or by drilling by machines. Whereas the pile bends under a horizontal load, a well or caisson rotates under a horizontal load. 'Caisson' literally mean "a large watertight case open at the bottom and from which water is kept out by pneumatic pressure." Caissons are installed with pneumatic pressure.

Since the methods of calculation of the bearing capacities, settlements, etc. of different types of ordinary piles 300–600 mm diameter are described in books dealing with geotechnical aspects of foundation engineering [1], we will not repeat them here. The method of calculating bearing capacities of large diameter piles 600 mm and above is discussed in Chapter 18. In this chapter, we confine ourselves the design of pile caps that are necessary to join a group of ordinary piles to act together. Large diameter piles do not need pile caps.

We should remember that the friction piles should generally have a spacing of at least three times the diameter of the piles and bearing piles at least 2.5 times the diameter so that their active zones do not overlap. (The British Code of Practice suggests a minimum spacing equal to the perimeter of piles for friction piles. IS 2911 (Part I) recommends a spacing of not less than 2.5 times the shaft diameter for bearing piles in loose sands.[2] When the piles are bearing on rock a minimum spacing can be two times the diameter of the shaft. IS uses the symbol D_p for the diameter of the piles. In British practice, it is usually represented as h_p.)

17.2 DESIGN OF PILE CAPS

Pile caps are used to transmit the column load to the pile foundation. The plan dimension of the pile cap should be based on the assumption that the actual final position of piles in construction can be up to 10 cm out of line from the theoretical centre lines. Pile caps should, therefore, be made very large to accommodate these deviations. In practice, pile caps are extended as much as 15 cm beyond the outer face of the piles with this objective. The design objectives of pile caps are that they should be capable of safely carrying the bending moment and shear force and *they should be deep enough to provide adequate bond length for the pile reinforcements and the column starter bars*. (These are fundamental requirements.) The important parameters in the design of pile caps are as follows:

- Shape of pile cap
- Depth of pile cap
- Amount of steel to be provided
- Arrangement of reinforcement.

To standardize the pile cap design, the following recommendations are commonly used in the British practice.[3]

17.3 SHAPE OF PILE CAP TO BE ADOPTED

The standard shapes and types of arrangements to be used in the layout of the piles should depend on the number of piles in the foundation. In this context, the following requirements should be taken into account:

1. The minimum spacing of piles (kh_p in Figure 17.1) permitted from soil mechanics depends on the type and diameter of pile as well as the soil conditions. CP 2004 requires a minimum centre-to-centre spacing of twice the diameter of the pile for end-bearing and three times the diameter for friction piles. (IS 2911: Part I, Secs. 1 and 2 recommend a minimum spacing of 2.5 times the diameter of the pile for both driven cast-in-situ and bored cast-in-situ piles.) For accommodating deviations in driving of piles, the size of the pile cap is made 300 mm (150 mm on each side) more than the outer-to-outer distance of the exterior piles. A cover of 75 mm is also usually provided for the pile cap surfaces in contact with earth and 60 mm against blinding concrete of 75–80 mm thick. In marine situations, the cover should be increased to a minimum of 80 mm.
2. Another requirement in arriving at the shape of the pile cap is that the centre of gravity of the piles and the pile cap should coincide so that all the piles are equally loaded with gravity loads.
3. In arriving at the final layout, the need to provide suitable reinforcement is a major consideration.

Based on these three requirements, the recommended shapes of the pile cap for 2 to 9 piles are as shown in Figure 17.1 and Figure 17.2.

No. of piles	*Shape of pile cap*	*Tension by truss theory* [Eq. 17.2a]
2	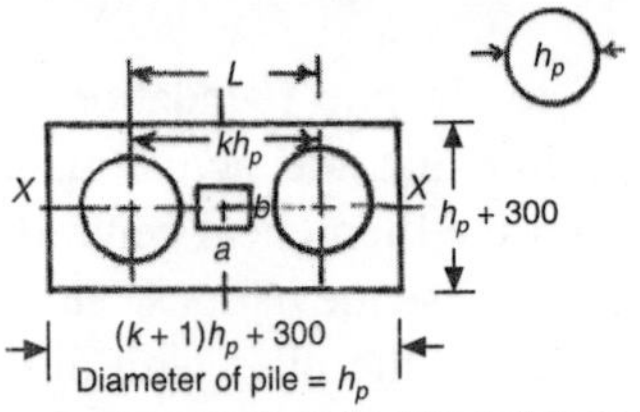 L; kh_p; h_p; X X; a; b; $h_p + 300$; $(k+1)h_p + 300$; Diameter of pile = h_p	(d = depth of pile cap) Along $XX = \dfrac{P}{12Ld}(3L^2 - a^2)$
3	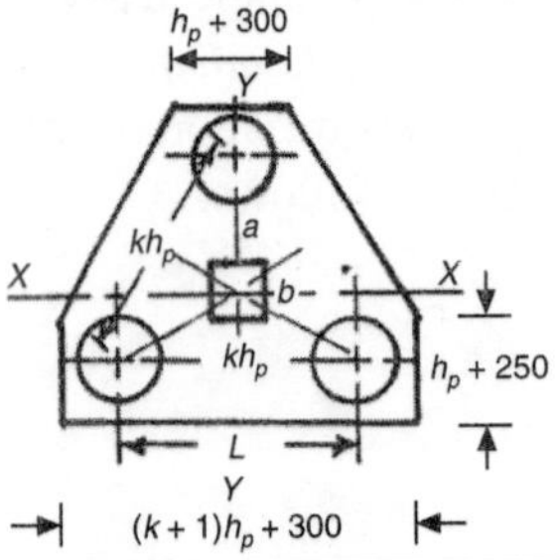 $h_p + 300$; Y; a; kh_p; X X; b; kh_p; $h_p + 250$; L; Y; $(k+1)h_p + 300$	Along $XX = \dfrac{P}{36Ld}(4L^2 + b^2 - 3a^2)$ Along $YY = \dfrac{P}{18Ld}(2L^2 - b^2)$
4	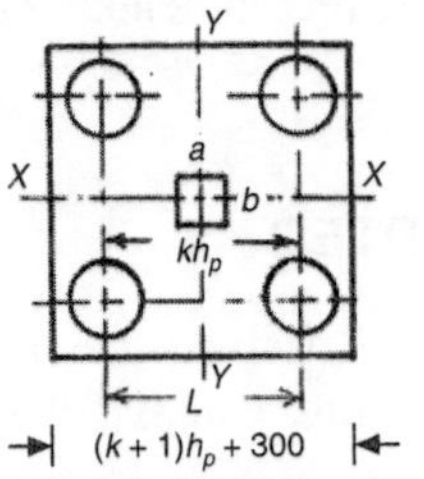 Y; a; X X; b; kh_p; Y; L; $(k+1)h_p + 300$	Along $XX = \dfrac{P}{24Ld}(3L^2 - a^2)$ Along $YY = \dfrac{P}{24Ld}(3L^2 - b^2)$
5	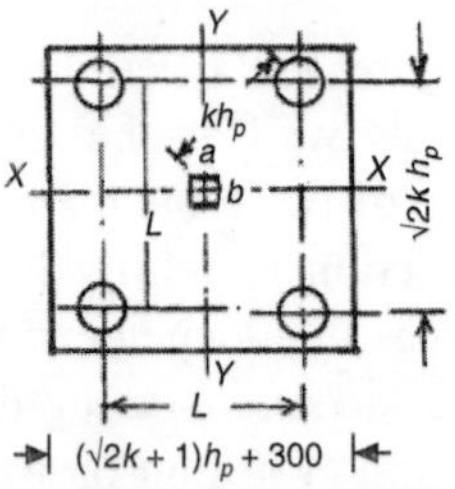 Y; kh_p; a; X X; b; L; $\sqrt{2}k\,h_p$; Y; L; $(\sqrt{2}k + 1)h_p + 300$	Along $XX = \dfrac{P}{30Ld}(3L^2 - a^2)$ Along $YY = \dfrac{P}{30Ld}(3L^2 - b^2)$
6	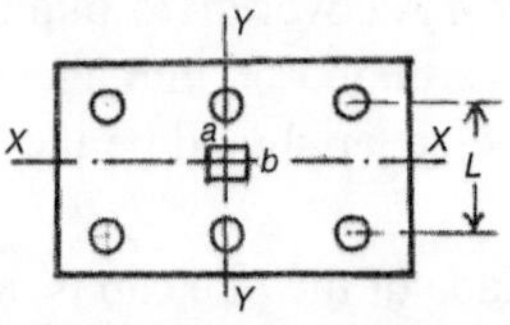 Y; X X; a; b; L; Y	Along $XX = \dfrac{P}{18Ld}(3L^2 - a^2)$ Along $YY = \dfrac{P}{36Ld}(3L^2 - b^2)$

Figure 17.1 Shape of pile caps for 2 to 6 piles and formulae for tension by truss theory for arrangement of steel. (Extend pile cap 150 mm beyond edge of pile. Dimensions given are in mm.)

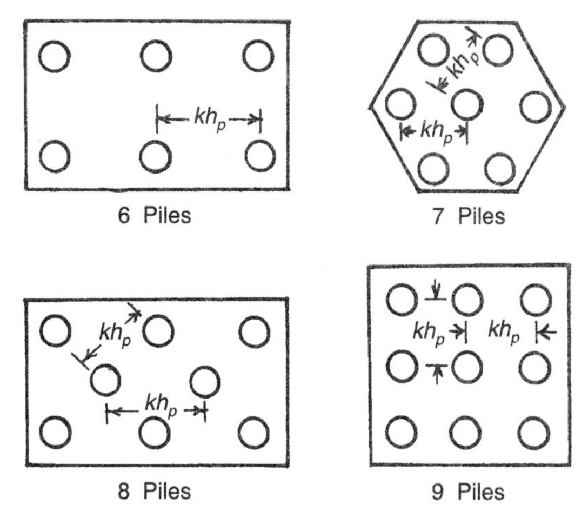

Figure 17.2 Shape of pile caps for 6 to 9 piles.

For example, as a pile cap for a three-pile group, any one of the shapes A, B and C as shown in Figure 17.3 can be considered. However, type A was rejected because of doubtful shear capacity, and B was rejected because of difficulty in setting out of steel. But shape C has been adopted as it can have a more practical layout of reinforcement and is also strong in shear.

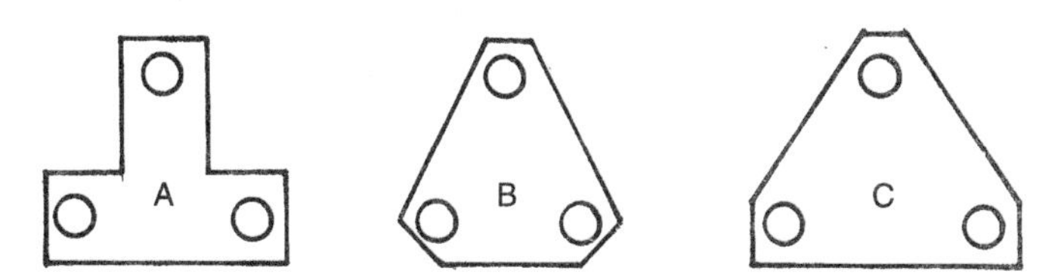

Figure 17.3 Pile cap for 3 pile group.

17.4 CHOOSING APPROXIMATE DEPTH OF PILE CAP

The pile cap depth is usually not less than 600 mm. The commonly taken depth is as follows:

Pile size (mm)	300	350	400	450	500	550	600
Pile cap depth (mm)	700	800	900	1000	1100	1200	1400

The study of the cost of pile caps for pile diameters 40–60 cm showed that the cost per kN of the load varied with the depth of the pile cap. Accordingly, one may generalize the relationship for an economical depth and arrive at the following relation: Assuming (h_p) as the diameter of the pile and (h) as the *total thickness of the pile cap*, we get

$$d = (2h_p + 100) \text{ mm} \quad \text{for } h_p \not> 500 \text{ mm} \tag{17.1a}$$

$$d = 1/3(8h_p + 600) \text{ mm} \quad \text{for } h_p \geq 550 \text{ mm} \tag{17.1b}$$

17.5 DESIGN OF PILE CAP REINFORCEMENT AND CAPACITY AND CHECKING THE DEPTH FOR SHEAR

The pile cap should be deep enough for the pile reaction from below the pile caps to be transferred to the pile cap. Similarly, considering the loads from the column at its top, the pile cap should have sufficient capacity to resist the bending moment and shear forces as well as the punching shear requirements to transfer the load to the piles through the pile cap.

There are two alternative theories on which pile caps can be assumed to transfer the loads from the columns to the pile foundations. They are (a) the truss theory and (b) the beam or bending theory.

Truss theory: Figure 17.4 shows the truss action for a four-pile group. Even in conventional designs when the angle of dispersion of load θ is less than 30° (tan 30° = 0.58), (i.e. the value of (a_v/d) ratio as shown in Figure 17.4 is less than 0.6), we may assume the load to be transferred to the pile by strut action. AB being in compression and BC in tension. This is called the *truss action*. Experiments show that this action (as in deep beams and corbels) can be predominant even up to (a_v/d) ratio equal to 2 or $a_v = 2d$. In this truss action, the tensile force between pile heads is assumed to tie the ends of the reinforcements at its ends as needed in the case of an arch. This is known as the *truss theory*.

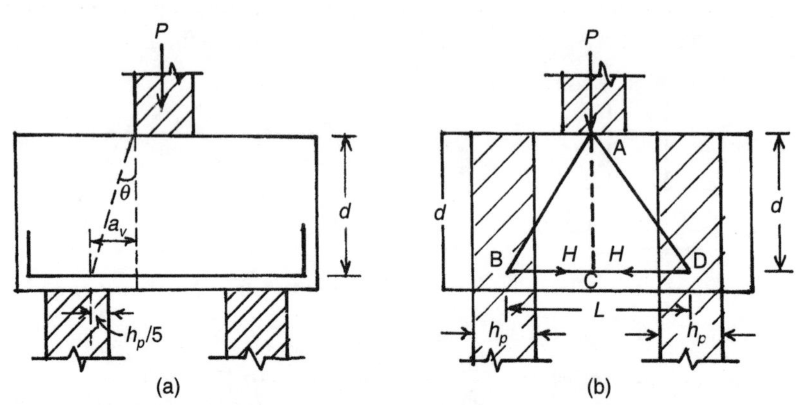

Figure 17.4 Design of pipe cap reinforcement by truss theory: (a) Transmission of load from column to pile by truss action, (b) Truss action in 2 pile group and use of tension steel. (*Note:* Value of a_v.)

Bending theory: The action of the whole pile cap as a beam is easy to imagine. When the shear span/depth of (a_v/d) is 2 or more, the bending action is more predominant than the truss action. Thus, the tensile reinforcement at the bottom acts like the tension reinforcement in an ordinary beam. Then the pile caps are to be designed by the *beam theory*.

In some cases, the design by beam theory may require far less steel than that by truss theory. It should, however, be remembered that the realization of these different actions is not so much for the determination of the amount of steel and the savings that can be done in steel quantity as for the appreciation of the real behaviour of the pile cap in the field and for the method of detailing of steel in pile caps. *The necessity of anchoring the main steel at their ends should be fully appreciated when conditions are favourable for truss action.* With truss action, the ends of the steel should be given full anchorage by providing the full development length by suitable bending of the steel inside the pile cap.

It should also be remembered that, theoretically, one can assume higher allowable anchorage bond stresses on top of the pile than specified for beams due to the effect of the compression present at the junction of the pile head and the pile cap.

17.5.1 Design for Steel

***Method 1* (By truss theory):**

(a) *Calculation of tensile force.* The truss action can be visualized in groups of piles up to six piles. For a pile cap over two piles, taking H as the tensile force and P as load from the column, from the triangle of forces of Figure 17.4, we have

$$\frac{H}{P/2} = \frac{L/2}{d}$$

The tensile force is thus given by the equation

$$H = \frac{PL}{4d} = F\left(\frac{PL}{d}\right), \qquad F = 1/4 \tag{17.2}$$

The area of steel required is given by

$$A_s = \frac{H}{0.87 f_y} = \frac{PL}{4(0.87 f_y) d}$$

If the size of the column (a along L) is also taken into account, the magnitude of the tensile force is obtained as

$$H = \frac{P}{12Ld}(3L^2 - a^2) \tag{17.2a}$$

These values of H for pile caps up to No. 6 are tabulated in Figure 17.1.

The ideal arrangement of reinforcement obtained from the truss theory is to place the steel in bands joining the top of piles with enough extension of the steel *for full anchorage.* In practice, this concentration of steel over piles is rarely possible and even distribution across the section with some concentration along the piles is resorted to, as described in Sec. 17.6.

***Method 2* (By bending theory):** When (a_v/d) ratio is more than 2, as in shallow pile caps or with the arrangement of 6 or more piles, the bending action is more predominant than the truss action. In this case, the pile cap is designed as a normal beam for bending moment and shear. The reinforcement is evenly distributed over the section.

For the bending action, the load on the pile cap may be considered to be uniformly distributed or concentrated. IS 2911 specifies the load from the column to be dispersed at 45° from the top *to the middle of the pile cap*. The reaction from the pile is also taken as distributed at 45° from the edge of the pile up to the mid-depth of the pile cap. The maximum bending moment and shear force are calculated on this basis. However, it is much easier to consider the loads as concentrated loads and calculate the BM and SH. The depth to be provided should be such that no extra shear reinforcement is necessary for the section.

17.5.2 Designing Depth of Pile Cap for Shear

***Method 1* (Using ordinary bending theory):** IS 456, Cl. 34.2.4.2 states that for checking shear in a footing supported on piles of diameter D_p (or h_p), we take a vertical section at a *distance equal to the effective depth* of the footing located from the face of the column pedestal or wall. All piles (of diameter D) located at $D/2$ or more outside this section are assumed as producing shear on the section. The reaction from any pile whose centre is located at $D/2$ or more inside the section shall be assumed as producing no shear on the section. For intermediate position of the pile centre, the portion of the pile reaction to be assumed as producing shear on the section shall be based on straight line interpolation between full value at $D/2$ outside the section to zero value at $D/2$ inside the section.

***Method 2* (Using enhanced shear theory (h_p = diameter of the pile)):** It has been pointed out that truss actions can be easily visualized for pile caps up to five piles. In such chases, the shear may be considered as in deep beams and corbels. The procedure for checking shear stress when using this theory is as follows.

The critical section for shear is taken as shown in Figure 17.5 along the piles, as also along the face of the piles, in square piles and at $h_p/5$ inside the pile for round piles. All piles with centres outside this line are considered for calculating shear across this section in the pile cap.

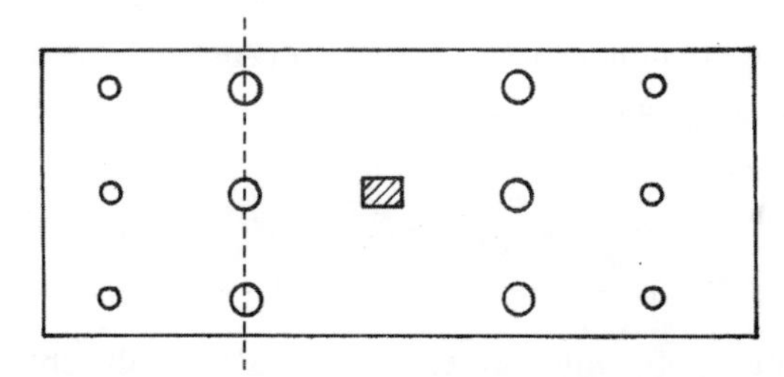

Figure 17.5 Checking shear in pile cap by enhanced shear theory.

No enhancement of shear is allowed if $a_v > 1.5d$. This enhancement is also allowed only for strips of width $3h_p$ immediately around the piles. If piles are spaced at more than $3h_p$ for the balance of the areas, only the normal shear value is allowed. The condition to be satisfied with the normal and enhanced shear values is as follows:

$$\text{Shear capacity along B.B} = \tau_c\, Bd > \Sigma P$$

where

ΣP = Sum of pile reactions from the edge to the section considered

B = Width of pile cap

d = Effective depth

The increased shear value is

$$\tau_c' = 2\tau_c (d/a_v) \tag{17.3}$$

where

τ_c' = Increased shear value
τ_c = Normal shear value in Table B.4
a_v = As shown in Figures 17.4

(*Note:* The increased value is applicable to a distance $a_v = 2d$ only. Critical sections are taken at $h_p/5$ for circular piles and along the face of the pile for rectangular piles.)

17.5.3 Checking for Punching Shear

The allowable punching shear is usually high. As a *general rule, if the piles come within the load spread of 45° of the supporting column then punching shear calculations are not normally necessary*. Otherwise, punching shear should be checked just as in other cases of footings around the columns.

17.6 ARRANGEMENT OF REINFORCEMENTS

The detailing of reinforcement in pile caps depends on their shape. In general, seven types of bars are used as pile cap reinforcements (Figure 17.6):

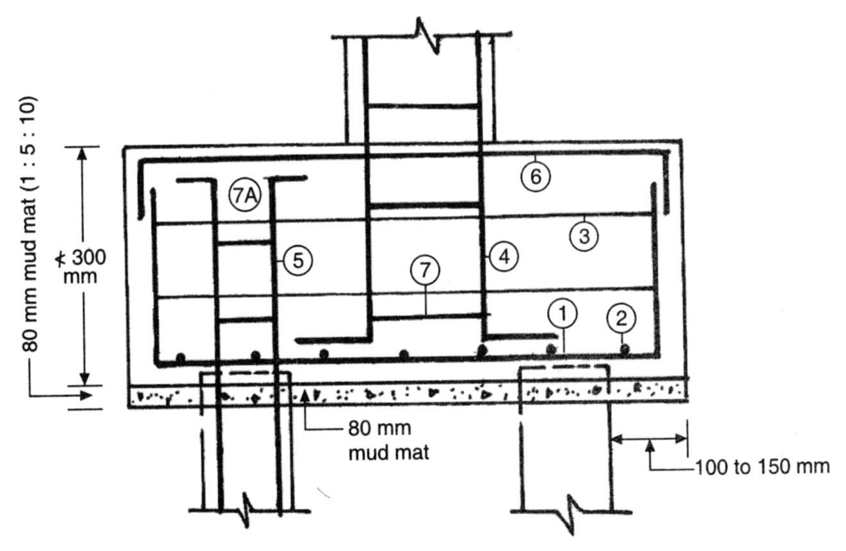

Figure 17.6 Reinforcement detailing of pile caps. ((1)–(7): Bar marks. (1), (2), Main bars; (3), Horizontal ties to resist bursting (T12–150); (4) Starter bars; (5) Pile bars; (6) Top bars; (7), (7A) links). (We generally design pile cap to be safe in shear without shear reinforcement.)

1. Type (1) steel main bars placed at the bottom in the *XX*-direction at 90° and extending to the top of the pile cap bent up at their ends to increase anchorage.

2. Type (2) steel main bars placed at the bottom in the *YY*-direction also bent up at their ends.
3. Type (3) steel consisting of two or three layers of 16 mm diameter horizontal ties (lacer bars) fixed to the upstands of the main bars as secondary steel to resist bursting.
4. Type (4) bars, the column starter bars, which are *L* shaped and turned back at the level of the bottom reinforcement. They are held together by links at two or more levels.
5. Type (5) bars are the reinforcements from the pile, which are extended into the pile cap for its full development length in compression.
6. Type (6) bars which are the top steel provided as compression steel in the slab if required by calculation. These reinforcements are tied together to form a cage before casting the pile cap.
7. Type (7) bars are the links to the column bars and 7A links to the pile reinforcements.

The arrangement of reinforcements for various types of pile caps is shown in Figure 17.7. In detailing of steel, special attention should be given to the anchorage length of these reinforcements.

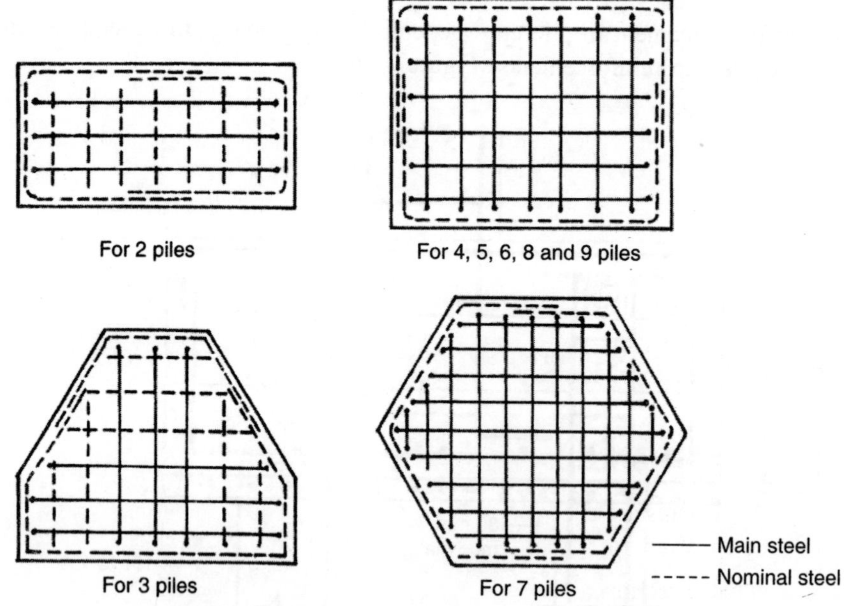

Figure 17.7 Arrangement of steel for pile caps.

17.7 ECCENTRICALLY LOADED PILE GROUPS

If the pile group is supporting a column subjected to moments due to vertical loads as in Figure 17.8, then the loads on the individual piles of the pile group have to be adjusted to take care of these moments also. In such cases, we can proceed as in the case of a footing subjected to eccentric loads.

1. If the moment is acting all the time, we can set the column at an eccentricity with respect to the CG of the pile group and see that all the piles are not loaded beyond their capacity.
2. If the moments act only temporarily, the change in pile loads due to the moment should be calculated and the resultant loads should be within the safe capacity of the piles.

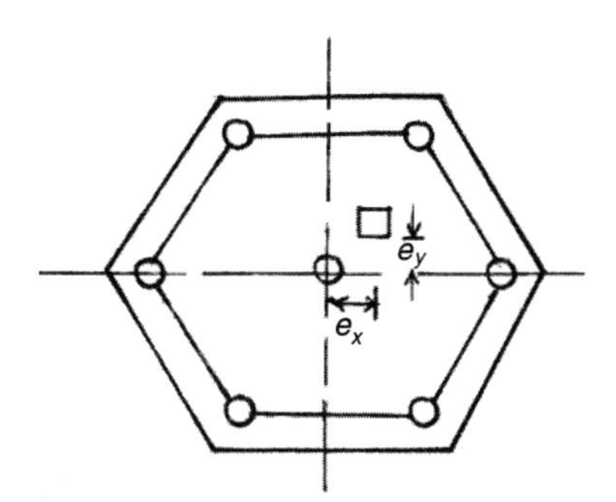

Figure 17.8 Eccentrically loaded pile cap.

The load on a typical pile will be given by the relations ($M_X = Pe_x$ and $M_Y = Pe_y$)

$$P = \frac{W}{N} \pm M_X \left(\frac{d_1}{\Sigma d_1^2}\right) \pm M_Y \left(\frac{d_2}{\Sigma d_2^2}\right) \tag{17.4}$$

where M_X and M_Y are the moments about CG of the group and $\frac{\Sigma d_1^2}{d}$ is the reciprocal of the modulus of the pile of the pile group.

We must also remember that if the moment caused is due to the external horizontal forces (in contrast to gravity loads), the group has to resist these horizontal forces also.

(*Note:* Moment due to e_x is designated as M_{XX}.)

17.8 CIRCULAR AND ANNULAR PILE CAP

Circular and annular pile caps are used to support circular or annular rafts in bad soils. The piles can be arranged in two ways:

- Along a radial direction at various specified diameters
- Along various specified diameters at equal spacing along the diameter but not radially.

(The method of analysis of the circular or annular raft on piles has already been discussed in Sec. 15.6 to 15.8.)

17.8.1 Analysis of Forces on Vertical Piles

When vertical loads W and moments M (like moment due to wind loads on a chimney) act, the loads are calculated as follows:

Maximum and minimum loads on pile in outer ring $= \frac{W}{N} \pm \frac{M}{I} y$ (17.5)

where W = Vertical axial load

N = Number of piles

$I = \Sigma (r \cos \theta)^2$

r = Radial distance of the pile

θ = Inclination of the pile from the direction of M

There should be no negative load on the pile.

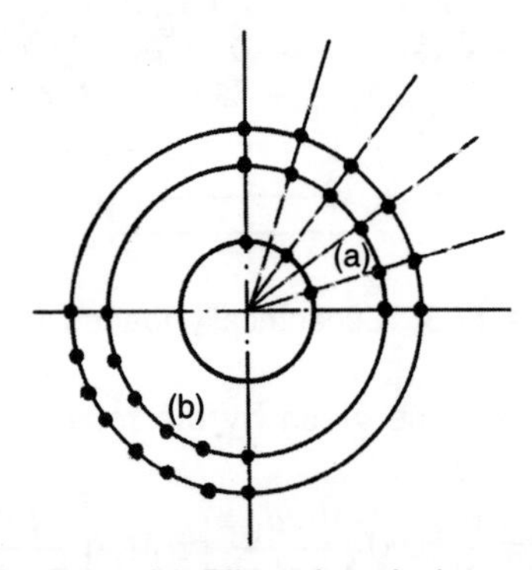

Figure 17.9 Circular pile caps: (a) Piles placed along circles in radial directions, (b) Piles placed at equal spacings along concentric circles.

17.8.2 Analysis of Raked Piles (Inclined Pile)

When lateral loads are very large, raking piles can be introduced, especially along the outer periphery. The pile rake should be neither greater than 1 in 2 nor less than 1 in 12. Usually, 1 in 5 is used as a value greater than 1 in 4 requires special field equipment. There are two methods of analysing this system as follows:

(a) The first method is the Culman's method as described in books on foundation engineering [1]. Here, vertical and raking piles take the load together.

(b) In the second method, it is assumed horizontal loads are resisted by raking piles only. In this method, we assumed that the maximum loading on the pile, as calculated by Eq. (17.5), is in the raked direction. The sum of the horizontal components (+ve on windward and –ve on the leeward side) resists the horizontal force. These forces can be determined by graphical construction on a large representation of the pile layout [3].

17.9 COMBINED PILE CAPS

As in the case of footings, it may be necessary in some cases to combine the pile caps for carrying two separate columns to a combined pile cap. The same principles, as used in combined footings to have the centre of gravity of loads to coincide with the centre of gravity of pile groups, can be used here also. Piles can also be used to carry an external column as in balanced footings [see Example 15.5].

17.10 SUMMARY

Pile caps should be designed with care. Their depths should satisfy the transfer of forces from columns to piles. The shape, size and reinforcements to be provided depend on the number of piles and the loads to be transferred. These aspects have been briefly discussed in this chapter.

EXAMPLE 17.1 (Checking for shear as per IS 456, Cl. 34.2 in pile caps)

A reinforced concrete column 500 × 500 mm is supported on a pile cap with 4 cylindrical piles of 300 mm diameter spaced at 1100 mm centres as shown in Figure E17.1. The column is placed with a pedestal 700 × 700 mm all around. The forces acting on the column are a gravity load of 800 kN and a moment of 600 kNm on the *YY*-axis. Check for shear if the effective depth of the cap is 530 mm.

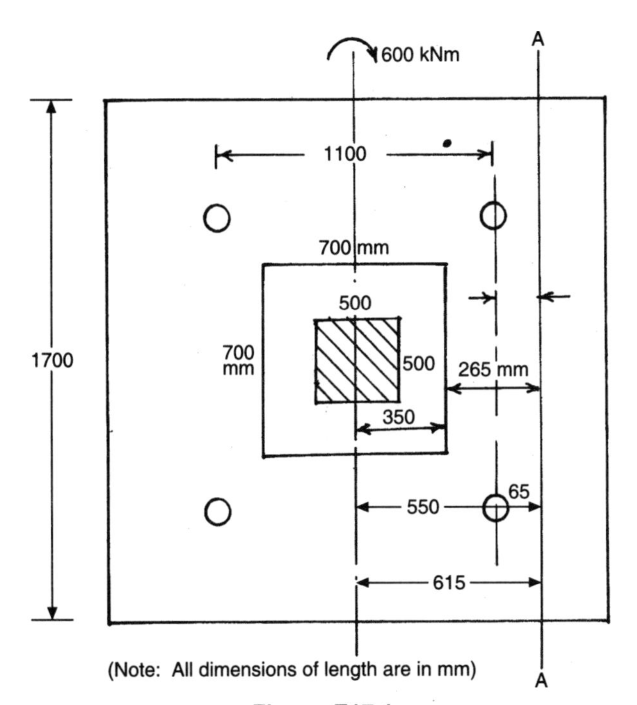

Figure E17.1

Reference	*Step*	*Calculation*
	1	*Find plan area of pile cap (with* 150 mm *extension)*
Sec. 17.3		Length L = 1100 + 300 + 150 + 150 = 1700 mm square
	2	*Calculate forces on piles (Pile diameter h_p)*

Reference	Step	Calculation
Figure E17.1		Let moment be taken left to right (clockwise) as shown in Figure E17.1
		Max load on *a row of two piles* on *the right*
		$= \frac{P}{A} +$ load due to moment (lever arm 1.1 m)
		$= \frac{800}{2} + \frac{600}{1.1} = 400 + 545 = 945$ kN
		Factored load = 1.5(945) = 1416 kN on 2 piles (approx.)
	3	*Find section AA along Y-axis for shear* (d = 530 given)
		Though the IS specifies the section for checking shear to be taken at a distance d for the footing, we will check shear at a distance $d/2$ from the pedestal to illustrate the principle of design.
		Section for shear at $\frac{d}{2} - \frac{530}{2} = 265$ mm from pedestal
		= 265 + 350 = 615 mm centre of pile cap.
		Location of the pile 550 mm from the centre line.
		Pile is 65 mm inside the section of AA
Sec. 17.5.2		According to the IS, shear is zero for piles located at $h_p/2$ = 150 inside AA and the effect is full for piles located at $h_p/2$ = 150 outside AA. For other positions, we take proportionately in the length
		$(h_p/2 + h_p/2) = h_p$
Step 2		Shear $= \frac{1417 \times (150 - 65)}{300} = 401$ kN
		Shear stress $= \frac{V}{bd} = \frac{401 \times 10^3}{1700 \times 530} = 0.44$ N/mm^2
		This is safe for a percentage of steel of 0.4% for M_{20}.
	4	*Find depth by thumb rule*
		$d = 2h_p + 100$ for $h_p < 500$
Sec. 17.4		= 600 + 100 = 700 mm

EXAMPLE 17.2 (Design of pile caps using truss theory)

A column 550 mm square has to carry a factored (design) load of 2600 kN to be supported on 4 piles each of 450 mm diameter and spaced at 1350 mm centres. Design a suitable pile cap assuming f_{ck} = 25 N/mm^3 and f_y = 415 N/mm^2.

Reference	Step	Calculation
	1	*Arrangement of pile cap*
Figure 17.1		With extension of the pile cap equal to 150 mm on all sides from the outer side of piles
		Size of pile cap = 1350 + 450 + 300 = 2100 mm
	2	*Depth of pile cap*
Sec. 17.4		Empirical formula = $2h_p$ + 100 = 1000 mm (too high)
		Adopt effective depth d = 1/2 spacing of the pile
		$= \frac{1350}{2} = 675$ mm and 75 mm cover.
		Assuming 20 mm rods, D = 675 + 75 + 10 = 760 mm
		Pile cap 2100 × 2100 × 760 mm
	3	*Check for truss action*
Figure 17.4		Shear span = a_v
		$$a_v = 675 - 275 - 225 + \frac{450}{5} = 265\text{ mm}$$
		$$\frac{d}{a_v} = \frac{675}{265} = 2.55 \text{ or } a_v/d = 0.39\text{ (only)}$$
		Truss action exists (a_v/d < 0.6).
Figure 17.1	4	*Tension steel*
		$$T = \frac{P}{24Ld}(3L^2 - b^2)$$
		$$= \frac{2600}{24 \times 1350 \times 675}[3(1350)^2 - (550)^2]$$
		= 614 kN
		$$A_s = \frac{614 \times 10^3}{0.87 \times 415} = 1701\text{ mm}^2$$
		Provide 16T 12, giving 1809 mm^2
	5	*Percentage of steel provided*
IS 456 Cl. 26.5.1.1		$$p = \frac{1809 \times 100}{2100 \times 675} = 0.127\% > 0.12\%$$
		(This is less than the 0.2% in IS and 0.33% in BS of steel usually provided for beams in which the bending action is predominant.)
	6	*Check for shear (f_{ck} = 25) using enhanced shear theory*
		Spacing of piles $3h_p$ = (3 × 450)

Reference	Step	Calculation
		Hence, enhanced shear is allowed in the section as per IS 456.
		$\tau_c' = \tau_c\left(\frac{2d}{a_v}\right) = \frac{0.36 \times 2 \times 675}{265} = 1.83$ but < 3.1 N/mm^2 (Enhanced shear)
		$\text{Shear} = \left(\frac{V}{2100 \times 6.75}\right) = \frac{2600 \times 1000}{2(2100 \times 675)}$
		$= 0.92$ N/mm^2 $< \tau_c' = 1.83$ N/mm^2
	7	*Detail steel*
		Detail steel according to the standard practice.
		[*Note*: If the depth of the pile cap provided is less than that given for truss action, then the pile cap is designed as a beam and checked for bending and shear.]

EXAMPLE 17.3 (Design of pile cap using bending theory)

Design a pile cap for a system of 3 piles supporting a column 500 mm square and carrying an axial load of 600 kN. Assume that the diameter of the pile is 400 mm, $f_{ck} = 20$ N/mm^2, $f_y = 415$ N/mm^2.

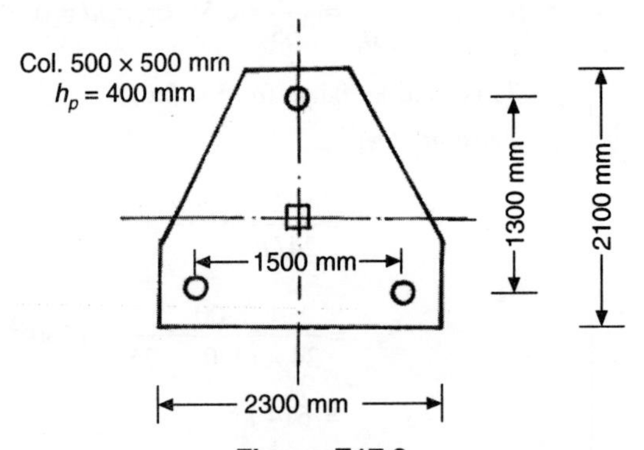

Figure E17.3

Reference	Step	Calculation
Fig. 17.1	1	*Layout of pile cap*
		Assume the standard layout with tolerance of 200 mm for the cap.
Eq. (17.1)		Spacing of pile $\not< 2h_p$.
		Adopt 1.5 m as size of pile cap; assume depth of cap $D = 600$ mm
		Bending theory as in a slab becomes applicable when the depth of pile cap to be provided is small, i.e.
		$d = 600 - 75 - \frac{16}{2} = 517$ mm

Reference	Step	Calculation
	2	*Loads on pile cap*
		Volume of the pile cap = Plan area × Depth
		Base length = 1.5 + 0.4 + 0.2 + 0.2 = 2.3 m
		$\text{Height} = \left(\frac{1.5 \times \sqrt{3}}{2}\right) + 0.4 + 0.4$
		= 1.3 + 0.8 = 2.1 m
		$\text{Plan area} = (2.3 \times 2.1) - (2.1 - 0.8)\left(\frac{2.3 - 0.8}{2}\right)$
		= 3.86 sq.m
		Wt. of pile cap = 3.86 × 0.6 × 24 = 55.5 kN
		Assume wt. as concentrated load at CG.
		$\text{Load on each pile} = \frac{1}{3}(600 + 56) = 219 \text{ kN}$
		Factored load on each pile = 1.5 × 219 = 328 kN
	3	*Transfer of load from piles to pile cap*
		First consider load transferred in the *YY*-direction and then in the *XX*-direction.
		According to IS 2911, Part I, Cl. 5.12.1, loads and reactions can be dispersed to the mid-depth of the pile cap. Taking loads as concentrated loads is a conservative estimate for thin slabs and we will adopt the same.
	4	*Bending of one pile loaded in the YY-direction*
Figure E17.3		(P × dist to CG of 3-pile loads)
		$M_1 = 328 \times \frac{2}{3} \times 1.3 = 284 \text{ kNm}$; Breadth ≈ 800 mm
		$\text{Required } d_1 = \sqrt{\frac{284 \times 10^6}{0.138 \times 800 \times 20}} = 359 \text{ mm} < 517 \text{ mm (available)}$
SP 16 Table 2		$\frac{M_1}{bd^2} = \frac{284 \times 10^6}{800 \times 517 \times 517} = 1.33,\ p = 0.4$
Table B.1		$A_s = \frac{0.4 \times 800 \times 517}{100} = 1654 \text{ mm}^2$
	5	*One-way shear in the YY-direction (enhanced shear theory)*
IS 456 Table 19		$v = \frac{\text{Pile load}}{\text{Section}} = \frac{328 \times 10^3}{800 \times 517} = 0.79 \text{ N/mm}^2$
		τ_c for 0.4% steel = 0.39 less than 0.79 N/mm²

Reference	Step	Calculation
		$\tau_c' = \tau_c \left(\frac{2d}{a_v}\right)$ (due to enhanced shear) (Taking at $h_p/5$)
Figure 17.4		$a_v = \frac{2}{3}(1.3) - 0.2 - 0.25 + \frac{0.40}{5} = 0.5$ m
		$\tau_c' = 0.39\left(\frac{2 \times 0.517}{0.5}\right) = 0.80 \text{ N/mm}^2$
	6	*Bending in the XX-direction*
		$M_2 = 328 \times \frac{1.5}{2} = 246 \text{ kNm} < M_1$
	7	*Bending shear in the XX-direction*
		$a_v = \frac{1.5}{3} - 0.2 - 0.25 + 0.08 = 0.13$
		$\tau_c' = 0.39\left(\frac{2 \times 0.517}{0.13}\right) = 3.012 \text{ N/mm}^2$
	8	*Punching shear around column to be supported*
		Section at $d/2$ from the column face
		$d = 517$ mm
		Perimeter for punching shear = $2(2 + b + 2d)$
		= 2(1000 + 1034) = 4068 mm
		$\tau_p = \frac{1.5 \times 600 \times 1000}{4068 \times 517} = 0.42 < 1.12 \text{ N/mm}^2$
	9	*Arrangement of steel*
		(a) Arrange main steel as beams in the *YY*- and *XX*-directions.
Figure 17.8		(b) Provide 5 Nos. of 10 mm rods as circumferential steel around the projected pile reinforcements which extend to top of the pile cap.
IS 2911 Sec. 3.4.2.4		[*Note*: Two-way shear or punching shear also has to be considered at $d/2$ from the column face where such action can develop, as in the case of a pile cap over a large number of piles. For this example, such action may not develop and hence, it is not considered.]

EXAMPLE 17.4 (Design of pile cap with truss and bending action)

Design a pile cap for a group of piles consisting of 6 piles of 350 mm diameter to support a 450 mm square column carrying a factored load of 280 tons. Assume f_{ck} = 25 N/mm^2 and f_y = 415 N/mm^2.

Reference	Step	Calculation
	1	*Layout of pile* (6 pile group)
Figure 17.1		Spacing $\nless 3h_p$ = 3 × 350 = 1050, let s = 1.1 m
		Using a close tolerance of 100 mm all around
		Length of the cap = 2.2 + 0.35 + 0.20 = 2.75 m
		B of the cap = 1.1 + 0.35 + 0.20 = 1.65 m. Let D = depth
		$D \nless 2h_p$ + 100 = (2 × 350) + 100 = 800 mm; cover = 40 mm
		$$d = 800 - 40 - \frac{20}{2} = 750 \text{ mm}$$
	2	*Load on pile*
		(γ_m)DL = 1.5(2.75 × 1.65 × 0.8 × 24) = 130 kN
		Load on each pile = (2800 + 130)/6 = 488 kN
	3	*Transfer of load*
		Assume the following:
		Transfer of the middle pile by truss action
		Transfer of the end pile by cantilever bending
	4	*Bending in the XX-direction as cantilever*
Table B.1		M = 2 × 488 × 1.1 = 1074 kNm
		$$d = \left(\frac{1074 \times 10^6}{0.138 \times 1650 \times 25}\right)^{1/2} = 434 \text{ mm} < 750 \text{ mm}$$
		$$\frac{M}{bd^2} = \frac{1074 \times 10^6}{1650 \times 750 \times 750} = 1.15; \ p = 0.39$$
		A_s = (0.39 × 1650 × 750)/100 = 4826 mm^2; $\frac{\text{Use 16 T 20}}{5026 \text{ mm}^2}$
		$$\text{Actual } p = \frac{5026 \times 100}{1650 \times 750} = 0.40$$
	5	*Shear along section YY*;
Table B.4		$$v = \frac{2 \times 488 \times 10^3}{1650 \times 750} = 0.79 \text{ N/mm}^2; \text{ (Load on pile 488 kN)}$$
		τ_c for 0.4% steel = 0.45 N/mm^2
		Taking section at $h_p/5$ of pile, we have
Figure 17.5		$$a_v = 1100 - \frac{1}{2}(450 + 350) + \frac{350}{5} = 770 \text{ mm}$$
		Enhanced shear is,

Reference	Step	Calculation
Eq. (17.3)		$$\tau_c' = \tau_c 2\left(\frac{d}{a_v}\right) = 0.45 \times 2 \times \frac{750}{770} = 0.88 \text{ N/mm}^2$$
		Similarly, check in the *XX*-direction as shown in Figure 15.4.
	6	*Steel in the YY-direction*
		Tension for truss action over the central two piles
		$$T = \frac{P(3L^2 - a^2)}{12Ld}$$
Figure 17.1		$L = 2.2$, $a = 0.45$, $d = 0.75$
		$$T = \frac{(2 \times 488)(14.32)}{12 \times 2.2 \times 0.75} = 705.9 \text{ kN}$$
		$$A_s = \frac{706 \times 10^3}{0.87 \times 415} = 1955 \text{ mm}^2$$
		Provide
		(a) 4 T 25 over central piles (b) 4 T 20 over outer piles (c) 4 T 20 in space in-between.
		(Total number 10 rods. Maximum spacings less than those allowed for distribution steel.)
	7	*Arrangement of steel*
Figure 17.7 Figure 17.8		(a) Main steel (16 Nos. 20 mm, ends bent up to the full depth) in the *XX*-direction along the length
		(b) Steel in the *YY*-direction along the breadth, as in Step 6.
		(c) Horizontal ties T12 at 300 mm
		(d) Starter bars, pile bars, links, etc. as per the standard specification.
	8	*Punching shear*
		This may be checked at *d*/2 from the column edge.

EXAMPLE 17.5 (Design of a balanced cantilever foundation on piles)

A 400 × 400 mm column A, as shown in Figure E17.5, carrying 1000 kN has to be placed very near an existing building. As the soil is bad 400 mm piles have been adopted for the foundation. It is to be carried by two piles and the minimum distance the pile can be installed is 1 m inside from the end column line. To counteract the moment due to eccentricity of the load, the pile cap has to be tied back to an internal column B, carrying 2000 kN to form a continuous pile cap. The interior column has four piles symmetrically placed around the column at 2 m centres. The distance between the columns is 4 m. Determine the loads in each pile and also explain how to design the cantilever pile cap.

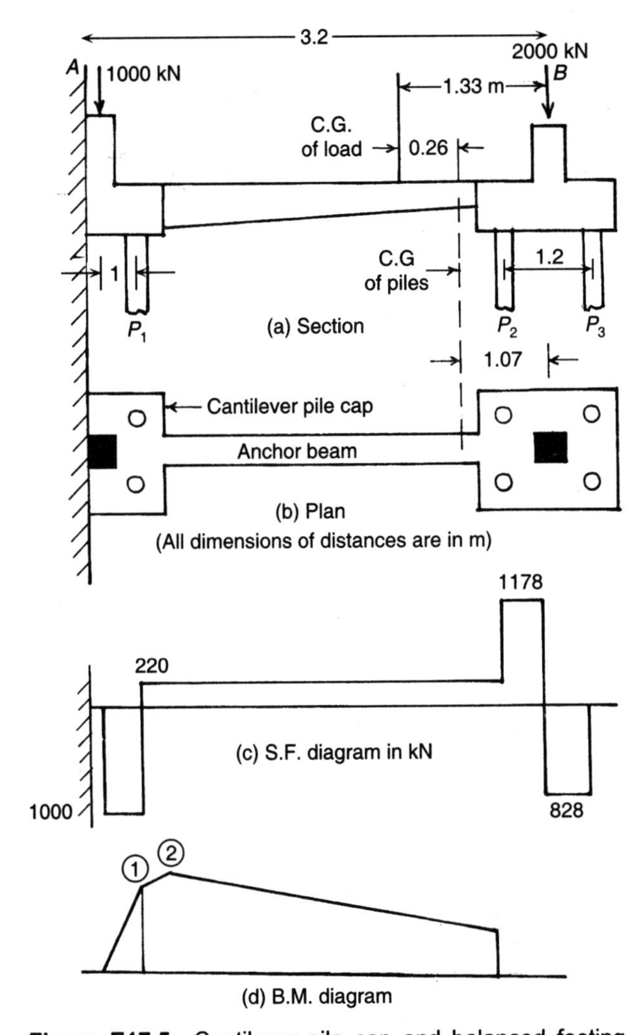

Figure E17.5 Cantilever pile cap and balanced footing.
(*Note:* The S.F. diagram can be refined by assuming the column load and the pile load is distributed through the pile cap.

Reference	*Step*	*Calculation*
		(*Note:* Let the external column be A and the internal one B. We have to distinguish between CG of loads and CG of piles.)
	1	*Find CG of load from column B*
		$3000 \times \bar{x}_1 = 1000 \times 4$ or $\bar{x}_1 = 1.33$ m from col. $B = \bar{x}_c$
	2	*Find CG of pile*
		Distance of 2 piles from B = (4 – 0.8) = 3.2 m
		$6\bar{x}_2 = 2 \times 3.2$ or $\bar{x}_2 = 1.07$ m from col. $B = \bar{x}_p$

Reference	Step	Calculation
	3	*Find moment of vertical loads about CG of piles* $M = 3000 \times (1.33 - 1.07) = 780$ kNm The foundation has $W = 3000$ and $M = 780$ kNm counter clockwise.
	4	*Calculate loads on each set of piles at* x_1, x_2 and x_3 $P_1 = \dfrac{W}{\text{No. of piles}} \pm \sqrt{\dfrac{Mx_1}{x_1^2 + x_2^2 + x_3^2}}$, where $I = \Sigma x_1^2$ where x_1, x_2 and x_3 are distances of piles (2 Nos.) on the left and right of the CG of piles. $x_1 = 3.2 - 1.07 = 2.13$ m (distance of 2 piles at boundary) $x_2 = 1.07 - 0.60 = 0.47$ m (distance of 2 piles under col. B) $x_3 = 1.07 + 0.60 = 1.67$ m (distance of 2 piles under col. B) $\Sigma x_1^2 = 2\left[(2.13)^2 + (0.67)^2 + (1.67)^2\right] = 15.1 \text{ m}^2$ Load on 2 of $P_1 = \dfrac{3000}{6} + \dfrac{780 \times 2.13}{15.1} = 610$ kN Load on 2 of $P_2 = 500 - \dfrac{780 \times 0.47}{15.1} = 476$ kN Load on 2 of $P_3 = 500 - \dfrac{780 - 1.67}{15.1} = 414$ kN $\Sigma = 2(1500) = 3000$ kN
	5	*Design of pile cap* (Refer Example 17.1 with moments) Based on the column loads and pile loads, we can draw the SF and BM diagrams along the length of the system and design the necessary dimensions and reinforcement details of the pile cap by using the bending theory. (Reference 4 deals with these type of problems in more detail.) (*Note:* In these types of foundations, it is usual to provide a continuous length of steel both on top and bottom faces of the pile cap with the lateral reinforcement in the form of stirrups all tied into a cage.)

EXAMPLE 17.6 (Analysis of circular pile cap)

An annular raft 8 m in outer radius and 3.2 in inner radius is loaded over a ring of 70 piles arranged along the radii of 7.5 m, 6.3 m, 5.0 m and 3.7 m, as shown in Figure E17.6. Assuming the system is subjected to a vertical load of 54000 kN and moment of 58000 kNm from a chimney, determine the maximum load on the pile system. Also, indicate how these loads are used to analyse the circular raft.

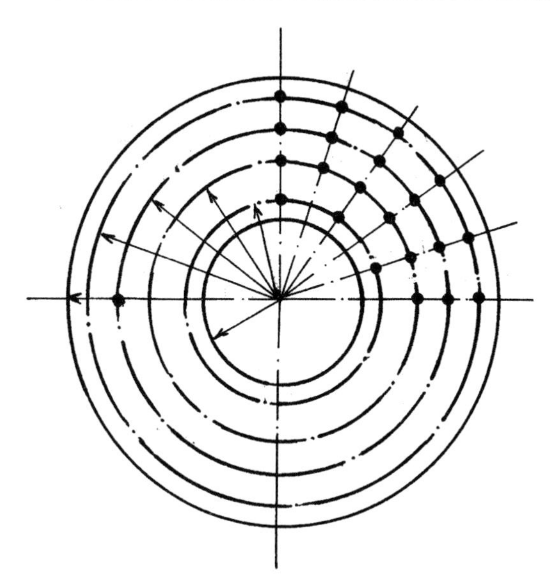

Figure E17.6

Reference	Step	Calculation
	1	*Find uniform pile reaction due to vertical load*
		$q_1 = \dfrac{54000}{70} = 771 \text{ kN}$
		Piles arranged at 18°, 36°, 54°, 72° and 90° radii around the circle.
	2	For finding the effect of moment,
		(a) *Assume moment in XX-direction (wind to right).* Find forces in piles
		$I_{YY} = r^2 \Sigma \cos^2 \theta$
		$= 7.5^2 \lfloor 2 + 4(\cos^2 18 + \cos^2 36 + \cos^2 54 + \cos^2 72) \rfloor$
		$+ 6.3^2 \lfloor 2 + 4(\cos^2 18 + \cos^2 36 + \cos^2 54 + \cos^2 72) \rfloor$
		$+ 5.0^2 \lfloor 2 + 4(\cos^2 18 + \cos^2 36 + \cos^2 54 + \cos^2 72) \rfloor$
		$+ 3.7^2 \lfloor 4(\cos^2 18 + \cos^2 54) \rfloor$
		$= 1278 \text{ m}^2$
	3	*Pressure due to moment (at outer radius = 7.5 m)*
		$q_1' = \pm \dfrac{58000}{1278} \times 7.5 = \pm 340 \text{ kN}; \ [M = 58000 \text{ kNm}]$
	4	*Total pressure on extreme piles*
		$q_1 \pm q_1' = 771 \pm 340 = 1111 \text{ kN and } 431 \text{ kN}$
		(No tension on the pile.)

Reference	*Step*	*Calculation*
	5	*Use of these loads to analyse the raft* For the analysis of the raft, we consider the 20 piles on the outer ring. Total max load = 1111 × 20 = 22220 kN. Assume it as a ring load on this circle of radius 7.5 m. Ring load $P_1 = \frac{22220}{2\pi \times 7.5} = 472$ kN/m along 7.5 m radius. Similarly, ring loads P_2 along 6.3 m radius, P_3 along 5.0 m radius and P_4 along 3.5 m radius are taken as ring loads on the raft for the analysis of the raft. Reference [5] gives more data on this subject.

REFERENCES

[1] Varghese, P.C., *Foundation Engineering*, Prentice-Hall of India, New Delhi, 2005.

[2] IS 2911, *Code of Practice for Design and Construction of Pile Foundation*, BIS, New Delhi.

[3] Varghese, P.C., *Limit State Design of Reinforced Concrete*, 2nd Ed., Prentice-Hall of India, New Delhi, 2002.

18

CHAPTER

Pile Foundations—Design of Large Diameter Socketed Piles

18.1 INTRODUCTION

Conventional piles are those ranging from 300–600 min in size. Piles larger than 600 mm in diameter are usually called *large diameter piles*. Lower diameter piles, 300–150 mm in size, are called *mini piles* and are used for slope protection and other uses. Those below 175 mm are called *micro* or *pin piles* (grouted piles) and they are commonly used for underpinning of building and seismic rehabilitation of bridges. As large diameter piles are generally placed at sites as drilled boreholes, they are also known as *drilled shafts*. The differences in the principles of design of each of these types should be clearly understood by a foundation engineer.

IS 2911 deals with the design and construction of conventional piles. The methods of design by static and dynamic approach are described in all textbooks on Soil Mechanics and Foundation Engineering [1]. Therefore, in this chapter, we confine ourselves to the design of large diameter piles.

In India, large diameter piles are commonly used for flyovers and tall buildings. They are also used extensively in India for bridges over rivers when rock deposits are available at reasonable depths. If rock is not available at reasonable depths for bridges, a well foundation or a number of conventional piles will be more economical. (The design of conventional piles and well foundations is dealt with in Ref. [1].)

There are a number of advantages in using large diameter piles instead of a number of regular piles to carry large loads. Large diameter piles do not need a pile cap. It is also easier to install a drilled shaft in dense sand and gravel than to drive a pile of any diameter. A drilled shaft also takes lateral loads more efficiently.

The drilled shafts of large diameter can be one of the following:

1. Shafts in cohesionless soils
2. Shafts in cohesive soils
3. Shafts bearing on solid rock
4. Shafts in *argillaceous* (clayey) weathered rock
5. Shafts in *granular-based* (like granite) weathered rock

The design of the first three types of piles is carried out as in the case of conventional piles. However, we find (see Sec. 18.4) that the most efficient way of taking the full strength of a large diameter pile is (a) to enlarge (or under-ream) its base dimension to a larger diameter, (b) to end it in good rock, or (c) socket it in a weathered rock. In this chapter, we primarily focus on the design of large diameter piles in weathered rock by socketing. [Well foundations used extensively in India for bridges over rivers need not end in rock. They consist of a concrete well plugged with concrete at the bottom (which is the bearing area) with inside sand filling and a top R.C. plug. The allowed bearing pressures of such wells are small. As per IRC specifications 78-1983, normally, the allowable bearing pressure should not exceed 2 N/mm^2 or 200 kN/m^2.]

Two ways in which the full capacity of large diameter piles can be obtained are:

- Under-ream the base of the piles up to three times the shaft diameter.
- To end the piles in rock or socket it in weathered rock (In weathered rock we may extend steel casing into rock and then socket with casing).

In this chapter, we deal with the second type only.

18.2 LOAD TRANSFER MECHANISM IN LARGE DIAMETER PILES

It is well known that the total load carrying capacity of a conventional pile can be expressed by static formulae with separate FS for friction and bearing as

$$Q_{\text{allowed}} = \frac{Q_{\text{ult. friction}}}{(\text{FS})_f} + \frac{Q_{\text{ult. bearing}}}{(\text{FS})_b}$$

In all types of piles with friction and bearing, the transfer of load is as shown in Figures 18.1 and 18.2. The load is first taken by friction and then only by bearing.

We may have different factors for safety for friction (1.5) and bearing (2.5) and an overall FS of 3 on friction plus bearing. When we load a pile, the load is taken first by the upper parts of the pile by friction. Gradually, the lower layers get stressed and, finally, the base resistance comes into action. This progressive transfer of load is shown in Figure 18.1. In granular soils, the lower layers, up to a certain level, offer more friction than the upper layers.

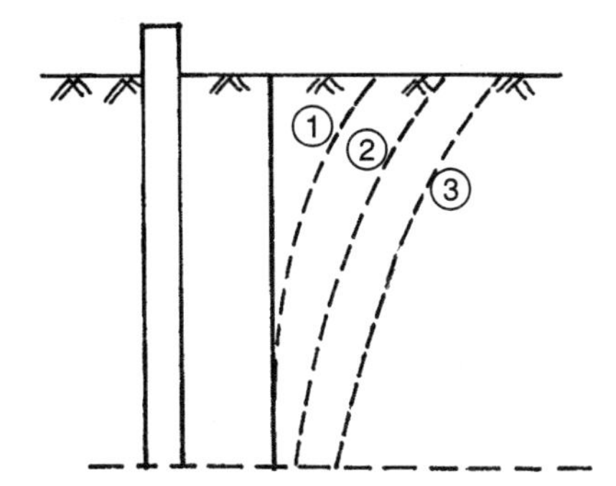

Figure 18.1 Progressive transfer of load from pile to soil by friction starting from transfer to top soil, (1) Loading at 10% frictional capacity, (2) Transfer at 30% frictional capacity, (3) Transfer at 80%.

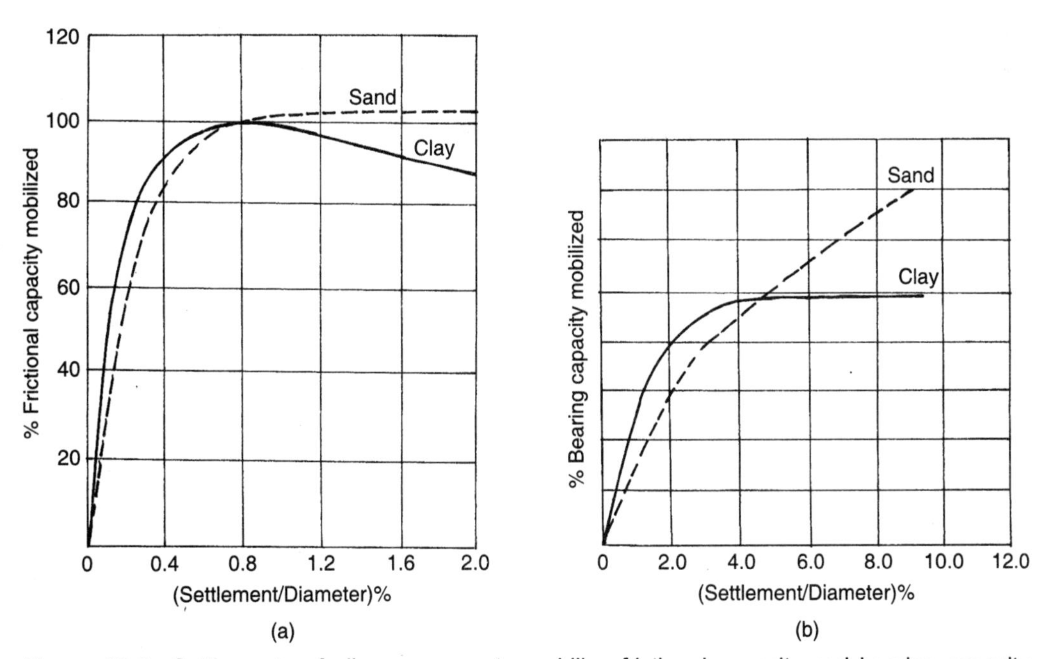

Figure 18.2 Settlements of pile necessary to mobilize frictional capacity and bearing capacity.

The difference between the magnitudes of the vertical displacement of the top of the pile for developing the full friction as against developing the full base resistance is not appreciated by many. Figure 18.2 shows the typical relation between frictional resistance and settlement for cohesionless and cohesive soils. Whereas frictional resistances is fully mobilized by as little as 0.4–0.6% of the diameter of the pile, it requires a movement of 4–10% of the diameter (as much as 10–16 times than that required for full mobilization of friction) to mobilize the full bearing capacity. This is evident from the basic relation we have studied in elastic settlement Δ of a plate in soil, which is

$$\Delta = \frac{p \times B(1-\mu^2) \times I_f}{E_s} \tag{18.1}$$

where p = Pressure
B = Breadth of plate
E_s = Modulus of elasticity of soil and μ its Poisson's ratio
I_f = Influence factor depending on the shape of the plate
Δ = A function of the diameter of the pile.

IS 2911, Part 4 specifies the safe capacity of piles as 2/3 the load that produces 12 mm settlement. Because of this restriction, the following discussions will show that to take full advantage of the large diameter, it is better to under-ream the pile or to end the piles in good rock if it is available or at least socket it in weathered rock, where the E value is large so that downward deformations due to loading will be small. [E is of the order of 10 N/mm^2–30 N/mm^2 in sands, whereas it is of the order of 25×10^3 N/mm^2 even in concrete. E of rock will be much larger. (see Table 14.2).]

18.3 ELASTIC SETTLEMENT OF PILES AND NEED TO SOCKET LARGE DIAMETER PILES IN ROCK

An approximate method for the calculation of settlement of piles has been described in Sec. 14.4.1 and, therefore, is not repeated here. We may, in addition, use Vesic's formula [2] for settlement.

Vesic proposed the following simple empirical expressions for S_b (settlement in bearing) and S_f (settlement in friction) in terms of the single value of *ultimate point resistance of the pile* q_{up}:

For point load $S_b = \dfrac{Q_p C_p}{D q_{\text{up}}}$ for base load Q_p (18.2)

For frictional load $S_f = \dfrac{Q_f C_s}{L q_{\text{sp}}}$ for side friction Q_f (18.3)

where $C_s = 0.93 + 0.16\sqrt{(L/D)C_p}$.

The recommended values of C_p are given in Table 18.1.

TABLE 18.1 Vesic's Value of C_p

Type of soil	*Bored piles*	*Driven piles*
Sand	0.09–0.18	0.02–0.04
Clay	0.03–0.06	0.02–0.03
Silt	0.09–0.12	0.03–0.05

[*Note:* Q_p is proportional to D^2 and Q_f is proportional to D.]

18.3.1 Example for Calculation of Deformations

Let us take an example to calculate the deformations of 300 mm and 1000 mm piles. Assume that *square piles* 300 mm in size are to be installed in a sand stratum (ϕ = 35) to a *depth of* 12 m. The allowable bearing capacity at that depth is estimated as 100 t/m^2 (1000 kN/m^2) and the allowable average friction as 16.5 kN/m^2. Let us compute the total settlements of 300 mm square and 1 m square piles installed in this stratum. Assume E of concrete = 20 × 10^6 kN/m^2 and E of soil = 30 × 10^3 kN/m^2 and μ = 0.3. [Refer Sec. 14.4.1.]

1. Settlement of 0.30 m pile (Let S denote settlement)

   Allowable frictional load (Q_f) = 0.3 × 4 × 12 × 16.5 = 238 kN (approx)

   Allowable bearing load (Q_p) = 0.3 × 0.3 × 1000 = 900 kN

   $S_t = S_p + S_b + S_f$. (Note concrete in pile also deforms = S_p)

   Assuming friction k = 0.6, we have

$$\text{Settlement } S_p \text{ of pile} = \frac{(Q_p + kQ_f)L}{AE_p} = \frac{900 + (0.6 \times 238) \times 12}{0.3 \times 0.3 \times 21 \times 10^6} = 1.48 \text{ mm}$$

   Taking I_{f1} = 0.85 and breadth of plate B = 0.3 m, we have for bearing

$$S_b \text{ of soil bearing } = \frac{q_p B}{E_s}(1-\mu^2)(I_{f1}) = \frac{1000 \times 0.3}{30000}(1-0.09) \times 0.85 = 7.73 \text{ mm}$$

   Again, taking $I_{f2} = 2 + 0.35\sqrt{12/0.3} = 4.2,$ we obtain

$$S_f = \frac{16.5 \times 0.3}{3000}(0.91)(4.2) = 0.63 \text{ mm}$$

   Total settlement = 1.48 + 7.73 + 0.63 = 9.84 mm

   This is less than 12 mm or 10% diameter of the pile as allowable settlement specified by the IS.

2. Now, let us calculate the settlement of 1.00 m square pile (larger pile): (B = 1 m).

$$Q_f = 1 \times 4 \times 12 \times 16.5 = 792 \text{ kN}$$

$$Q_p = 1 \times 1 \times 1000 = 1000 \text{ kN}$$

$$S_p = \frac{(1000 + 0.6 \times 792)2}{1 \times 21 \times 10^6} = 0.85 \text{ mm}$$

$$S_b = \text{Proportional to } B = \frac{7.73}{0.3} = 25.7 \text{ mm}$$

$$S_f = \text{Proportional to } B = \frac{0.63}{0.3} = 2.1 \text{ mm}$$

$$s_e = 0.85 + 25.7 + 2.1 = 28.65 \text{ mm}$$

According to IS 2911, Part 4, dealing with load test on piles, the safe bearing capacity of a pile is taken as only 2/3 load at a settlement of 12 mm, or 50% load at which the settlement is 10% the diameter, which is smaller. *It is clear that the full bearing capacity of large diameter piles cannot be mobilized in that limited movement allowed.*

[*Note:* This is the reason that for bridges, instead of large diameter piles resting on soils, we resort to large well foundations with large bearing area and low bearing pressure. But on land where there are space restrictions and we have to use large diameter piles, *for taking full advantage of large diameter piles*, we have to extend them to rocks or weathered rock. These facts are illustrated by Figure 18.2, which shows the types of load settlement curves we get in the testing of normal piles.]

18.4 SUBSURFACE INVESTIGATION OF WEATHERED ROCK AND ROCK

Conventional soil investigation can be used for the investigation of soils. The investigation of rocks and weathered rocks has to be different. The following different methods are commonly used for the investigation of the characteristics of rocks and weathered rocks when we are to end the piles in rocks.

[*Note:* It is most important that borelog should identify the type of rock as granite, basalt, sandstone, shale, etc. The common practice of many soil investigating firms of simply designating a hard situation as rock or weathered rock should not be allowed. We should also conduct a 48-hour water immersion test to test its stability under water.]

18.4.1 Method 1: Core Drilling

Investigation by core drilling can be used for all types of rocks and should be always carried out. It is done by core drilling machines capable of taking cores not less than 54 mm in diameter. The following properties of rock have to be determined:

1. **Rock recovery ratio.** This is the percentage of cores recovered in one drilling operation.
2. **Rock Quality Designation (RQD).** This is an important property that describes the quality of the rock deposit. It is defined as follows:

$$\text{RQD} = \frac{\text{Sum of length of cores recovered of lengths} \geq 100 \text{ mm}}{\text{Total length of drilling}}$$

   Breaks caused by drilling should be ignored.
3. **Laboratory tests for cylinder strength q_t and for modules of elasticity E_t.** In standard test, the height of the cylinder is twice the diameter (54 × 108 mm). For lesser height, diameter ratios of sample correction factors are to be used. With lesser heights, we get higher strengths. Thus, the cylinder strength of concrete is taken as 0.8 times the cube strength. Therefore, we may assume a linear reduction ratio of 0.8 for *H*/*D* ratio of one varying linearity to 1.0 for *H*/*D* = 2. [See also Sec. 18.10]

Tables 18.2 and 18.3 give the statistical values of the probable values of q_{mass} (unconfined compression of the rock mass) and E_{mass} (modulus of elasticity of the rock mass) to be derived from RQD and laboratory tests.

TABLE 18.2 Average Unconfined Compression Strength of Rocks (Compare these values with M20 concrete)

Type of rock	*Strength* (N/mm^2)
Good rock	> 80
Moderately weak	40
Weak	20
Very weak	8

TABLE 18.3 RQD and q_{mass}/q_{core} and E_{mass}/E_{core}
(This table gives the probable value of rock mass from test on core samples)

RQD	*Rock quality*	$\frac{q_{mass}}{q_{core}}$ *and* $\frac{E_{mass}}{E_{core}}$
< 25	Very poor	0.15
25–50	Poor	0.20
50–75	Fair	0.25
75–90	Good	0.3–0.7
90–100	Excellent	0.7–1.0

Estimation of shear resistance from tests on core drilling data and laboratory tests

From laboratory tests and Tables 18.2 and 18.3, we can estimate the properties of in-situ rock mass. As discussed in Sec. 18.6, the Reese and O'Neill's method of determination of the strength of drilled shafts use these q_{mass} values for calculating the frictional strength of sockets and also the bearing strength of piles. It also uses the value of E_{mass} to determine the deformation of the socket. In a socketed pile, the resistance in shear is between concrete and rock. The socket rock may get weakened by the boring. Consequently, a reduction factor α is used for the strength. The strength of the socket will depend on the lesser value of the punching shear strength of the rock mass and the concrete. The ultimate punching shear can be taken as a *function of cylinder strength* as follows:

$$\tau_p \text{ is lesser of } 0.28\sqrt{q_{mass}} \text{ or } 0.28\sqrt{f_{concrete}} \tag{18.1}$$

[*Note:* In the Cole and Stroud method, this ultimate punching shear strength is taken as 0.05 times concrete strength, i.e. $\tau_p = 0.05 f_{ck}$. This is similar to punching shear in concrete $= 0.25\sqrt{f_{ck}}$.]

18.4.2 Method 2: Cole and Stroud Method of Investigation of Weathered Rock

In the Cole and Stroud method, such rocks are classified on the basis of SPT values as shown in Table 18.4. This method is more suited for argillaceous weathered rock such as shale than granitic rocks. In this method, the strength of rock is to be extrapolated from N values for full penetration. Hence, they are useful for medium to hard rock (shale, siltstone, etc.) only. For hard rocks, such as granite and basalt which usually contain boulders, the method seems to give high values and consequently high pile capacity values. It has been suggested that for granular rocks, the number of blows for 10 cm be taken as the index rather than for full standard 30 cm penetration. Table 18.4 gives the estimated value of the shear strength based on N values.

TABLE 18.4 Scale of Strength and SPT *N* Values for Rocks (Cole and Stroud; IS 2911)

Value of N	*Shear strength* (kg/cm^3)	*Description of strength*	*Grade*	*Breakability*	*Penetration with knife*	*Scratch test*
	400	Strong	A	Difficult to break against solid object with hammer		Cannot be scratched with *knife*
	200					
600						
	100	Moderately strong	B	Broken against solid object by hammer		Can just be scratched with *knife*
	80					
	60					
400						
	40		C	Broken in hand by hitting with hammer		Can be just scratched by thumb *nail*
200	20	Moderately weak	D	Broken by leaning on sample with hammer		Can be scratched by thumb *nail*
	10	Weak			None	
	8		E	Broken by hand	2 mm	
100 80	6					
60	4	Very weak	F	Easily broken by hand	5 mm	

(N less than 60 are considered as soil)

[*Note:* In all Soil Reports, where we specify rock or weathered rock, we should give also its classification, e.g. granite, basalt, gneiss, and limestone etc.]

18.4.3 Method 3: Chisel Energy Method [3] for Classification of Rocks

The Chisel energy method was suggested by Datye and Karandikar [3] in 1988 for the Bombay region for granular weathered rocks such as weathered braccia (with unconfined compression strength of 1.2 to 4 N/mm^2) and weathered basalt (with UCS 9.4 to 54 N/mm^2). The core recovery is usually 20 to 70% and RQD Nil to 40%. The method uses a *one or two ton chisel to cut out the rock and is called the* **chisel energy method**.

The chisel energy method uses a drop chisel of weight W tons falling through H metres. The penetration d in centimetres achieved in half an hour of chiseling in the rock during which N chisel blows is measured and related to the area A of cross-section in square metres. The chisel energy is expressed as:

$$E = \frac{WH}{A(d/N)} = \frac{WHN}{Ad} \text{ in t m/m}^2\text{/cm} \tag{18.2}$$

where W = Drop weight in tons
A = Area in m^2
H = Drop in metres
d = Penetration in cm

This value of E is reduced by a factor of 0.8 when the ground water is present in the bore hole. Based on the above results, rocks are classified as follows: Rock suitable for socketing should have an energy level of 50 t/m^2/cm, and piles should be terminated at an energy of 10 t/m^2/cm only. Empirical values of 12% energy level are taken in t/m^2 as *safe capacity in socket frictions* and 3 times the energy level is taken as *safe capacity in bearing*. Thus, for energy level 50, the friction value is 50 × 12/100 = 6 t/m^2 and the bearing value is 3 × 50 = 150 t/m^2.

18.5 CALCULATION OF BEARING CAPACITY OF SOCKETED PILES

We have seen that the settlement necessary for developing the full capacity of piles (including both friction and bearing) up to 400 mm diameter in ordinary soils is small and will be within allowable limits. For large diameter piles, the amount of settlement necessary is much more than that allowed by the Code of Practices (IS 2911). In such cases, socketing in rock becomes necessary. Thus, allowable deformation, a major problem in socketed the full capacity large diameter piles in soils, is removed if they bear on rock or shocked in weathered rock.

In all such cases, the method of design by socketing in weathered rock described below should be used only after a full understanding of the load transfer mechanism, conditions at the site, and the quality of workmanship. Thus, for example, if sufficient care is not taken in cleaning the bottom of the hole, the base resistance will be practically nil, and if the socket strength is not enough, the pile will settle considerably.

18.5.1 Estimation of Total Pile Capacity of Large Diameter Piles

Broadly, the following assumptions are made for the estimation of the pile capacity. The socket capacity and the base resistance are usually calculated separately.

1. Design capacity based on shaft resistance only
2. Design capacity based on end bearing only
3. Design based on ultimate shaft resistance with a FS = 3 and the end bearing resistances with a FS = 4
4. Design based on full shaft resistance plus the end bearing resistance. (The end bearing is to be based on computed settlements obtained from E values.)

It is also common to divide weathered rock into two categories: argillaceous (clay-based such as shale) and decomposed granular rock such as granite. SPT values can be taken as a measure of the strength of the argillaceous type and unconfined strength as an indication of the strength of granular types of rocks. (In Bombay, with granitic rock, it is related to the chisel energy.)

We may assume that if the calculated settlement is more than 10 mm, the bond in the socketing will be broken and the entire load has to be taken by the bearing. Both the side friction and the bearing values are estimated in crystalline rock from its unconfined strength of the rock as shown in Table 18.1. The maximum friction in the socket will depend on the shear strength of the weaker of the two materials involved, namely, concrete and rock.

The allowable bearing pressures usually vary considerably, depending on the state of weathering of the rock. The bearing pressure is to be estimated from the unconfined compression strength and degree of weathering. The movement for full bearing action also depends on the nature of weathering. Though we may estimate these values by theory, the real values can be obtained only by field tests. A summary of the methods used to estimate the pile capacity is given in the following sections.

18.6 ESTIMATING CARRYING CAPACITY OF LARGE DIAMETER PILES

The commonly used methods of estimation of the capacity of socketed piles are:

1. Energy level test method
2. Cole and Stroud method
3. Reese and O'Neill method
4. IRC recommendations

18.7 ENERGY LEVEL TEST METHOD (By Datye and Karandikar)

The principles of energy level test method have been dealt with in Sec. 18.4.3 and the procedure is given in Example 18.1.

18.8 COLE AND STROUD METHOD [1], [4]

In the Cole and Stroud method, the values of the shear strength are derived from the penetration tests given in Table 18.4. A reduction of strength $\alpha = 0.3$ is usually made for shear resistance of the rock due to softening of the rock due to drilling (see Ref. [4], Sec. 1.12.2 for more details).

The socket bond strength will be lesser of the punching shear strength of the rock or concrete. We assume the ultimate shear to be $\alpha\tau_c$ as determined by the SPT test or $0.05c_u$ (where c_u is the cylinder strength of concrete.) Hence, the strength of the shaft of length L of diameter D is as follows:

$$\text{Ultimate shaft strength in friction} = (\alpha\tau)\,\pi DL$$

As regards the full bearing strength, we use only the bearing capacity in bearing will be as follows:

$$\text{Ultimate bearing capacity} = c_u N_c \left(\frac{\pi D^2}{4}\right)$$

$$\text{Total safe capacity} = \left(c_u N_c \frac{\pi D^2}{4}\right)\frac{1}{\text{FS}} + \frac{\alpha\tau\pi \text{DL}}{\text{FS}} \qquad (18.3)$$

Usually, an overall factor of safety of 3 or separate factors of 1.5 in friction and 2.5 in bearing are used in design. This method is illustrated by Example 18.2.

18.9 REESE AND O'NEILL METHOD[5]

The Reese and O'Neill method is used in USA for the design of drilled shafts. It recommends that the pile capacity should be based on either the side resistance between the shaft and the rock or the base resistance at the bottom only *but not on both of them added together*. The following steps can be used to estimate the pile capacities in friction and point bearing:

Step 1: From RQD and laboratory core test data, estimate the unconfined cylinder compression strength q_u of rock mass. Let the cylinder strength of concrete be f_c.

Step 2: Estimate the controlling shear strength which will be lesser of the values of the concrete and rock. We assume friction $= f =$ punching shear $= 0.28\sqrt{\text{cylinder strength}}$ or $0.28\sqrt{f_{ay}}$ but less than $0.15f_c$.

Step 3: Calculate the ultimate frictional capacity of socket strength as

$$Q_f = f \times (\pi DL)$$

(Check that the lesser of the above two values is f.)

where

D = Diameter of the pile

L = Length of the socket

Safe value = $Q_f/3$ with FS = 3.

Next check deformation.

Step 4: To calculate the deformation of the socket, find I_f, the factor affecting deformation of the socket. This is given by the "Reese and O'Neill curve" shown in Figure 18.3.

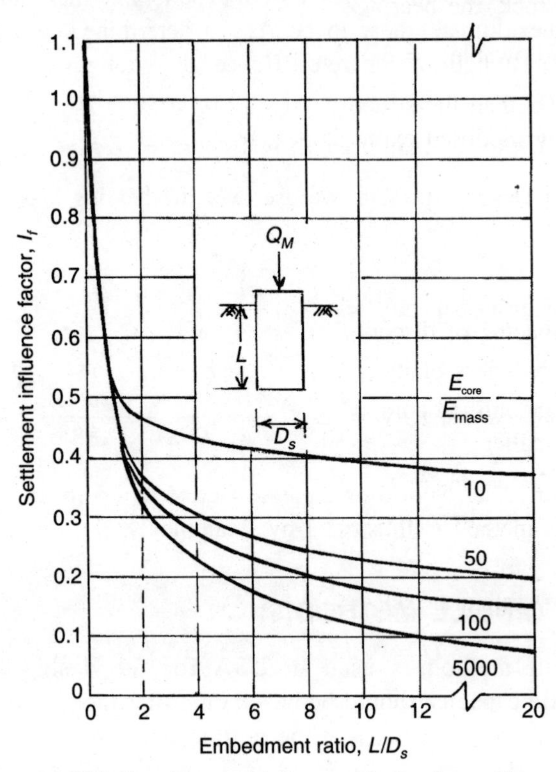

Figure 18.3 Reese and O'Neill settlement influence factor for settlement in socketed piles.

Estimate E_{mass} of rock from Table 18.3 and calculate E_{core}/E_{mass}.
Read-off settlement influence factor I_f with L/D value shown in Figure 18.3.

$$\text{Settlement of socket} = \frac{Q_f}{(D/I_f)E_{mass}} = \frac{QI_f}{DE_{mass}}$$

Step 5: Estimate the total movement in concrete and rock socket. If it is more than 10 mm, the friction is broken.

$$\Delta = \frac{Q_f L}{AE_{concrete}} + \frac{Q_f I_f}{DE_{mass}} \not> 10 \text{ mm}$$

If settlement is less than 10 mm, take the capacity of the pile as that due to socket friction only.

Step 6: If settlement is greater than 10 mm, we assume that the socket resistance is broken. Calculate the point resistance as follows.

Step 7: Calculate the bearing capacity. (If deformation of the socket is more than 10 mm, the bearing capacity controls the design as the socket should be assumed to have lost its resistance.)

For uniform rock, the bearing capacity is as follows:

$$Q_p \text{ (Allowable)} = 3 \times \text{Unconfined strength of rock} \times \text{Area}$$

but not greater than the structural capacity of the pile.

The following modified expression is given for Q_p by Reese for stratified strata:

$$Q_p = 3\left[\frac{3 + e/D}{10(1 + 300\,\delta/e)^{0.5}}\right]q_u \tag{18.4}$$

where

e = Spacing of discontinuities in same units as D

δ = Thickness of individual discontinuity

q_u = Unconfined compression strength or rock or concrete at tip, whichever is smaller

(This is applicable where $e < 300$ mm and $\delta < 5$ mm.)

[*Note:* We may alternatively use the simple expression, Ultimate point resistance = Area × q_v.]

Step 8: Calculate the structural capacity

$$Q = 0.25 f_{ck} \text{ (Area of the pile)}$$

Safe load on the pile capacity should not be more than the structural capacity of the pile.

18.10 IRC RECOMMENDATIONS

IRC–78–2000 (Rev., Aug 2005) gives data for the socket length to diameter ratio (*L*/*D*) and unconfined strength of the rock to calculate the socket strength.

18.11 SUMMARY

Sound engineering judgement must be used when deciding whether the *capacity of large diameter piles* should be based on combination or individual values of side resistance and end bearing. We should understand that the action of socketed piles in weathered rock is very much different from the action of piles in soils. Also, we must be aware that the settlement at which full base resistance will be developed should not be in excess of the allowable settlements. The investigations to be made in argillaceous rocks such as shale are usually different from the granular-based rocks. In both cases, cores of samples and their unconfined strengths can give us a good idea of their shear resistance and ultimate bearing capacity. In all cases, in our soil investigation reports, we should give the identification of the rock on which the pile rests.

EXAMPLE 18.1 (Determination of socketed pile capacity from energy level test)

A 600 mm pile is founded in weathered granite rock with 5D socketing, of which 3D socketing is with energy level 50 t/m^2/cm and 2D is with 100 t/m^2/cm. It is stopped at a layer with 100 t/m^2/cm energy level. Estimate its safe bearing capacity.

Reference	Step	Calculation
	1	*Estimate frictions and bearing capacities at given energy level*
		Friction at 50 t/m^2/cm = $\frac{12}{100} \times 50 = 6$ t/m^2
		Friction at 100 t/m^2/cm = $\frac{12}{100} \times 100 = 12$ t/m^2
		Bearing at 100 t/m^2/cm = 3 × 100 = 300 t/m^2
	2	*Calculate the capacity in friction*
		For 3D socketing = (π × 0.6) × (3 × 0.6) (6) = 20 t
		For 2D socket length = (π × 0.6) × (2 × 0.6) (12) = 27 t
		Total friction = 47 t
	3	*Calculate capacity in bearing*
		$Q_b = \frac{\pi \times (0.6)^2}{4} \times 300 = 85$ t
	4	*Total capacity*
		Friction + Bearing = 132 t

EXAMPLE 18.2 (Determine the socketed pile capacity by Cole and Stroud method)

Estimate the allowable load of the pile of 900 mm in diameter socked to 3D in an argillaceous shale deposit. Assume f_c of concrete 20 N/mm^2 and the SPT value of the weathered rock is 150 for 30 cm penetration.

Reference	Step	Calculation
	1	*Estimate shear strength from SPT value*
Table 18.1		N = 150, shear strength = 10 kg/cm^2 = 1 N/mm^2
		Ultimate shear with reduction factor due to weakening by drilling = 0.3 × 1 = 0.3 N/mm^2
		Ultimate punching shear in concrete = 0.05 × 20 = 1 N/mm^2
		Lesser of the two = 0.3 N/mm^2 = 300 kN/m^2
	2	*Calculate side friction for 3D socket length*
		Ultimate Q_f = ($\alpha\tau$) (πDL) = 300(π × 0.9)(3 × 0.9) = 2289 kN
		Allowable (FS = 3) = 2289/3 = 763 kN = 76 t

Reference	Step	Calculation
	3	*Calculate point bearing (shear strength = 1 N/mm²) (no reduction)*
Step 1		Allowable $q_b = \dfrac{9 \times c}{FS} = 3c = 3 \times 1 = 3\ \text{N/mm}^2$
		$= 3 \times 10^3\ \text{kN/m}^2$
		Allowable $Q_b = q_b\ (pD^2/4)$ (Using the above value)
		$= 3 \times 10^3 \times (\pi \times 0.9 \times 0.9)/4 = 1907$ kN
		= 190 t
	4	*Total capacity assuming friction + bearing acts*
		$\Sigma Q_f + Q_b = 76 + 190 = 266$ t
	5	*Structural capacity (allowable) of pile with factor of safety*
		$Q_c = 0.25 f_{ck} \times \text{area} = 0.25 \times 25 \times 10^3\ \dfrac{(\pi \times 0.9 \times 0.9)}{4}$
		= 3974 kN = 397 t
		Hence allowable structural capacity = 266 t
		[*Note:* It is reported that the results of this method are usually on the higher side.]

EXAMPLE 18.3 (Socketed pile capacity by Reese and O'Neill method)

Estimate the allowable load of a socked pile of diameter 900 mm socketed 3D *into soft granite rock*. RQD of rock is 80%. UCS of rock is 70 N/mm², concrete of the pile is M25. The laboratory value of E of rock is 2×10^3 N/mm² and E of concrete is 20×10^3 N/mm². Allowable settlement is 3.8 mm.

Reference	Step	Calculation
	1	*Calculate ultimate side resistance of socket*
		q_u of rock = 70×10^3 kN/m² > q_u of concrete
		Hence, friction in concrete controls the strength of the socket.
	2	*Estimate controlling shear strength*
		In this case, concrete controls as cylinder strength of concrete is less.
Sec. 18.4.1		Cylinder strength = $0.8 f_{ck} = 0.8 \times 25 = 20\ \text{N/mm}^2 = f_c$
		Friction = $0.28\sqrt{20} = 1.25\ \text{N/mm}^2$
		However, friction < $0.15 f_{ck} = 0.15 \times 20 = 3\ \text{N/mm}^2$
		Adopt 1.25 N/mm² = 1.25×10^3 kN/m²
		[*Note:* In the Cole and Shroud method, adhesion is taken as 0.05 × cylinder strength]
	3	*Calculate frictional capacity of pile*
		$Q_f = \pi DLf = (3.14 \times 0.9)(3 \times 0.9)(1.25 \times 10^3)$ kN = 9540 kN

Reference	Step	Calculation
	4	*Estimate E values*
		$E_{concrete}$ = 20 × 10⁶ kN/m²
Table 18.3		RQD = 80%; E_{mass}/E_{core} = 0.55
		E_{mass} = 0.55 × 2 × 10³ = 1.1 × 10³ kN/mm² = 1.1 × 10⁶ N/mm²
		$\frac{E_{core}}{E_{mass}} = \frac{20}{1.1} = 18;\ L/D = 5;\ I_f = 0.35$
	5	*Calculate movement in socket and check* < 10 mm
		Total movement allowed $\not> 10$ mm. Let $s_f = \Delta$.
		$s_f = \frac{Q_f L}{AE_c} + \frac{Q_f I_f}{DE_{mass}}$
		where I_f = (Influence factor) for L/D = 3
Step 3		$s_f = \frac{9540 \times 2.7}{(\pi/4)(0.9 \times 0.9) \times 20 \times 10^6} + \frac{9540 \times 0.35}{0.9 \times 1 \times 10^6}$ m
		$= 2025 \times 10^{-6} + 3710 \times 10^{-6} = 5735 \times 10^{-6}$ m
		= 5.7 mm < 10 mm.
		Therefore, Q_u = 9540 kN
		Thus, socket capacity controls.
	6	*Find allowable load with FS*
		$Q_{allowable} = Q/3$ = 9540/3 = 3180 kN = 318 t
	7	*Find safe structural capacity of the pile on concrete strength*
		$Q_s = 0.25 \times f_{ck} \times$ (area) = 5 (area) N
		= 0.25 × 25 × (3.14) × (900)²/4 = 3974 kN
		= 397 t
	8	*Find end bearing capacity on rock strength*
Table 18.1		For RQD = 80%; q_u field = $0.55 q_{tert}$
		q_u = 0.55 × 70 = 38.5 N/mm² = 38.5 kN/m²
		$c = \frac{q_u}{2} = \frac{38.5}{2} = 19.25\ \text{N/mm}^2$
		Allowable stress in bearing = 3c = 3 × 19.25 = 57.75 N/mm²
		This is more than that allowed in concrete of 6.25 N/mm².
		Hence, the structural capacity on concrete strength controls.
	9	*Find safe capacity of pile*
		Structural capacity = 397 t
		Socket capacity = 318 t

REFERENCES

[1] Varghese, P.C., *Foundation Engineering*, Prentice-Hall of India, New Delhi.

[2] Raja, B. and M. Das, *Principles of Foundation Engineering*, 5th Ed., Thomson Asia, Singapore.

[3] Karandikar, D.V., *Failure of Undersocketed Bored Cast in-Situ Piles in Weathered Rocks of Mumbai*, Proceedings, Vol. I, Indian Geotechnical Conference ICG, 2006, Chennai.

[4] Thomlinsion, M.A., *Foundation Design and Construction*, ELBS with Longman, Singapore.

[5] Reese, L.C., F.T. Touma and M.W. O'Neill, 'Behaviour of Drilled Piers under Axial Loading', *Journal of Geotechnical Engineering Division*, ASCE 102, 1976.

CHAPTER 19

Design of Cantilever and Basement Retaining Walls

19.1 INTRODUCTION

Earth slopes are stable when their slope angles satisfy their stability requirements. In many places, due to lack of available space, the slopes will have to be made steeper. To some extent, this can be accomplished by revetments or by inclined walls, etc. as shown in Figure 19.1. The factors of safety of revetments or inclined walls can be determined by the slip circle method, and as the forces involved are positive gravity forces, a factor of safety of 1.2–1.5 is considered satisfactory for such slopes. Another modern method to build steeper slopes is to use reinforced earth, using geo-fabrics. Such steep slopes and even vertical walls are commonly used to conserve space in approach roads for flyovers in cities. These are described in books on Soil Mechanics and Foundation Engineering [1].

However, when we have vertical earth face, traditionally we use rigid or flexible walls as shown in Figure 19.2. In places where proper foundations can be built, we use rigid walls. In places where we cannot have a proper foundation, for example, by the side of a waterway such as the sea, we may have to adopt *sheet piles* or *sheet piles with tieback arrangements* (called *bulk heads*). We may also use *diaphragm walls*. In mountainous regions where stones are in plenty and the ground is stable, we may adopt crib walls, also called *Gibon walls*. Vertical faced *reinforced earth* walls are also nowadays commonly used in such places. The theory and design of these walls are described in books on Soil Mechanics [1].

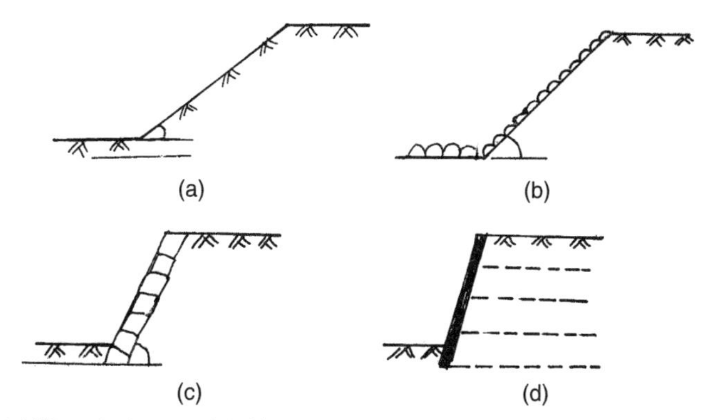

Figure 19.1 Stability of slopes. (a) Naturally stable slope, (b) Slopes with stone revetment, (c) Inclined walls with rubble masonary, (d) Reinforced earth slopes.

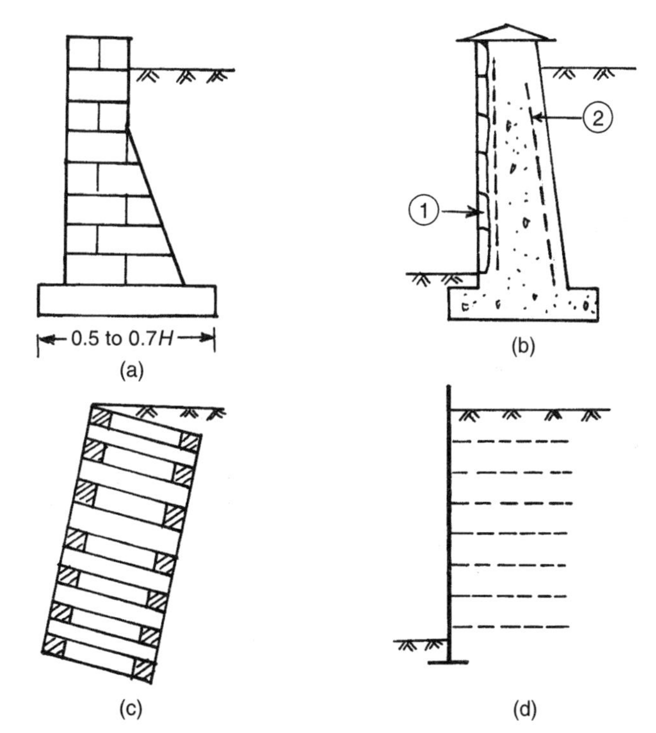

Figure 19.2 Vertical non R.C.C. walls: (a) Masonary gravity walls, (b) Semigravity mass concrete walls which allow some tension in concrete: (1) Usually provided with masonary or brick facing, (2) Nominal temperature steel in the front and back, (c) Crib walls filled with stones, (d) Reinforced earth walls.

Rigid walls can be of the gravity type or semigravity (partly reinforced) type or of the fully reinforced concrete cantilever or R.C. counterfort walls as shown in Figures 19.3 and 19.4. Very high retaining walls are built with relieving platforms, where the horizontal platforms take the vertical load of earth, thus considerably reducing the lateral earth pressure effects in very high walls as shown in Figure 19.4. The following is a list of the commonly used earth retaining structures [2].

Type I: **Gravity type walls**

1. Gravity retaining walls made of masonry (brick, stone, block, etc.)
2. Semigravity concrete retaining walls with temperature steel
3. Abutment walls of bridges
4. Crib walls and Gibon walls made of stones in a concrete crib or weld mesh cage

Type II: **Reinforced earth walls**

5. Reinforced earth (reinforced with geo-fabrics) walls.

Type III: **Reinforced concrete low walls**

6. Basement and cantilever walls
7. Counterfort walls
8. Buttress walls

Type IV: **Reinforced concrete very high walls**

9. Walls with relieving platforms (For very high walls or for very heavy traffic on top of the earth as in quuy walls, we build relieving platforms which reduce the earth of pressure considerably (Figure 19.4)).

Type V: **Walls without foundations (Figure 19.5)**

10. Cantilever sheet pile wall
11. Anchored bulk heads (These are cantilever walls with tie buck arrangements)
12. R.C. concrete diaphragm walls (which can be constructed by slurry method and if necessary, tied back).

A detailed treatment of these various types of walls is given in books on retaining structures [2]. In this chapter, we briefly deal with the general conditions to be checked for the overall safety of retaining walls as well as with the design of R.C. cantilever walls and basement walls that we came across in the construction of buildings.

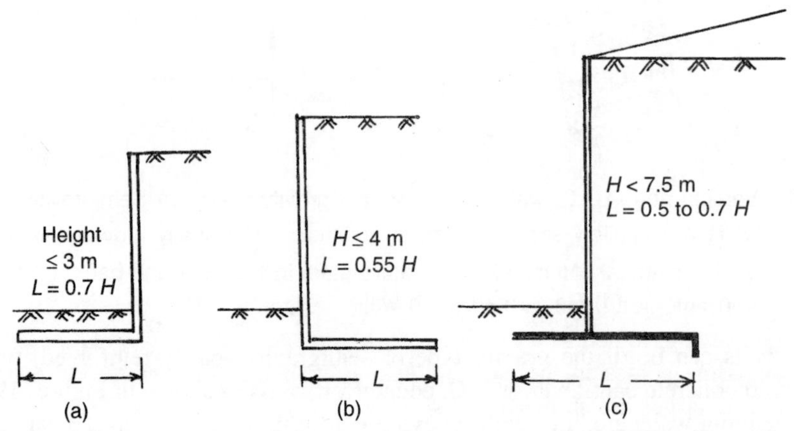

Figure 19.3 Reinforced concrete cantilever vertical earth retaining walls, (a) and (b) L shaped walls, (c) T shaped walls.

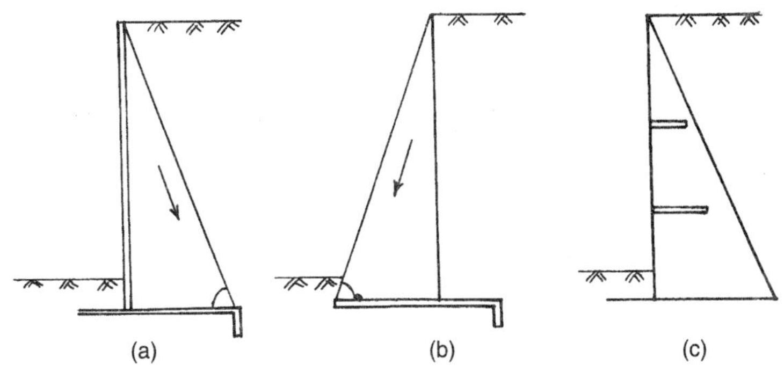

Figure 19.4 Reinforced concrete counterfort vertical earth retaining walls: (a) Counterfort at the back, (b) Counterfort in the front (Buttress walls), (c) Counterfort walls with relieving platforms for very high walls.

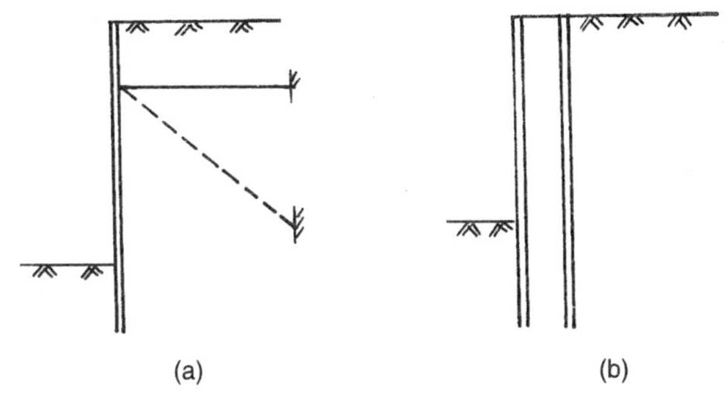

Figure 19.5 Retaining walls in weak soils: (a) Sheetpile walls without or with tie backs (bulkheads), (b) Diaphragm walls.

19.2 EARTH PRESSURE ON RIGID WALLS

As we have seen, rigid walls have foundations whereas flexible walls such as sheet pile or bulk heads are structures penetrating into the soil for stability. The way we design them is different and can be studied from books dealing with Foundation Engineering [1]. In this section, we deal with earth pressures on rigid walls with foundations. The front portion of the foundation of a retaining wall facing the air is called the *toe* and the back facing the soil is called the *heel.*

Terzaghi has shown that the *earth pressure at rest* acts where the wall cannot rotate and it is reduced to active earth pressure with rotation at the base [3]. The passive pressure comes into action when the earth is pushed in. The difference in failure surfaces is shown in Figure 19.6. There are many formulae and constructions, such as Rebhan's construction to find the resultant earth pressures on retaining walls. However, in the design of walls with sloping backs, cantilever walls, etc., we take the pressures on a vertical line at the end of the heel, and the earth pressure is assumed to act in the same direction as the top slope. Some of the special cases that we come across in the calculation of pressures on retaining wall, such as sloping backlifts and surcharges are shown in Figures 19.7 and 19.8. These are dealt in detail in books on Retaining Walls and Soil Mechanics [2][4].

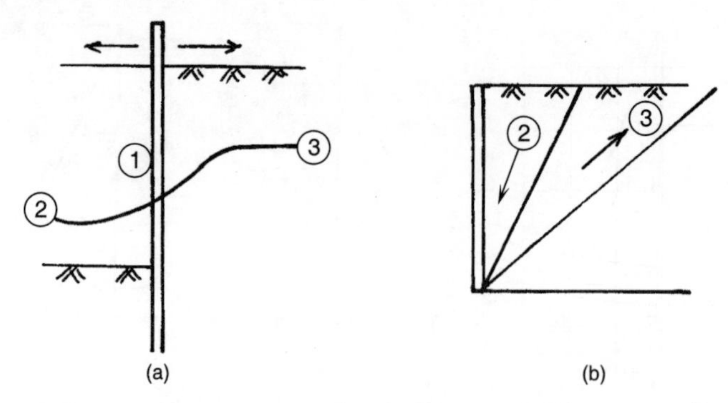

Figure 19.6 Earth pressure theory: ① Earth pressure at rest, ② Active earth pressure, ③ Passive earth pressure.

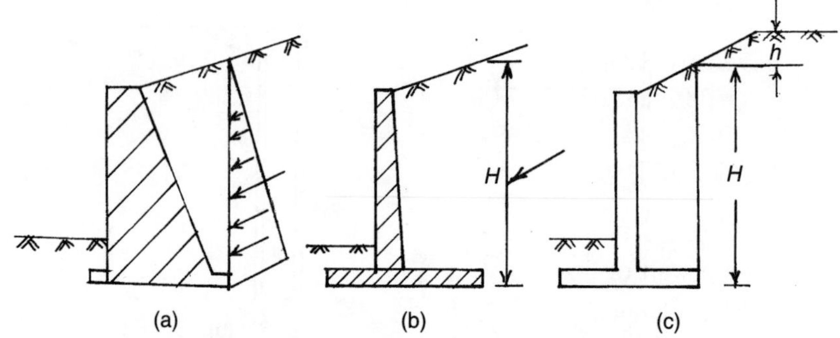

Figure 19.7 Active earth pressure against retaining walls: (a) Gravity wall with sloping back, (b) Cantilever wall with sloping back, (c) Cantilever well with slope for a limited distance.

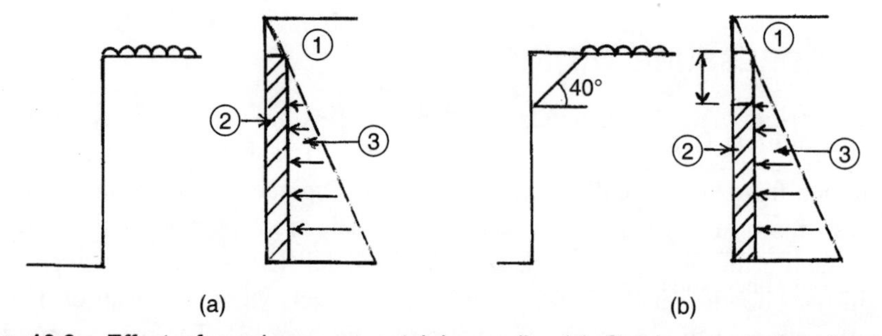

Figure 19.8 Effect of surcharge on retaining walls: (a) Surcharge starting from wall, ① Effect of surcharge, ② Pressure due to surcharge, ③ Earth pressure.

19.2.1 Calculation of Earth Pressure on Retaining Walls

1. Empirical procedure

The most practical method to calculate the earth pressures on rigid retaining walls is the use of the values recommended by Terghazi and Peck for different types of soils [4], as shown in

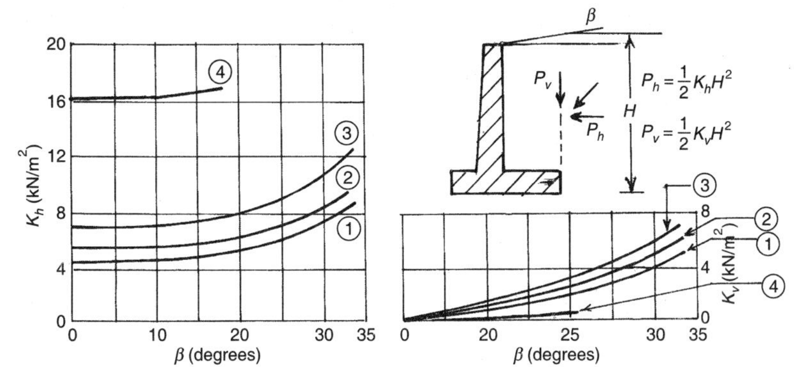

Figure 19.9 Terzaghi's design charts for active pressures on cantilever retaining walls for typical backfills. (Coefficients for calculation of pressures per metre length of wall K_h and K_v include weight of backfill also. Description of soils are given in Table 19.1.)

Figure 19.9. They give horizontal and vertical earth pressures on a retaining wall with a level or inclined backfill for different types of soils. The symbols used are the following:

$$\text{Coefficient } k_h = \frac{1-\sin\phi}{1+\sin\phi}; \quad K_h = k_h\gamma$$

Pressure at depth $h = p_h = k_h\gamma h = K_h h$

Total pressure on a wall of height H, $P_h = 1/2\ K_h H^2$

We calculate the horizontal and vertical pressures and find the inclination of earth pressures on a cantilever retaining wall, as shown in Figure 19.8.

$$P_h = 1/2\ K_h H^2 \tag{19.1a}$$

$$P_v = 1/2\ K_v H^2 \tag{19.1b}$$

The values of K_h for a lever backfill are given in Table 19.1 and those for various values of β are given in Figure 19.9.

TABLE 19.1 Empirical Design of Retaining Walls with Horizontal Backfills [Figure 19.9]

Numbers in Figure 19.9	*Description of soil backfill*	*Value K_h** kN/m²/m
①	Coarse-grained soils without fines very permeable like clean sand or gravel	4.8
②	Coarse-grained soil of low permeability	5.6
③	Residual soils with stones and soils such as fine silty sands, and granular materials with clay content	7.6
④	Very soft clays, organic silts, silty clays	16.0
5	Medium or stiff clay deposited in chunks in such a way that during floods and heavy rains negligible water enters through the spaces contained in them	20.0

*These are values of the chart for $\beta = 0$

(*Note:* For pressures on walls with sloping backfills of soil types 1 to 4, refer Figure 19.9.)

2. Earth pressure for clay soils with friction and cohesion

For clay soils with friction and cohesion, we may use Bell's theory and write the pressure coefficient as

$$p_h \text{ (Active)} = k_a \gamma h - 2c\sqrt{k_a} \times \sqrt{\frac{c+c'}{c}} \tag{19.2a}$$

where c is the apparent cohesion and c' is the wall adhesion.

$$p_h \text{ (Passive)} = k_p \gamma h + 2c\sqrt{k_p} \times \sqrt{\frac{c+c'}{c}} \tag{19.2b}$$

(We note that as cohesion *decreases*, the active pressure *increases* and hence omitting c gives us conservative values of earth pressures.) This aspect is further explained in Sec. 19.6.1.

19.3 DESIGN OF RIGID WALLS

There are two aspects in the design of rigid walls. The first is the layout for stability. The following are the six conditions regarding the overall stability of the retaining wall to be satisfied before we take up the second aspect of the structural design of the various components of the retaining walls.

Condition 1: The resultant of the horizontal and vertical forces acting on the wall should fall within the middle third of the foundation. This will ensure that there will be no tension at the base and the base will be always in contact with the ground.

Condition 2: The maximum base pressure should not exceed the safe bearing capacity of the soil.

Condition 3: As there is a horizontal force on the structure, it should be resisted by passive pressure and friction at the base. The friction developed at the base of the wall should be at least twice the horizontal force on the wall. If passive pressure is also to be mobilized, it should be given ample factor of safety. Passive pressure at toe cannot be depended on if there is any likelihood of it being excavated.

Condition 4: The wall and the earth fill should not fail by base failure (foundation failure), especially if there are soft layers within 1.5 times the height of the wall. The soil condition should be specially looked into and analysed for base failure.

Condition 5: There should not be undue settlement or differential settlement between the toe with respect to the heel. This can lead to forward tilting of the wall and final failure.

Condition 6: It is very necessary that full arrangements for drainage of retaining walls should be made, as described in books on Foundation Engineering [1], so that these walls do not have to act as dams instead of acting as simple earth retaining walls. Otherwise, the effect of water pressure also has to be taken into account in calculations.

19.4 DESIGN OF ORDINARY R.C. CANTILEVER WALLS (Figure 19.10)

In this section, we deal with cantilever walls and basement walls only. The terms toe and heel of retaining walls are distinguished by assuming that the wall carries the earth on its back.

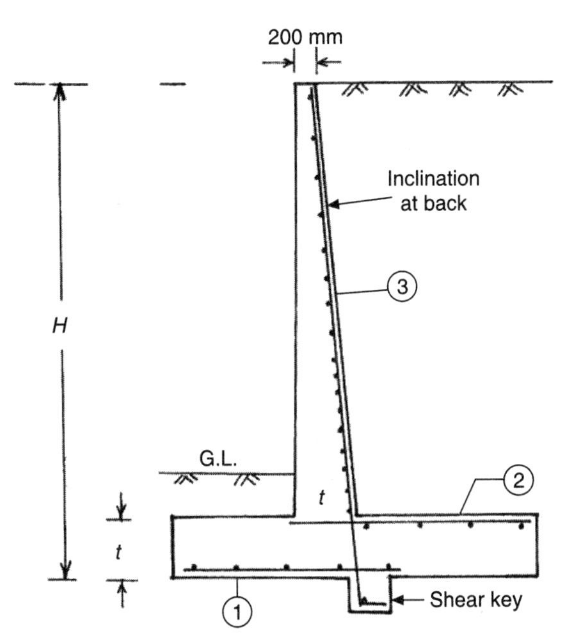

Figure 19.10 Detailing of main types of steel in cantilever retaining wall: (1) Bottom toe steel, (2) Top heel steel, (3) Main wall for cantilever action. In addition provide nominal steel also at top of toe, bottom of heel and in front of the wall.

As in the design of other civil engineering structures, we first arrive at workable dimensions of the structure by thumb rules and then check its stability. Then, we design various components of the wall. Accordingly, the following steps can be used for design of cantilever earth retaining walls:

Step 1: Decide the depth of the foundation from practical consideration. Usually, a depth not less than 1 m *above the top* of base slab is adopted. (The use of Rankine's formula has no relevance.)

Step 2: Taking H as the height from the top of the backfill to the bottom of the base slab as in Figure 19.9, the thickness of base slab is taken as 8–12% of H but not less than 300 mm. Assume a thickness of stem not less than the base thickness. The top width should not be less than 200 mm and the latter is usually given at the back to bring the front in line with the property line.

Step 3: Next we assume the length of the base L as 50–70% of H. Place $L/3$ in front of the wall as the toe. (Figure 19.3 also gives some guidance on preliminary dimensioning of cantilever walls.)

Step 4: Estimate the earth pressure coefficients either from the theory or from the curves given by Terzaghi and Peck [4] (Figure 19.9).

Step 5: Check stability condition. Find the eccentricity of resultant on the base of the wall by taking moments about the heel. Calculate the maximum and minimum base pressures. The maximum should not exceed the safe bearing capacity and the minimum should be positive. In clays, the pressure at the toe should not produce excessive settlement and tilting of the wall.

Step 6: Check for sliding. The friction at the base should be at least two times the horizontal force acting on the wall. If it is deficient, provide a key (or nib) as given in the base (see Example 19.1). It is better to give a key below the heel or under the stem. If the key is provided at the toe, the soil there can get disturbed.

Step 7: Proceed with the structural design of the toe slab.

Step 8: Take up the structural design of the heel slab.

Step 9: Take up the structural design of the stem. If necessary, provide a splay at base junction to increase the thickness of the stem and reduce the effect of sharp corners.

Step 10: Detail the steel. The principal steel *in the toe is at the bottom of the toe, in the heel, at the top of the heel* and in the stem, in the *backside of the stem*. Provide the other areas with nominal steel as shown in Figure 19.11. We should remember that retaining walls are exposed to all weather conditions of the site or exposed to soil. Clear cover of 50–70 mm is normally used in practice.

These aspects and the structural design of cantilever walls are illustrated by Example 19.1.

19.5 DESIGN OF CANTILEVER WALLS WITHOUT TOE

As shown in Figure 19.3 for walls less than 3 to 4 m we need to provide only a heel or a toe in the base length. In such cases, the base length will have to be made nearly equal to the height of the wall. The design principles are the same as in the conventional type of cantilever wall.

19.6 DESIGN OF BASEMENT WALLS

When constructing basements of buildings, the walls around the building have to retain the outside earth in addition to carrying the vertical loads acting on them. Depending on the site and the project, there can be two types of constructions as shown in Figure 19.11.

1. Free standing basement retaining walls
2. Embedded basement retaining walls

The free standing type is more common for ordinary, not too deep basements. Embedded walls can be diaphragm walls first built to facilitate excavation and then incorporated into the building. In the free standing system, we have to determine the base length required for stability. In the embedded type, we have to find the length of embedment for stability. Many

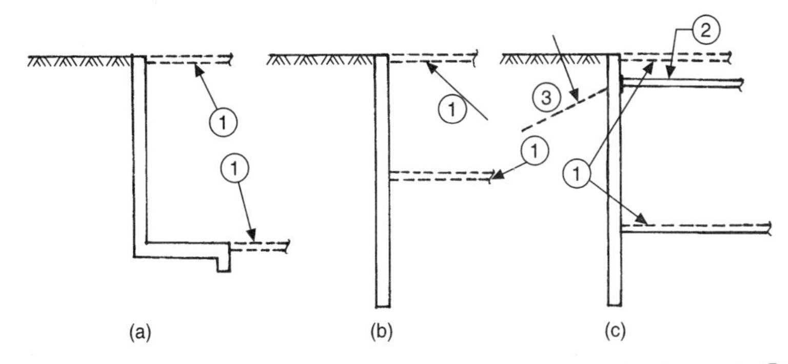

Figure 19.11 Type of basement wall: (a) Free standing, (b) and (c) Embedded, ① Final support by permanent flows, ② Temporary support, ③ Ground anchor.

methods of calculation can be used to find the required length of embedment. The theory is similar to finding the depth of penetrating in sheet pile walls. In this section, we briefly deal with the design of free standing basement walls. References [5] and [6] can be used for a detailed study of both types of walls.

19.6.1 Calculation of Earth Pressures in Clays (see also Sec. 19.1)

The calculation of earth pressures due to clay deposits can be done in three ways:

1. Use c' and ϕ' values obtained from drained laboratory tests.
2. Using $c = 0$ and assuming a ϕ value to give satisfactory results, found to be good from experiences. Values given in Table 19.2 can be used for this purpose.
3. Using empirical values given in Table 19.1 (as recommended by Terzaghi and Peck [4]) or values given in Table 19.3.

TABLE 19.2 Recommended Equivalent ϕ Values for Earth Pressures Due to Clay Deposits

P.I. value (%)	ϕ (degrees)
15	30
20	28
25	27
30	25
40	22
50	20

19.7 DESIGN OF FREE STANDING BASEMENT WALLS

Free standing basement walls can have their legs inside or partly outside the building line, depending on the sequence of construction. The floor slab props the cantilever wall and prevents it from tilting forward to produce active earth pressure. It acts as a propped cantilever. If earth is packed after construction of the wall, there will be no rotation of the wall. Thus, earth pressure at rest will be acting on the wall. The usually recommended values of coefficient of earth pressure at rest k_{rest} are given in Table 19.3.

TABLE 19.3 Values of "Coefficient of Earth Pressure at Rest" for Soils

S. No.	*Type of soil*	k_{rest}
1.	Dry sand loose	0.64
2.	Dry sand dense	0.49
3.	Saturated sand loose	0.46
4.	Saturated sand dense	0.36
5.	Clay compacted (P.I.9)	0.42
6.	Clay compacted (P.I.30)	0.60

The reaction of a fully propped cantilever at the cantilever end is $F/5$, where F is the total triangular load. The maximum bending moment is less than that in a free cantilever and it reverses in sign. In terms of the total load F. The bending moment at x, will be as follows:

$$M_x \text{ (at } x \text{ from the fixed end below)} = \frac{F}{15}(L - x)\left(\frac{5x}{L} - 2\right)$$

At $x = 0$, $\quad M_0 = -2FL/15$

Also BM at $x = 2/5L = 0$ and for lesser and more values, it reverses in sign. Hence, it is important to reinforce both faces of the wall as shown in Figure 19.12.

It is also important that we should provide some nominal face reinforcement continued from the floor slab to the wall. Generally, U bars are provided from the wall into the floor slab onto which the floor slab bottom steel is lapped. The vertical steel in the wall is the main steel and the horizontal distribution steel is placed inside the main steel. Detailing of steel is made as shown in Figure 19.12.

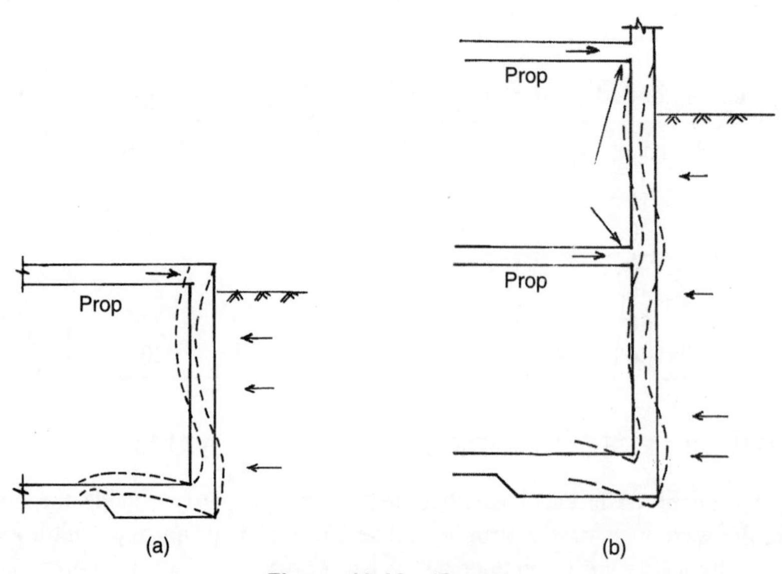

Figure 19.12 (Contd.)

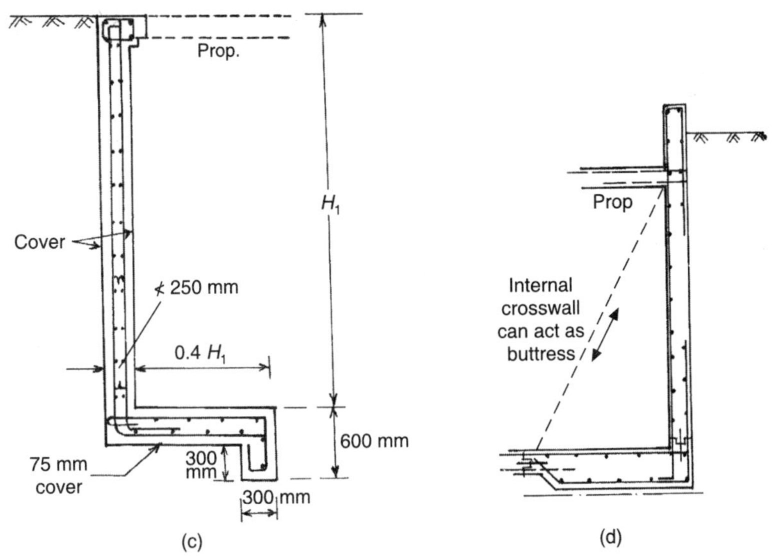

Figure 19.12 Propped cantilever retaining walls as basement walls. (a) and (b) Deformations of walls—Reverse bending requires steel on both sides of the wall, (c) and (d) Detailing of steel in basement walls.

19.8 SUMMARY

This chapter gives a brief account of the different types of retaining walls and their features. It also discusses in detail the procedure for the design of medium sized reinforced cantilever walls and basement walls.

EXAMPLE 19.1 (Preliminary dimensioning and design of an ordinary cantilever retaining wall)

Give the layout and design of a conventional cantilever retaining wall with level backfill to retain 4 m of earth ($\phi = 30°$) of unit weight 19 kN/m^2.

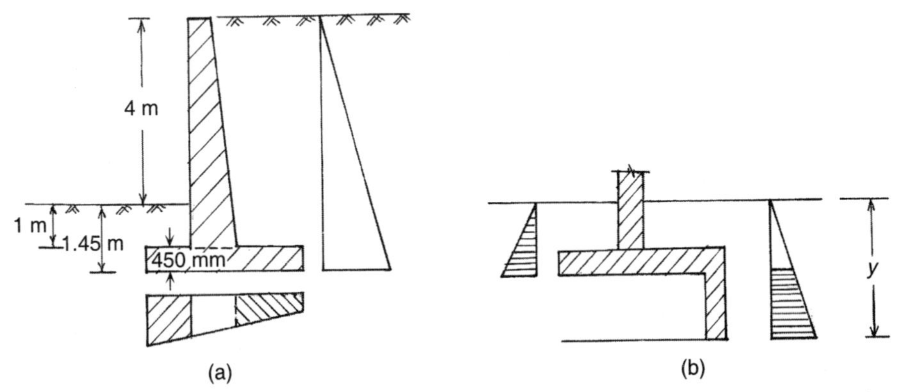

Figure E19.1 (a) Design of toe and heel slabs and stem, (b) Design of nib below the wall.

Reference	Step	Calculation
	1	*Estimate total height*
		Assume a depth of foundation = 1.45 m (bottom of base slab)
Sec. 19.4		Total height = 4 + 1.45 = 5.45 m
	2	*Estimate base thickness* (8–12% H)
		Try 8% H = 0.08 × 5.45 = 436 mm
		Adopt thickness of base slab 450 mm.
	3	*Estimate thickness of stem*
		Adopt 450 mm as the thickness of the stem at base (as above)
		Let the top width be = 200 mm
		Adopt a wall sloping at back from 450 to 200 mm at the top on the earth side.
	4	*Estimate length of base* (L = 0.5 to 0.7 H)
		Assume $0.5H = 5.45/2 = 2.73$ mm
		Adopt 3 m as base length
Sec. 19.4		Put $L/3$ = 1 m as toe in front of the wall
		(These steps give us the dimensions of a wall to design.)
	5	*Estimate earth pressure coefficients for active pressure*
		Assume $\phi = 30°$, $w = 19$ kN/m²
		$k_h = \dfrac{1-\sin\theta}{1+\sin\theta} = \dfrac{0.5}{1.5} = \dfrac{1}{3}$, and $k_p = \dfrac{1+\sin\theta}{1-\sin\theta} = 3$
Figure 19.9		$K_h = \gamma k_h = 19 \times \left(\dfrac{1}{3}\right) = 6.33\ (\text{kN/m}^3)$
		(Taking value of K from Terzaghi chart = 4.6)
		Adopt K = 6.33
	6	*Check stability conditions with characteristic loads and find base pressures*
		Find the point of intersection of the resultant pressure at the base to check
		(a) Maximum and minimum base pressures, and
		(b) No tension at base and also moment of forces of the bottom of the heel

Item	*Weight* (kN)	*Length above base*	*Moment*
Stem	0.20 × 5 × 25 = 25.0	1.65 m	41.25 kNm
Stem (slope)	(0.25/2) × 5 × 25 = 15.6	1.83	28.60
Base	3 × 0.45 × 25 = 33.8	1.50	50.70
Earth	1.55 × 5 × 19 = 147.3	0.775	114.16
Total	= 221.7 kN		= 234.71 kNm

Reference	Step	Calculation
		Lateral pressure = 1/2 KH^2 = 1/2 × 6.33 × $(5.45)^2$ = 94 kN
		Moment = 94(H/3) = 94.0(5.45/3) = 170.77 kNm
		Total moment = 234.71 + 170.77 = 405.48 kNm
		Resultant from O = M/W = $\dfrac{405.48}{221.70} = 1.83$ m from O
		Total length of the base = 3 m
		e = Eccentricity = 1.83 – 1.50 = 0.33 m
		Eccentricity towards toe, where pressure is maximum is
		$L/6 = 3000/6 = 500$ mm > 330 mm
		Hence, no tension in the base.
		Max and min pressure $= P/A \pm \dfrac{Pe}{Z}$ and $Z = L^2/6$
		$= \dfrac{221.7}{3} \pm \dfrac{(221.7 \times 0.33) \times 6}{9} = 122.7$ and 25.1 kN/m^2
		p max = 122.7 kN/m^2 < SBC 175 kN/m^2
		p min = 25.1 kN/m^2 (No tension at base)
	7	*Check for sliding (with characteristic load and FS)*
		P = Lateral pressure = 94 kN
		Total weight = 221.7 kN
		Let friction coefficient be 0.55.
		Frictional resistance = 0.55 × 221.7 = 121.94 kN
		Factor of safety $= \dfrac{121.94}{94} = 1.29 < 2.0$ (prefered)
Figure E19.1		We will provide a nib at the end of the heel as shown. Calculate the depth of the nib required.
		Lateral resistance required = 2P = 2 × 94 = 188 kN
		Resistance required from nib = 188 – 122 = 66 kN
		1. *Design of nib (projection below slab)*
		Nibs are not generally provided at toe as that region may be excavated after construction. Nib design is made by different designers using different assumptions as indicated below.
		(a) The nib only increases the soil riction by converting friction as soil to soil instead of concrete to soil friction, which can be taken as 0.35 for silt to 0.60 for rough rock. A factor of safety of 1.4 is used. Tan 30 for soil is taken as 0.58.)
		(b) Passive resistance of the soil below the rib only contributes to the resistance.

Reference	Step	Calculation
		(c) The nib resistance is the difference between the passive resistance at the front of the nib and the active resistance at its back.
		(d) The failure surface is a plane and circle starting at ϕ degrees at the foot of the nib as shown in Figure E19.1.
		2. *Let us assume (b) and find depth of nib required*
		(Assume $k_p = 3$ and $\gamma = 19$). First consider front.
Step 4		Let us consider the depth up to the bottom of the slab 1.45 m depth. Passive pressure at 1.45 m = $k_p \gamma h$
		$P_1 = 3 \times 19 \times 1.45 = 82.7$ kN/m
		Total pressure $= \frac{1}{2} \times 82.7 \times 1.45 = 60$ kN
		Consider the back. Let the nib project be y metres from the top of earth.
		Passive pressure at y depth = $3 \times 19 \times y = 57y$
		Total pressure $P_2 = 1/2 \times 57y \times y = 28.5y^2$
		The resistance by nib only = $P_2 - P_1 = 66$ kN
		Hence, $28.5y^2 - 60 = 66$ gives $y = 2.12$ m
		Depth of nib below slab = 2.12 – (1.45 + 0.45) = 0.22 m = 220 mm
		Make the nib at least equal to the slab thickness = 450 mm
		[For the nib to be rigid, the depth should not be more than four times its thickness. If such a condition arises, redesign the base length L to give more friction.]
	8	*Design of toe part of the slab (steel to be provided at bottom) as base pressure is the main force*
		Both SF and BM have to be taken at junction of the stem, pressure varies from 122.7 to 25.1 kN/m^2
		Ground pressure at stem junction
		$= 25.1 + \frac{(122.7 - 25.1)}{3} \times 2 = 90.2 \ (\text{kN/m}^2)$
		(a) *Design for moment*
		Moment of base reaction $= \left(\frac{122.7 + 90.2}{2}\right)(1) \times \frac{1}{3}(90.2)$
Sec. 19.4		$= \left(\frac{122.7 + 90.2}{2}\right)(1) \times \frac{1}{3} \frac{[90.2 + (2 \times 122.7)]}{[122.7 + 90.2]} = 55.93$ kNm
		We may deduct the moment due to self-weight of the slab and the earth above. But for a safe design, we omit these.
		$d = 450 - 50 - 10 = 390$ mm
		M_U = Factored design moment = 1.5 × 55.93 = 84 kNm

Reference	Step	Calculation
		$\dfrac{M_U}{bd^2} = \dfrac{84 \times 10^6}{1000 \times 390 \times 390} = 0.55$ gives $p_t = 0.158\%$
		Provide at least 0.2% steel for foundation.
		A_s = (0.2/100) × 1000 × 390 = 780 mm^2/m
		Provide 16 mm @ 250 mm spacing = 804 mm^2/m
		(b) *Check for shear (Neglect wt. of earth above)*
		Upward reaction – wt. of the slab
		Shear = $\left(\dfrac{122.7 + 90.2}{2}\right)(1) - (1 \times 0.45 \times 25) = 95.2$ kN
		$\tau = \dfrac{V}{bd} = \dfrac{95.2 \times 10^3}{1000 \times 390} = 0.244 \text{ N/mm}^2$
		This is less than 0.33 allowed for 0.2% steel.
		Provide nominal temperature steel at top (see Step 9).
	9	*Design of heel (Steel to be provided at top of the slab) (The weight from earth is the major force)*
		Forces acting are: (a) weight of earth, (b) weight of the slab and (c) ground reaction at base
		Ground pressure at junction with stem
		$= 25.1 + \dfrac{(122.7 - 25.1) \times 1.55}{3} = 75.5$ kN
		Find moment of forces (tension at top)
		(1) Earth = 1.55 × 5 × 19 × 0.775 = 114.12 kN
		(2) Slab = 1.55 × 0.45 × 25 × 0.775 = 13.52
		(3) Reaction
		$= \left(\dfrac{75.5 + 25.1}{2}\right)(1.55) \times \dfrac{[75.5 + (2 \times 25.1)]}{75.5 + 25.1} \times \left(\dfrac{1.55}{3}\right) = 50.33$ kNm
		Net M = 114.14 + 13.52 – 50.33 = 77.31 kNm
		M_U = 1.5 × 77.31 = 116 kNm
Table B.1		$\dfrac{M}{bd^2} = \dfrac{116 \times 10^6}{1000 \times 390 \times 390} = 0.76;\ p = 0.221\%$
Table C.2		A_s = (0.221/100) × (1000 × 390) = 862 mm^2/m
		Provide 16 mm @ 230 gives 873 mm^2/m. (as seen from Table C2)
		Minimum steel is provided at the bottom (0.12%) and along the wall (@ 0.12%). See Step 9.
	10	*Design of stem (height* 5 m)
		Depth $\nless$ that of base; adopt D = 450 mm, d = 390 mm
		$K_h = k_h \gamma$ = 0.33 × 19

Reference	Step	Calculation
		$$M = \frac{K_h H^2}{2} \times \left(\frac{H}{3}\right) = \frac{6.33 \times (5)^3}{6} = 131.87 \text{ kNm}$$
		M_U^n = Factored moment = 1.5M = 197.8 kNm
		$$\frac{M_U}{bd^2} = \frac{197.8 \times 10^6}{1000 \times 390 \times 390} = 1.30; \ p = 0.392\%$$
		$$A_s = \frac{0.392}{100} \times 1000 \times 390 = 1529 \text{ mm}^2$$
		Provide 16 mm bars at 130 gives 1547 mm^2.
		Check stem also for shear. Depth should be enough without shear reinforcement.
		Distribution steel
		$$\frac{0.12 \times 1000 \times 450}{100} = 540 \text{ mm}^2$$
Table C.2		Provide 10 mm bars @ 140 mm gives 560 mm^2/m.
		Provide distribution steel on both faces
		We have to provide 10 mm @ 280 mm only as steel is provided to both faces.
		Curtailment of steel
		Check depth of section and requirement of steel at mid-height for medium walls and 1/3 and 2/3 heights for high walls.
	11	*Detail the steel*
		Detail steel as shown in Figure 19.10.

REFERENCES

[1] Varghese, P.C., *Foundation Engineering*, Prentice-Hall of India, New Delhi, 2005.

[2] Huntington, *Earth Pressures and Retaining Walls*, John Wiley, New York, 1968.

[3] Terzaghi, K., Large Scale Retaining Wall Test, *Engineering News Record*, Vol. 112, 1934, pp. 136–140.

[4] Terzaghi, K. and R.B. Peck, *Soil Mechanics in Engineering Practice*, 2nd Ed., Wiley, New York, 1967.

[5] Peck, R.B., W.E., Hanson and T.H. Thornburn, *Foundation Engineering*, 2nd Ed., Wiley Eastern Ltd., New Delhi, 1980.

[6] Thomlinsion, M.J., *Foundation Design and Construction*, ELBS, Longman, Singapore.

Infilled Virendeel Frame Foundations

20.1 INTRODUCTION

Ordinary reinforced concrete strip wall footings (which are used for residential buildings), when built on soils of low bearing capacity, are subjected to large differential settlement. Therefore, short piles or under reamed piles with grade beams are commonly used in places where the foundation condition can enable sufficient load bearing *capacity of the pile to be mobilized at less than* 4 m *depth. However, in deep deposits of low bearing capacity (where short piles will not work) for ordinary buildings*, an economical choice will be the use of a rigid ground beams as foundations. The use of reinforced T or U beam for these foundations in fairly good soil is described in Chapter 9. Such a ground beam can be made more rigid by tying the reinforced concrete footing with the plinth beam by *stub columns* between them. This converts the foundation to a Virendeel frame as shown in Figure 20.1. The concrete footing and the top R.C. plinth beams form the bottom and top chords, respectively, of the frame. They are connected at intervals with *verticals or stub columns*. These form panels and these panels are infilled with bricks. Such Virendeel frame can be used in two ways. First, it can be used as *a frame* resting on the ground all along its length in the same way as a rigid beam (beam on elastic foundation). Secondly, it can be used as a *Virendeel truss girder* supported at the two ends on rigid supports like piers on long piles and loaded from top by the building and supported by very weak soil below, which may give only little resistance. The brickwork can be built in such a way that they are

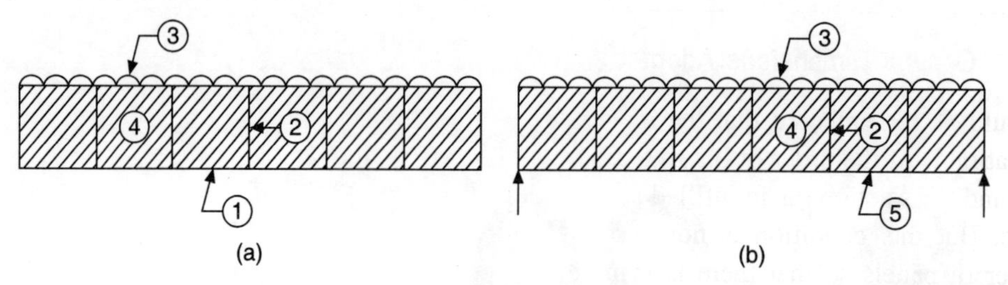

Figure 20.1 Virendeel frame: (a) As infilled beam under walls, ① Strip foundation, ② Stub columns, ③ Plinth beam, ④ Brick infill, (b) Virendeel girder, ⑤ Bottom beam.

all well bonded together. In deep deposits of weak soils, these will serve as a useful foundation for ordinary buildings. These frames are also useful in places of high water table where the foundation brickwork can get submerged under water and get weakened.

In this chapter, we briefly examine the approximate methods of design of the Virendeel frame as girders and beams. Their use as beams will be similar to the use of the less rigid T beams described in Chapter 9.

20.2 BEHAVIOUR OF VIRENDEEL GIRDERS WITHOUT INFILLS

Virendeel girders are highly indeterminate structures. Their elastic or plastic analysis is highly complex. Results of their approximate analysis should always be interpreted with a full appreciation of the limitations of the approximations made.

One of the most important factors that affect the behaviour of these frames is the depth/span ratio of the truss and the number of panels of the girder. It can be proved that if the number of panels is more than 4–6, the *depth/span ratio* is the most important factor that will affect the performance of the Virendeel girder.

The important stress resultants to be considered in the analysis of the Virendeel girder (see Figure 20.2) are:

- *Tension* and compression in the *bottom* and top chords, respectively
- Moments in the top and bottom chords
- Moments in the verticals
- Deflection of the truss

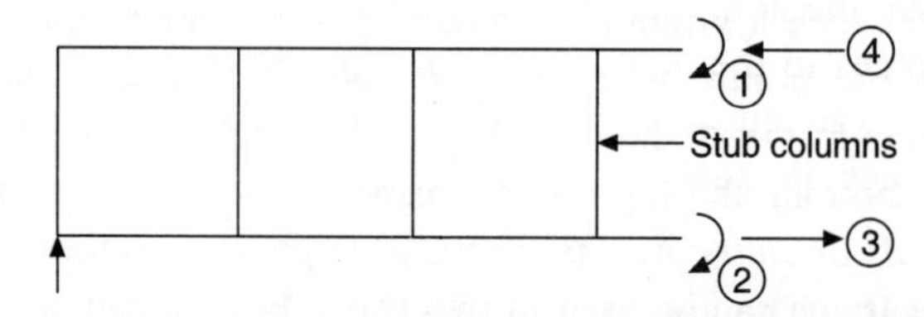

Figure 20.2 Analysis of Virendeel girder-stress resultants: ① Moment in top chord, ② Moment in bottom chord, ③ Tension in bottom chord, ④ Compression in top chord.

20.2.1 General Dimensions Adopted

It should be remembered that the most common depth/span ratio that we usually meet within foundation is of the order of 0.2 to 0.3 with 4–6 square panels. For better action of infill against shear and truss action (as in infilled shear wall for wind loads), it will be advisable to use square panels. But this condition is not absolutely obligatory. It is also desirable to have an even number of panels so that there is symmetry in action.

20.3 APPROXIMATE ANALYSIS OF VIRENDEEL GIRDERS

Symmetrically loaded Virendeel girders can be easily analysed for nodal loadings by using Naylor's method of moment distribution, where we allow joint and sway distribution, simultaneously but require modified set of distribution factors. (Stiffness of vertical members is multiplied by 6.) This method is described in books on the theory of structures.

A very approximate method of analysis of Virendeel girders can be carried out by converting them into a determinate structure by assuming that the points of contraflexure are at mid-points of horizontal chord members and also at the mid-points of vertical members as shown in Figure 20.3. The loads are carried to the nodes by beam action between the nodes. The method is illustrated by Example 20.1.

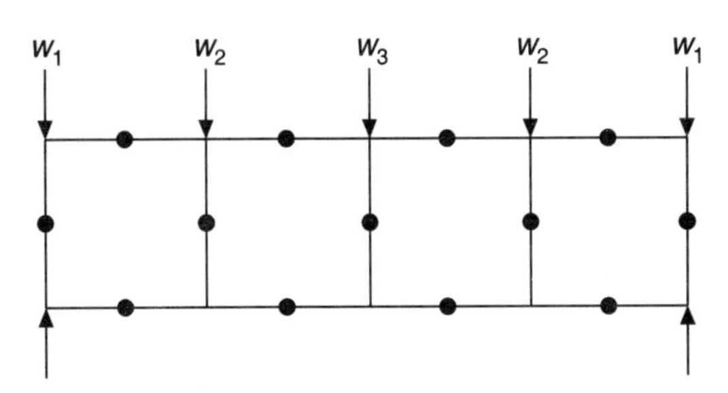

Figure 20.3 Approximate analysis of Virendeel girder assuming points of inflections.

20.4 RESULTS OF REFINED ANALYSIS

Other methods such as Kloucek's knotted cantilever method can also be used for more refined analysis. In approximate method, the chord moments will not vary with the depth/span ratio. However, the direct forces in the chord members are very high with small depth/span ratios, and the effect of these large tensions will be taken into account in the exact analysis. We can summarize the difference in analysis between the approximate and the exact method as follows:

1. **Chord moment.** The approximate method gives no variation in the value of chord moment with depth/span ratios. The exact analysis shows that the value of chord moment increases considerably with increase in the depth/span ratio as shown in Figure 20.4 [1].

2. **Direct force in chords.** The approximate method gives much higher values than those obtained by more refined analysis for direct forces in the top and bottom chords.

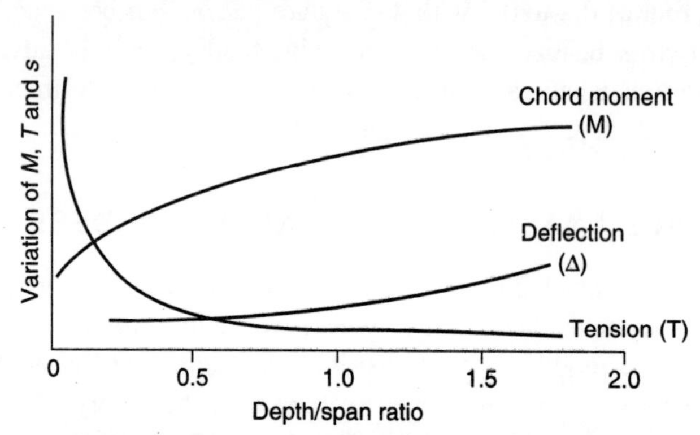

Figure 20.4 Results of theoretical analysis of a four bay Virendeel girder without infill showing variation of chord moment, deflection and chord tension (and compression) [1]. [*Note:* Laboratory tests on these girders with well bonded infilled brickwork showed that they were 8 to 10 times stiffer than infilled girders.]

3. **Maximum deflection.** Though for a given Virendeel girder *without infill* the deflection is reported to slightly increase with depth/span, this will not be true for infilled frames (Figure 20.4).

20.5 DESIGN OF VIRENDEEL FRAME AS A BEAM OR A GIRDER BASED ON SOIL CONDITION

The procedure for the design of Virendeel foundations can be described as follows:

Case 1: **Empirical design as a beam to reduce settlement** In cases where the soil under a load bearing wall can take the load as a continuous footing (but we need a rigid foundation to take care of the *differential settlement*), we proceed as follows. We suitably design an R.C. continuous strip footing to fully take the gravity loads. To take care of the differential settlement, we convert it into a Virendeel frame by adding R.C. stub columns connecting the foundation slab and the top plinth beam so that the foundation is made into a Virendeel frame. It is desirable to have the verticals as the same size of the plinth beam. The plinth beam is the compression chord and reinforced with about 1.5% steel, with equal steel on all corners and nominal ties.

Case 2: **Theoretical design as a rigid beam** Alternatively, where the foundation soil is fairly good but not as good as in case (1), we may consider the wall *resting on a Virendeel frame as a rigid beam* on elastic foundation and design it by more involved theoretical methods.

Case 3: **Empirical or approximate design as a girder between rigid supports** If the foundation soil is very poor to support the load of the building, it will be necessary to plan the *Virendeel frame as a girder* between strong pile or pier foundations at its ends spaced at fairly large intervals. The girder spans between the supports with very little support from below.

20.6 APPROXIMATE ANALYSIS OF VIRENDEEL GIRDER

When the frame acts as a girder between two rigid supports, we assume that the total moment at any section of a girder M_0 is carried as

$$M_0 = M_d + M_i + M_m$$

where

M_d = Moment due to direct force (i.e. due to compression and tension) in the chords
M_i = Moment taken by infill brickwork
M_m = Sum of moments in the chords

As the value of M_i depends on many factors, we neglect it and use the analysis of girders without infill for the design of the concrete members. This gives us a conservative design.

20.7 PROCEDURE FOR DESIGN OF VIRENDEEL FRAME FOUNDATION

As already stated, we can use Virendeel frames in two ways—as a beam on elastic foundation and as a girder between pillars.

Case 1: **(Infilled Virendeel frame used as a rigid beam and subjected to vertical loads)** We have two options as follows:

(a) First, as already indicated, we may design the bottom chord as a conventional continuous R.C. foundation, satisfying the bearing capacity consideration at a suitable depth. For providing rigidity and reducing differential settlement, we connect these foundations to the plinth beam by stub columns spaced at suitable intervals (so that the number of panels will be not less than four to six). These verticals and plinth beams are to be of the same dimension with not less than 1.5% steel and nominal shear reinforcement in the term of ties. The frame is to be infilled with brickwork.

(b) Secondly, we can consider the whole frame with infill as a beam on elastic foundation.

Case 2: **(Infilled Virendeel frame as girder spacing between rigid supports)** In this case, the load is first assumed to be transmitted *along the top chord as concentrated loads at the nodes*. The structural action of such an arrangement is very complex and exact and the design will be very lengthy. We may simplify the design by making the following assumptions:

(a) The loading on the top chord can be assumed to be taken to the nodes of the girder by the bending action of the two chords, assuming brickwork is only a filler.

(b) The Virendeel frame is then analysed and designed for shear bending moment and direct forces as shown in Example 20.1.

For the design as a girder, we have to check the following

(a) Assuming there is no infill, the chord top members should be strong enough to transfer loads on them to the adjacent nodes. (This is the first step to transfer uniformly distributed loads as nodal loads.)

(b) The Virendeel frame, when *used as a girder*, should be able to bear the foundation load without the help of the brick infilling as a frame. As the soil below is weak, we neglect the upward reaction from the soil. This assumption gives a safe design.

(c) Shear may be assumed to be taken as in infilled shear walls. Symmetrical even number of panels will be more convenient and infill can be assumed as struts between corners.

(d) The end supports must be piles or piers, which can safely take the reactions.

20.8 DETAILING OF STEEL

The detailing of steel in Virendeel frames is as shown in Figure 20.5.

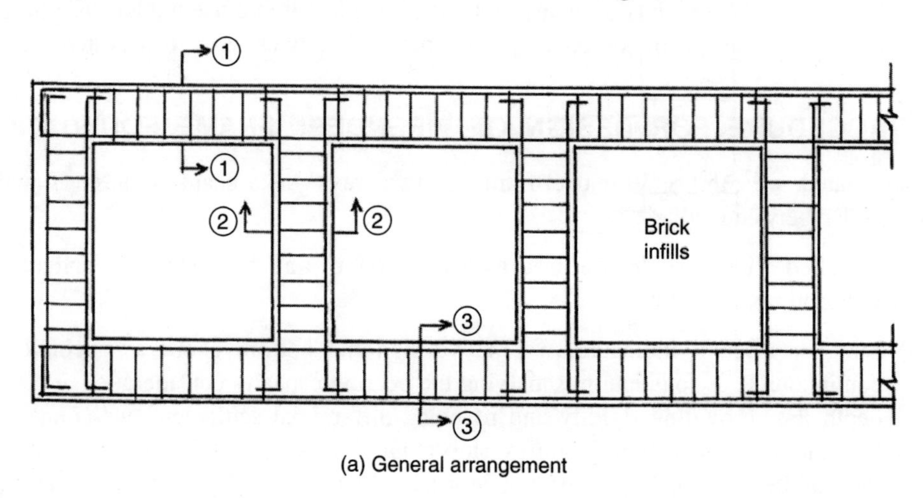

(a) General arrangement

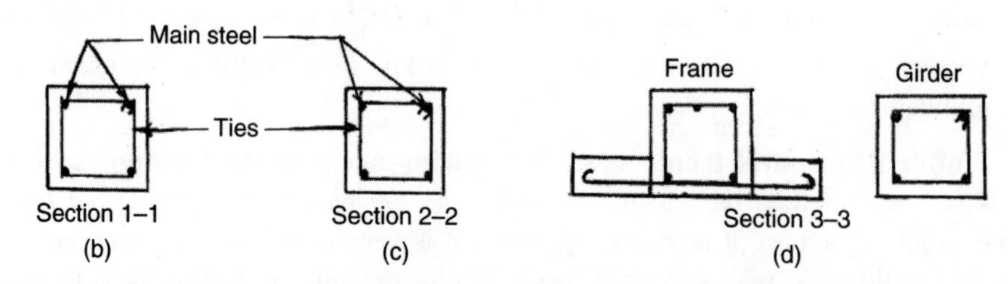

Figure 20.5 Detailing of Virendeel frame and girder. [*Note:* (d) In a frame we use strip foundation of the wall as bottom member. In a girder we provide a bottom beam section as bottom chord. To act as a girder it has to be supported at the ends on rigid supports like piers or piles.]

20.9 SUMMARY

Brick-infilled Virendeel frames have been successfully used in many places as *rigid beams and girders between supports* in foundations of ordinary buildings built on soils of low bearing capacity. They are also useful in regions where foundations are subject to high water table or flooding which can lead to weakness of foundations brickwork. This chapter deals with the empirical design of Virendeel frames as beams supported all along their length and also as girders supported on rigid supports only at their ends.

EXAMPLE 20.1 (Approximate analysis of Virendeel girder)

A Virendeel girder consists of six panels. Each panel is of 2.5 m in chord length and 2.5 m in height. A load of 15 kN acts on each interior nodal point on the top. Estimate, using the approximate method, the maximum design values of axial forces bending moment and shear in (a) the top chord, (b) the bottom chord and (c) the verticals.

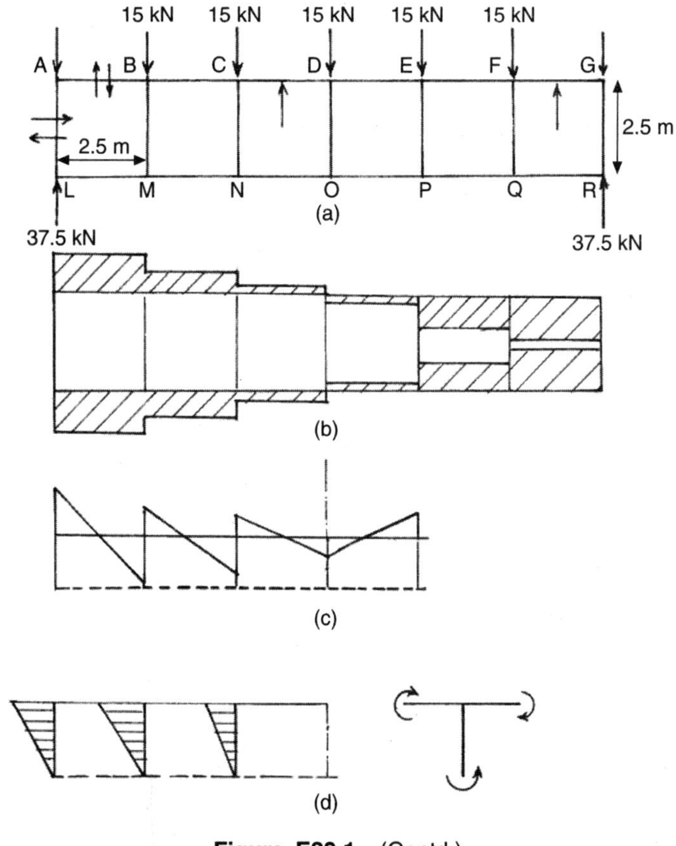

Figure E20.1 (Contd.)

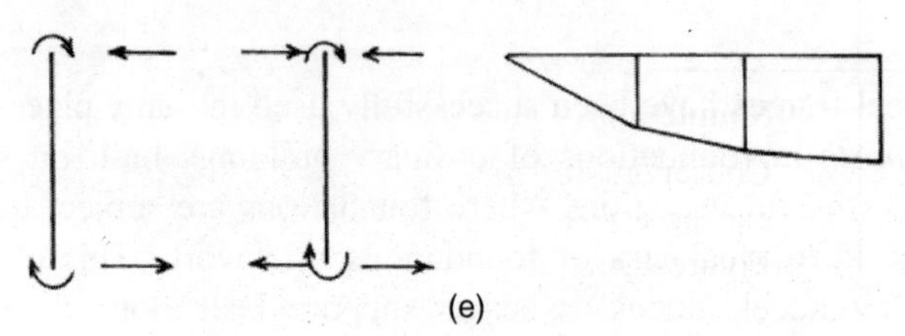

Figure E20.1 Details of approximate analysis of Virendeel girder.

Reference	*Step*	*Calculation*
	1	*Draw the shear forces in the panel [Figure E20.1(b)]*
Figure 20.2		(Assume the top and bottom chords share SF equally)
	2	*Find bending moments in chords [assuming point of inflection at mid-point of chords Figure E20.1(c)]*
		$M_{AB} = M_{BA} = 18.75 \times 1.25 = 23.44$ kNm (maximum)
		$M_{BC} = M_{CB} = 11.25 \times 1.25 = 14.06$ kNm
		$M_{CD} = M_{DC} = 3.75 \times 1.25 = 4.69$ kNm
	3	*Find moments in vertical members. It is the sum of beam moments at junction*
		[For an exact analysis of symmetrical girders, we may use moment distribution using Naylor's method of normal moment distribution on one half using (stiffness of beam × 1) and (stiffness of column × 6)]. However, for the approximate method, we use simple statics [see Figure E20.1(d)]
		$M_{AL} = 23.44$ kNm
		$M_{BM} = 23.44 + 14.06 = 37.05$ kNm (maximum value)
		$M_{CN} = 14.06 + 4.69 = 18.75$ kNm
		$M_{DO} = 0$
	4	*Find axial forces (compression and tension) in top and bottom chords (from equilibrium in verticals) Figure E20.1(e)*
		It is the sum of moment in top and bottom chords divided by height.
		Along $AB = \dfrac{2M_{AE}}{h} = \dfrac{2 \times 23.44}{2.5} = 18.75 \text{ kN} = 18.75 \text{ kN}$
		Along $BC = 18.75 + \dfrac{2 \times 37.5}{2.5} = 18.75 + 30 = 48.75 \text{ kN}$
		Along $CD = 48.75 + \dfrac{2 \times 18.75}{2.5} = 48.75 + 15 = 63.75 \text{ kN}$ (see Note below)
		(Max shear in vertical = (48.75 – 18.75) = 30 kN equal to 30 above)

Reference	Step	Calculation
		Simple approximate analysis
		Conservative values of tension and compression in the chords can be estimated as force from the bending moment concept. Figure E20.1(e)
Sec. 20.6		Compression = Tension = $\dfrac{\text{BM at section } XX}{\text{Distance between chords}}$
		In the above example for maximum values,
		Max tension = Max compression $= \dfrac{M_{max}}{h}$
		$= \dfrac{(37.5 \times 7.5) - 15(5 + 2.5)}{2.5} = 67.5 \text{ kN}$
		(This compares very favourably with 63.75 kN as obtained above. The difference is due to the fact that we neglect chord moments obtained from Step 2.)
	5	*Tabulate maximum values of axial forces, moments and shear in member*

Member/ Values	*Top member*	*Bottom member*	*Vertical*
Moment	23.44 kNm	23.44 kNm	37.5 kNm
Axial force	63.75 (comp.)	63.75 (tension)	18.75 kN (comp.)
Shear	18.75 kN	18.75 kN	30.0 kN

Reference	Step	Calculation
		(Use factored value (1.5 × above value) for limit state design)
	6	*Design of members*
		Design top chord for moment + compression
		Design bottom chord for moment + tension
		Design verticals for moment + compression
		[*Note:* Compression members should have a length/least dimension ratio not greater than 12.]
		[*Note:* If the upper chord has to carry the load along its length to transfer the loads to the D nodes, as in Figure E. 20.1(a), they have to be separately checked as beams. In infilled frames, the resistance may be assumed to be shared by top and bottom chords equally.]

EXAMPLE 20.2 (Transfer of top chord loads as nodal loads on Virendeel girder)

The top chord of a Virendeel girder foundation is 2.5 m span between nodes. If it carries 12 kN/m load on the top chord, determine the initial design of top chord. (Refer Figure E20.1.)

Reference	Step	Calculation
	1	*Find total load and moments* The total load on each panel = 12 × 2.5 = 30 kN Reaction on each panel = 15 kN
	2	*Preliminary design of top chord* Assuming the ends are fixed $M_{max} = wl^2/12$ Design the top chord as a beam to transfer these loads to the nodes.

Note: This example shows how the UDL on the top chords of the Virendeel girder between the nodes is transferred to the top nodal points. This is an approximation as it assumes that the girder is not in filled.

REFERENCE

[1] Varghese, P.C. and T.P. Ganesan, An Analysis of Virendeel Girders, *Journal of Institution of Engineers India*, July 1972.

Steel Column Bases

21.1 INTRODUCTION

In this chapter, we briefly deal with the design of foundations for steel columns. Details of design of connections of steel column to base plate by angles, channels, etc. are covered in textbooks on steel design. Bases are designed to transmit axial loads only or axial loads with moments and horizontal forces to the foundations. Moments are transmitted to the concrete foundation by the steel plate and the holding-down bolts. Horizontal loads are transmitted by the shear of the holding-down bolts as well as the friction between the base plate and concrete. The calculation of tension developed in the holding bolts and their design are important topics in steel design. In this chapter, we also deal with the design of simple concrete bases for steel columns and with the design of grillage foundations that are used for heavily loaded stanchions.

21.2 TYPES OF BASES

The following are the generally used types of foundation for the steel column (Figure 21.1):

1. Slab base over concrete footing (Figure 21.1)
2. Gusseted base over concrete footing (Figure 21.2)
3. Grillage foundation under steel bases (Figure 21.3)
4. Pocket base (Figure 21.4)

A brief description of each of these is given here.

Slab or bloom base: It is used for columns with axial load only and no bending. A slab base consists of a base plate welded or fixed by angles to the column and bolted to the foundation block. The column bears on the base plate which, in turn, bears on the concrete. The thickness of the plate can vary from 40 mm for 100 t to 120 mm for 1000 t column loads. For such heavy loads, the stanchion end and the slab are usually machined to have a close fit. In such cases, the weld connection between stanchions and base plate is only designed to hold them in position. For lighter columns, when machining is not done, the weld should be designed to transmit the load from the column to the base through these connections. The labour required for slab bases is less than that for a gusseted base. But they require more steel. These slab bases are easier to maintain than gusseted bases. Unlike the latter, the slab bases have no pockets in which water and dust can collect, thereby leading to easy corrosion.

The formula used for the design of these *slab bases* is fully described in textbooks on steel design. In short, the thickness (t in mm) for the slab base for an I section columns is calculated from the following formula:

$$t = \sqrt{\frac{3w}{p}\left(a^2 - \frac{b^2}{4}\right)} \tag{21.1}$$

where, w = (Load/Area), i.e. pressure on the underside of steel plate in N/mm^2 (When resting on concrete, it should not exceed $0.25 f_{ck} \sqrt{A_1 / A_2}$. Refer IS 456, Cl. 34.4.)

p = Permissible bending stress in steel in N/mm^2 (185 N/mm^2)

a = Greater projection of plate beyond the column faces in mm [Figure 21.1(a)]

b = Lesser projection of the plate beyond the column faces in mm

For steel round columns, the minimum thickness of the square base in mm is given by,

$$t = \sqrt{\left(\frac{9W}{16p} \times \frac{D}{D-d}\right)} \tag{21.2}$$

where W = Total load in Newtons

D = Diameter of base $\not< 1.5(d + 75)$ mm

d = Diameter of the column in mm

p = Permissible bending stress = 185 N/mm^2

Gusseted bases: Gusseted bases consist of gusset plates on the sides of the flanges, cleats, stiffeners and base plate all welded or riveted together to form a stiff assembly, which transmits column load uniformly to the concrete foundations. *They are very useful for columns with direct load and bending*. Part of the load is to be transmitted from the column through the gusset plates. As the gussets and stiffeners help the slab against bending, a much thinner base slab can be used, leading to economy of materials. However, these bases are difficult to maintain due to the presence of pockets where corrosion can set in. Gusset bases require less steel than slab base. We will not go into the details of their design. It can be found in textbooks on steel design.

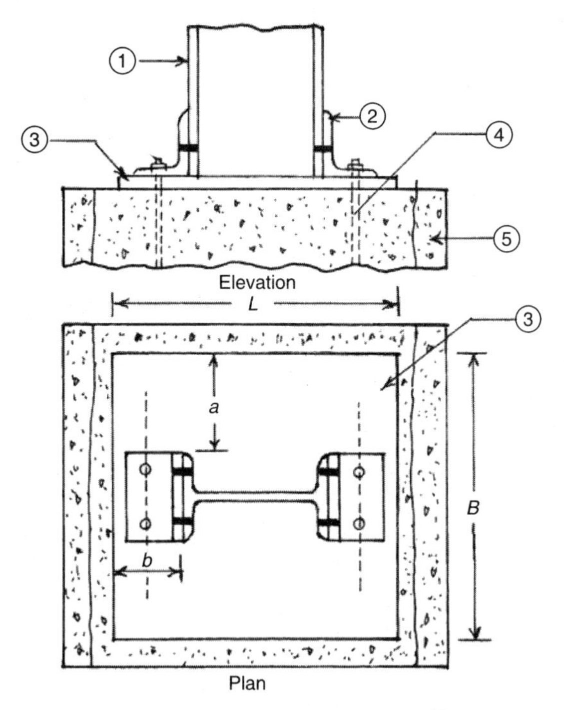

Figure 21.1 Slab or bloom base over concrete footing (1) Steel column, (2) Angles, (3) Base plate, (4) Bolts, (5) Concrete base.

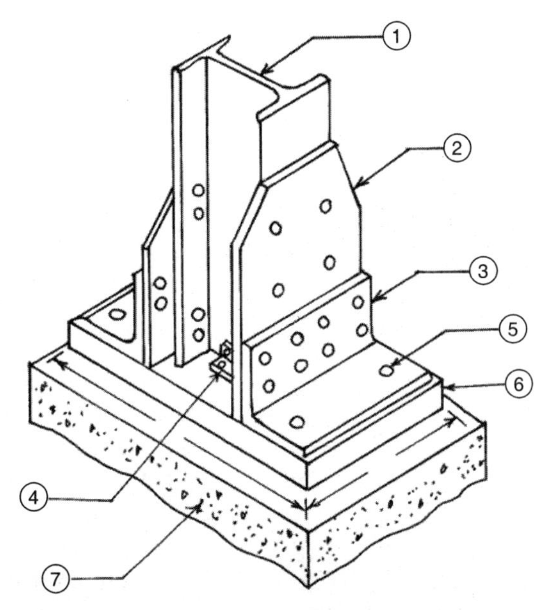

Figure 21.2 Gusseted base on concrete footing: (1) Column, (2) Gusset plate, (3) Gusset angle, (4) Web cleat, (5) Foundation bolts, (6) Base plate, (7) Concrete base.

Grillage foundations: Where very heavy loads, as in buildings, are to be transmitted to shallow foundations, a steel grillage can be used to transmit the load to the foundation. A simple R.C.C. base will be too deep for the area of foundation required. A grillage foundation consists of two or three tiers of steel beams placed under the column with a base plate. In a two-tier system, the base plate transmits the load first to three or four I beams under it with their length along the length of the foundation. These beams are placed, in turn, on a larger number (8–10) closely spaced steel I beams, placed over the foundation soil with proper care. The whole system is encased in concrete (of strength not less than M_{20}) and the cover specified is not less than 100 mm. The concrete is well compacted around the beams. The girders in each tier are kept in position by pipe separators (or spacers) threaded over the bolts placed in holes made in the webs of the girders. This arrangement keeps them in position during concreting.

The first type of failure that can occur in this set-up is buckling of the web of the first tier. It is the failure (instability) of the web as a column under a concentrated load on the flange. It is a sudden failure and has occurred in bridge construction at several places. The grillages were temporary grillage not covered with concrete. Another type of possible failure is the crippling failure of the web. Crippling (become like a cripple) of the web is a local failure just under the flange (at the root of the fillet) under concentrated loads. Such a failure is also called *crushing failure* (see Figure 21.6). Both buckling and crippling should be checked for temporary grillages used for erection of bridges and in all grillages if concreting is done after loading the column. In temporary constructions, stiffeners (vertical pieces) or diaphragm (horizontal pieces) can be provided to prevent buckling. Stiffeners can prevent crippling or crushing also. (These problems are further discussed in Sec. 21.6.)

[*Note:* More recently, another device called Rigid Steel Pedestals [2] has been evolved to spread the column loads to concrete foundations. It is a sort of steel pedestal into which the column is installed.]

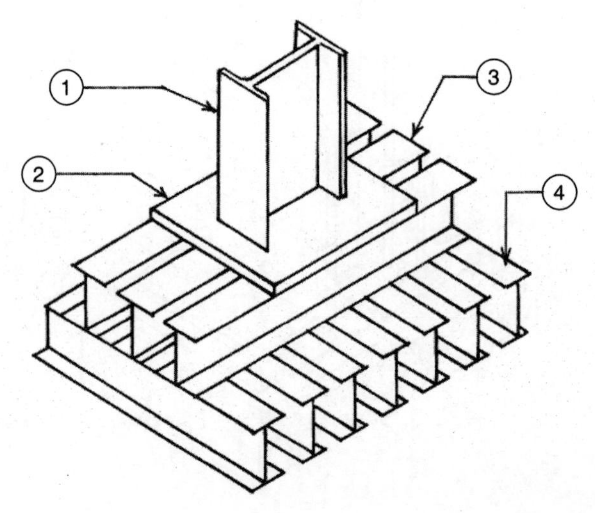

Figure 21.3 Steel grillage foundation: ① Column, ② Base plate, ③ Top tier steel beams, ④ Bottom tier steel beams.

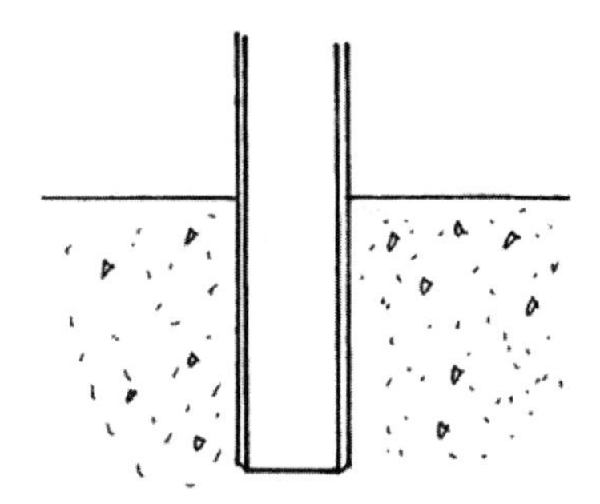

Figure 21.4 Pocket base with I section (pipes are also used) for advertisement posters.

Pocket base: This base, shown in Figure 21.4, is the most primitive type of foundation for light steel columns or posts. In pocket bases, the stanchion is grouted into a pocket in the concrete foundation. Pocket bases are used only for light columns. For example, all the *columns of steel frames for advertisement posters* are only pocket bases. The axial load is resisted by direct bearing on the concrete as well as by bond between steel and concrete. If there is a moment in the column, it is resisted by compression forces acting between the concrete and the flanges of the column as in I sections or the faces as in box sections.

21.3 DESIGN OF R.C. FOOTINGS UNDER STEEL COLUMNS

After the design of the base plate and anchor bolts, the design of R.C. footings is carried out in the normal way. IS 456, Cl. 24.2.3 specifies that the value of *shear* for design of footings under column with slab base or gusseted bases is to be taken halfway between the faces of the column or pedestal and the edge of the base plate instead at a distance equal to the effective depth of the footing from the face of the column for concrete columns built integral with the footings. Bending and punching shear considerations should be made, as in R.C. column, at the edge of the base plate.

21.4 DESIGN OF STEEL GRILLAGE FOUNDATION

We will consider the design of grillage foundations in more detail. As per IS 800 (1984), allowable stresses in steel grillage beams can be increased by $33\frac{1}{3}\%$ from that specified for normal beams. The allowable stress works out to 1.33×165 [for Fe 250] ≈ 220 N/mm^2. We assume that the load from the column is uniformly distributed over the base plate and that each tier of beams distributes the load uniformly to the lower layer.

21.4.1 Design Moments and Shears

The various suggested methods to calculate the bending moments and shears *in centrally* loaded and eccentrically loaded grillages are shown in Figures 21.5 and 21.6.

The most commonly used values of bending moments and shears in the members for design are the following. We first compute total BM and Z values for the upper tier taken together. After that, we determine the type of steel section and the number required. Similarly, we analyse the lower tier also.

Let,

P = Load in column in Newtons (without moments)

a, b = Length and width of base plate of column (a placed along L of upper tier)

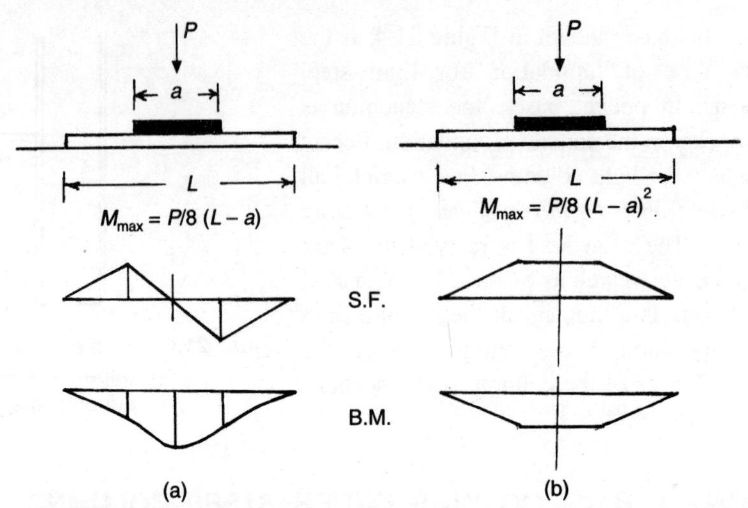

Figure 21.5A Two of the assumptions used for design of grillage foundation for centrally loaded column: (a) Assuming distribution of load through plate to girder, (b) Grillage beam acting as a cantilever.

L = Length of the *upper tier* in m (along the length of the foundation)

B = Length of the *lower tier* in m (along the breadth of the foundation)

(We find L and B from $L \times B$ or area = P/Safe bearing capacity.)

In the L-direction, the load per unit length of all the beams together is equal to P/L upwards.

Taking moments about the centre line of the column for the upper *cantilevers*, we get the cantilever moment from Figure 21.6 as

$$M = \frac{P}{L}\left(\frac{L}{2}\right)^2 \times \frac{1}{2} - \frac{P}{a}\left(\frac{a}{2}\right)^2 \times \frac{1}{2}$$

$$= (P/8)(L - a)$$

Similarly, total maximum moment in *beams* along length

$$L = P/8(L - a) \text{ (upper tier).}$$

Total maximum moment in *beams* along length

$$B = P/8(B - b) \text{ (lower tier)} \qquad (21.3)$$

Maximum shear for beams along

$$L = (P/2)\left(\frac{L - a}{L}\right)$$

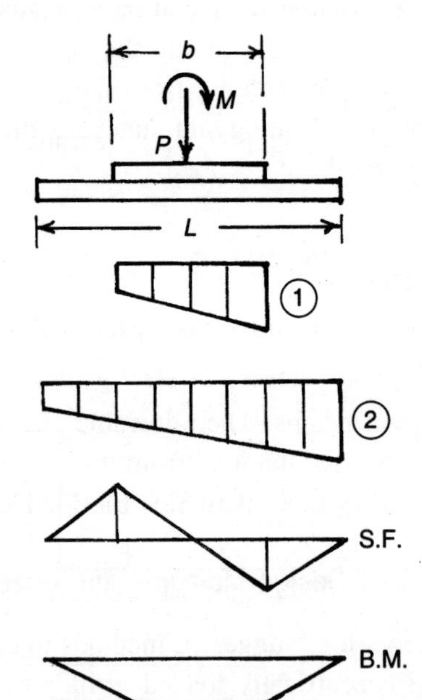

Figure 21.5B Design of grillage foundation for a column with vertical load and moment: ① Load distribution under plate, ② Load distribution under beam.

$$\text{Maximum shear for beams along } B = (P/2)\left(\frac{B-b}{B}\right) \qquad (21.4)$$

21.4.2 Steps in Design of Grillage Foundations

We work on characteristic loads and elastic design. The steps in the design of grillage foundations are as follows:

1. Find the area required of the foundation from the safe bearing capacity.
2. Consider the upper tier beam placed along the length L of the foundation.
   (a) Calculate the maximum BM for the upper tier
   (b) Compute the total section modulus using f_s = 220 N/mm² and select the number of beams required (3 or 4 in numbers). Check clearance for beams to be accommodated in the bearing plate size. Clear distance should be more than 75 mm.
   (c) Calculate the maximum shear and check for the shear. Max allowable shear = 100 N/mm².
   (d) Check for web buckling and web crippling.
3. Consider the lower tier beams placed along the breadth B of the foundation.
   (a) Calculate the maximum BM for the lower tier.
   (b) Compute the section modulus and find the number of beams required to cover the length B with clearance not less than 75 mm.
   (c) Check for the shear
   (d) Check for web buckling and web crippling.

Details of a steel grillage foundation are shown in Figure 21.6.

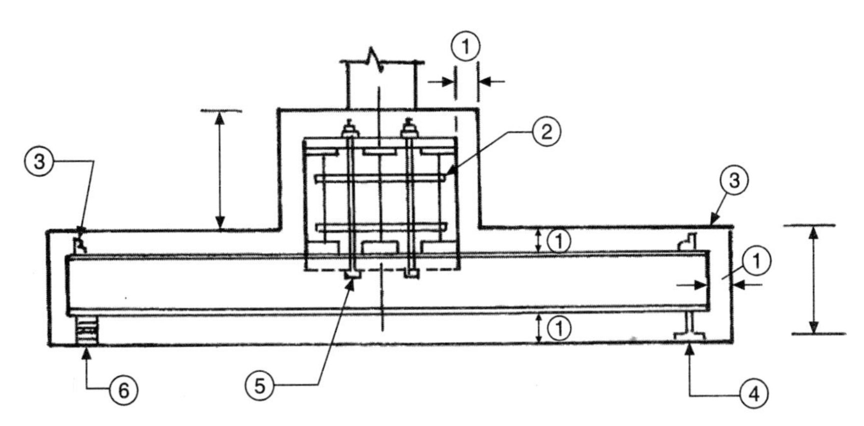

Figure 21.6 Details of grillage foundation: (1) Cover 100 mm, (2) First tier I beams with tubular distance pieces over threaded rod bolts, (3) Angle connecting bottom tier of I beams, (4) 100 mm T for levelling, (5) Anchor arrangement using 150 × 75 mm channels, (placed webside down to ensure filling with concrete) placed specially when both are subjected to tension, (6) Wedges for levelling I beams.

21.5 GRILLAGE FOUNDATION AS COMBINED FOOTING

For the design of a combined footing for two column loads, we make the CG of loads to coincide with the CG of the footing. We design the grillage system as an inverted beam, supported on the two columns with uniform pressure from the ground on the grillage.

21.6 WEB BUCKLING AND WEB CRIPPLING (CRUSHING) OF I-BEAMS UNDER CONCENTRATED LOADS [Fig. 21.7]

Web buckling is different from web crippling or crushing and both should be checked whenever we place concentrated loads on I-girders, and plate girders. Stiffeners should be provided if the calculations show that they are needed.

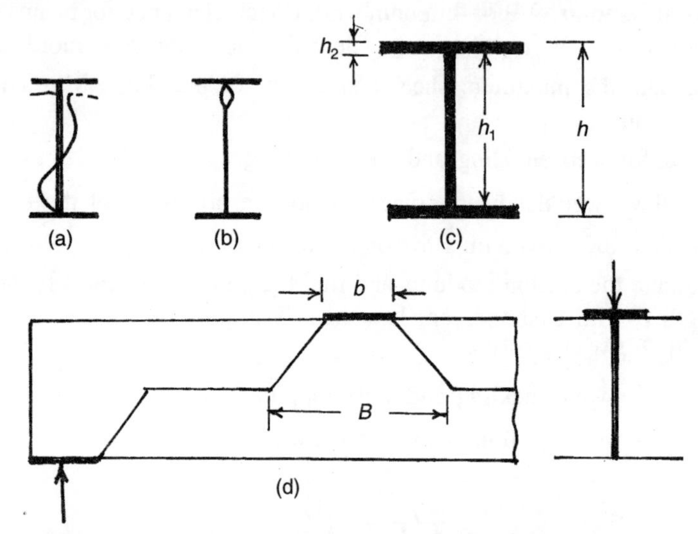

Figure 21.7 Web failures in I beams: (a) Web buckling, (b) Web crushing, (c) Dimensions of I beam in steel tables used for web buckling, (d) Calculation of web buckling (b is length of plate).

21.6.1 Web Buckling

The concentrated load from the bearing plate is assumed to be distributed in the B-beam to the middle of the beam. With this assumption, the slenderness ratios of the web behaving as a strut (see textbooks on Steel Design) are calculated.

1. Slenderness ratio for web buckling

$$\text{Slenderness ratios} = \frac{\sqrt{3}h_1}{t} \tag{21.5}$$

where h_1 = depth of the web = $(h - 2h_2)$(h, h_1 and h_2 are given in steel tables)
t = thickness of the web
h_2 = depth to the root of the fillet

2. Stress calculation for buckling. The stress at the mid-level is

$$\sigma = \frac{P}{B_1 t} \text{ (should not exceed allowable)} \tag{21.6}$$

where

B_1 = 45 degree dispersed width at the mid-depth of the web for $(b + h)$
P = Load on the plate
t = thickness of the web

The stress σ should not exceed the maximum allowed for the slenderness ratio mentioned before as given in steel tables) [1].

21.6.2 Web Crippling or Web Crushing

Crippling of the web occurs with large bearing stress just below the flange at the junction of the flange and the web, i.e. the root of the fillet. The thickness of the flange at the fillet is given in steel tables as h_2. We assume a 30° dispersion (with the horizontal) of the load in this case to the root of the fillet. The maximum allowable bearing stress is 187 N/mm^2.

If P is the load, then the bearing stress is equal to P/bearing area.

Bearing area for load from top = $B_2 t$

$$B_2 = b + \sqrt{3}\,(2h_2) \tag{21.7}$$

where b is the length of the bearing plate.

Thus,

$$\text{Bearing stress} = \frac{P}{t[b + \sqrt{3}\,(2h_2)]} \tag{21.8}$$

should not exceed 187 N/mm^2.

21.6.3 Checking for Web Buckling and Web Crippling

It can be found that in general if the beam section is safe on web crippling, it will be safe in web buckling also. Thus, we need to check only for web crippling in routine cases. The problem is illustrated by Example 21.1.

21.7 DESIGN OF POCKET BASES

We briefly deal with pocket bases also. The forces to be resisted are axial load P and moment M. In this type of bases, the stanchion is grouted into a pocket in the concrete foundation. The axial load is taken by the bond and the bearing. The moment is resisted by compression forces between the surfaces of steel and concrete as shown in Figure 21.8. For flanged beams for resisting moments, it is assumed that the compression forces can act on *both faces (outside and inside of flanges)* as concrete encases the whole section. In hollow sections, it is assumed that compression can act on one side (the outside) only as shown in Figure 21.8.

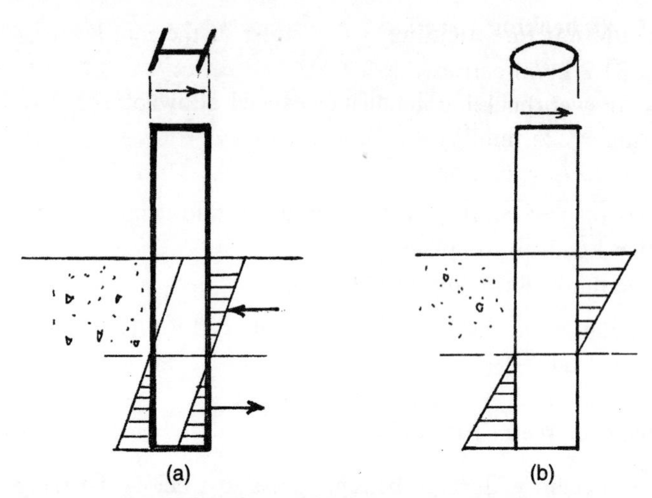

Figure 21.8 Design of pocket bases: (a) I section encased in concrete-reaction acting all along the length of embedment, (b) Circular (tubular) section-web reaction acts on compression face only.

Case 1: **(I section)** *For structural I section*, if d is the depth of the embedment on both faces, b is the breadth of the flange, and p_c the allowable bearing stress on concrete, as assumed at all contacts, then

$$F = (p_c/2) \times (d/2) \times b = p_c db/4 \quad \text{[on one flange (Figure 21.8)]} \tag{i}$$

where p_c is equal to the permissible stress in compression in concrete.

The moment of the total forces from each face will be,

$$= 2 \times F \times \left(\frac{2}{3}d\right) = (2) \times \left(p_c d \frac{b}{4}\right) \times \left(\frac{2}{3}d\right) = p_c bd^2/3 \tag{ii}$$

From the allowable stress in steel (p_s), the maximum moments the steel section can take,

$$M = p_s Z_{XX} \tag{iii}$$

Equating the moments (ii) and (iii), the necessary length of the embedment for developing full capacity is obtained as,

$$d = [3p_s Z_{XX}/p_c b]^{1/2} \tag{21.9}$$

Case 2: **(Hollow section)** A similar derivation will lead to d for a hollow section. For full bending capacity, the value of d will be double the value of Eq. (21.9) and equal to,

$$d = [6p_s Z_{XX}/p_c b]^{1/2} \tag{21.10}$$

21.8 SUMMARY

This chapter briefly dealt with the design of the different types of foundation of steel columns, namely, pocket bases for advertisement posters, slab bases, gusseted base and grillage foundations.

EXAMPLE 21.1 (Checking I section for web buckling and web crippling or (crushing))

An ISMB 225 at 31.2 kg/m carries a load of 250 kN through a 700 × 110 mm bearing plate. Check the beam for web buckling and web crushing (crippling). From steel tables, h = total depth of the section = 225 mm, h_2 = depth to the root of the fillet = 25.85 mm, h_1 = depth of the web = 173.3 mm, t_w = 6.5, and t_f = 11.8 mm (breadth of the flange = 110 mm).

Reference	*Step*	*Calculation*
		(a) Web buckling
IS Tables	1	*Find slenderness ratio and allowable stress*
Eq. (21.5)		Slenderness ratio $= \sqrt{3}h_1/t$
		$= (\sqrt{3} \times 173.3)/6.5 = 46.17$
IS 00(1984)		Allowable stress for slenderness ratio for grade 240 steel and l/r = 50 = 125.5 N/mm^2
	2	*Find stress after dispersion of load at 45° from base plate to mid-depth*
		B_1 = (Length of the plate + h) for dispersion at 45°
		= (700 + 225) = 925
Eq. (21.6)		Stress = $\dfrac{P}{B_1 t_w} = \dfrac{250 \times 10^3}{925 \times 6.5} = 41.5 < 125\ \text{N/mm}^2$
		Thus, safe against buckling.
		(b) Web Crippling
	1	*Find width of bearing at root of fillet and crushing stress at that lever with 30° dispersion to horizontal*
		P = 250 kN
Eq. (21.7)		B_2 = (Length of the base plate) + $2\sqrt{3}h_2$
		$= 700 + (2\sqrt{3} \times 25.85) = 789.5$ mm
Eq. (21.8)		Crushing stress = $250 \times 10^3/(789.5 \times 6.5) = 48.7\ \text{N/mm}^2$
		This is less than 187 N/mm^2. Thus, safe against crushing.
		[*Note:* Generally, when the load is safe against crushing, it will be safe against buckling also.]

EXAMPLE 21.2 (Design of base plate of a steel column on concrete footing)

Design the base plate for a steel column ISHB 350 @ 67.4 kg/m, carrying an axial load of 3000 kN and resting on a M_{20} concrete footing.

Reference	*Step*	*Calculation*
Steel Tables	1	*Find dimension of I-beam*
		B = 250 mm. Total length h = 350 mm
	2	*Find bearing area required on concrete and find w*

Reference	Step	Calculation
IS 456 (Cl.34.4)		Max bearing pressure allowed $= 0.25 f_{ck}\sqrt{A_1/A_2}$
		(Assume) $0.25f_{ck} = 0.25 \times 20 = 5$ N/mm^2
		Min. area required $= \dfrac{3000 \times 10^3}{5} = 60 \times 10^4$ mm^2
		Min. size $= \sqrt{60 \times 10^4} = 775$ mm square
		Adopt 870 × 870 mm plate as the slab base.
		Actual pressure $w = 3000 \times 10^3/870 \times 870 = 3.96$ N/mm^2
	3	*Find projections of I-beam on base plate and w*
		Place steel beams 870 × 870 mm plate centrally
		Greater projection $a = (870 - 250)/2 = 310$ mm
		Smaller projection $b = (870 - 350)/2 = 260$ mm
	4	*Calculate thickness of base plate required*
Eq. (21.1)		$t = \sqrt{\dfrac{3w}{p}[a^2 - (b^2/4)]}$. Assume $p = 185$ N/mm^2
		$= \sqrt{\dfrac{3 \times 3.96}{185}[(310)^2 - 260^2/4]} = 71$ mm
		Use a 870 × 870 mm size 72 mm thick plate as the slab base.

EXAMPLE 21.3 (Design of grillage foundation)

Design a two-tier grillage foundation to carry an axial load 1000 kN, with IS steel beams encased in M_{20} concrete. Assume a safe bearing capacity of 150 kN/m^2 and the size of the base plate under the column is 700 × 700 mm.

Reference	Step	Calculation
Sec. 21.4.2	1	*Find area of base from safe bearing capacity*
		Axial load + self wt. = 1100 kN, BC = 150 N/mm^2
		Required length of a square base $= \sqrt{\dfrac{1100}{150}} = 2.71$ m
		Adopt a square foundation 2800 × 2800 mm
	2	*Consider upper tier beam of length L*
Eq. (21.3)		(a) Compute maximum BM $= \dfrac{P}{8}(L - a)$ with $a = 700$ mm
		$= 1000 \times 10^3(2800 - 700)/8 = 263 \times 10^6$ N-mm
		(b) Find the total section modulus using 30% more allowable stress.

Reference	*Step*	*Calculation*
		Allowable stress = 1.33 × 0.66 × 250 = 220 N/mm^2
		$Z = 263 \times 10^6/220 = 1195 \times 10^3$ mm^3
		(c) *Select 3–4 beams to give* $Z = 1195 \times 10^3$ mm^3
IS Steel Tables		Choose 4 Nos., Z required for each beam $= \dfrac{1195}{4} = 298.75$
		Choose ISMB 225 @ 31.2 kg/m (height 225 mm)
		Width of flange = 110 mm
		h_2 = Depth to the root of the fillet = 25.85 mm
		h_1 = Height between the root of the fillet = 173.3 mm
		t = 6.5 mm
		$Z = 305.9 \times 10^3$ mm^3 × 4 = 1223.6×10^3 mm^3
		Check accommodation under the bearing plate.
		(700 – 4 × 110)/3 spaces = 86.67 mm > 75 mm
		(d) *Check maximum shear for each girder*
		Shear force $V = \dfrac{1}{n}\left[\dfrac{P}{2}\right](L-a)/L, \quad n = 4$
		$= \dfrac{1 \times 1000 \times (2800 - 700)}{4 \times 2 \times 2800} = 94$ kN (per girder)
		Shear stress $= \dfrac{V}{ht} = \dfrac{94 \times 10^3}{225 \times 6.5} = 64 < 100$ N/mm^2
		Thus, safe in shear.
		(e) *Check for web crushing (crippling)*
		Depth to the root of the fillet = 25.85 mm
		Width of the plate = 700 mm
Sec. 21.5.2		Width of dispersion at 45° to horizontal on both sides
Eq. (21.7)		$= (2\sqrt{3}h_2 + 700) = (2 \times \sqrt{3} \times 25.85) + 700 = 789.5$ mm
		Load on each beam = 1000/4
		Crushing stress $= \dfrac{1000 \times 10^3}{4 \times 789.5 \times t} = \dfrac{1000 \times 10^3}{4 \times 789.5 \times 6.5}$
		= 48.7 < 187 N/mm^2 (allowed)
		(As it is safe in web crushing, it will also be safe in web buckling.)
	3	*Consider lower tier (B = L for a square base)*
Eq. (21.4)		(a) Max BM $= \dfrac{P(B-b)}{8} = 263 \times 10^6$ N/mm (as in step 2)
		(As $B = L$ and $a = b = 700$ mm)

Reference	Step	Calculation
		(b) *Find Z required* = 1195 mm³ (as in 2 above)
		Choose 10 beams
		Z required = 119.5 mm³ for each
Steel Table		ISLB 175 @ 16.7 kg/m (total height = 175 mm)
		$Z = 125.3$ mm³
		Width of the flange = 90 mm
		$h_2 = 16.7$ mm, $h_1 = 141.6$ mm and $t = 5.1$ mm
		Check accommodation (100 mm cover of concrete)
		$\text{Spacing} = \dfrac{(2800 - 200) - 90 \times 10}{9} = 188.8 \text{ mm}$
		Spacing is more than 75 mm specified.
		(d) *Check for shear*
		$\text{Shear} = \dfrac{1}{10} \times \dfrac{1000 \times 10^3}{2} \left(\dfrac{2800 - 700}{2800} \right) = 37.5 \times 10^3 \text{ N}$
		$\text{Shear stress} = \dfrac{37.5 \times 10^3}{175 \times 5.1} = 42 \text{ N/mm}^2$
		This is less than 100 N/mm², the maximum allowed.
		(e) *Check for crushing (load transmitted by top beam with flange width* 110 mm)
		$\text{Stress} = \dfrac{1000 \times 10^3}{10\,(110 + (2\sqrt{3} \times 16.7)) \times 5.1} = 116.8 \text{ N/mm}^2$
Example 21.1		This is less than 187 N/mm², the maximum specified. Thus, safe against crushing. Therefore, safe for buckling also.

EXAMPLE 21.4 (Design of a pocket base for a steel column)

An I-section used as a column for a large advertisement poster has the following properties: $h = 305$ mm, $b = 125$ mm, $Z_{xx} = 611.1 \times 10^3$ mm³, sectional area, and $A = 6080$ mm². The maximum allowable bending compression in concrete p_c is 7 N/mm² and permissible bending stress in steel p_s is 165 N/mm². If the stanchion is embedded to develop its full bending strength, find the depth at which it should be placed. Also, find the safe axial load it can take (Figure 21.5).

Reference	Step	Calculation
	1	*Find depth of embedment of the I-section for full bending capacity, assuming compression on outside and inside faces of I-section*
Figure 21.8		Full bending strength = $p_s Z_{xx} = M$, $p_s = 165$
		$\text{Required depth for full capacity} = d = \left[\dfrac{3M}{p_c b} \right]^{1/2}$

Reference	*Step*	*Calculation*
Eq. (21.9)		$d = [(3p_s) \times (Z_{xx})/p_c b]^{1/2}$
		$= [(3 \times 165 \times 611.1 \times 10^3)/(7 \times 125)]^{1/2}$
		= 585 mm
	2	*Find the axial load capacity*
		Axial load = 1/2 [surface area in bond × bond stress] + full bearing
Section 21.7		The bond is assumed to act on both sides of one flange and full web.
		Perimeter length = 2 × (125 + 305) = 860 mm (approx.)
		Allowable bond stress = 0.83 N/mm^2 (assumed)
Eq. (21.2)		Axial bearing allowable = 5.5 N/mm^2
		Area of I-section = 6080 mm^2
		Bond area in 585 mm depth = 585 × 860 = 5.0 × 10^5 mm^2
		Safe axial load = Bond strength + Bearing strength
		= (5.0 × 10^5 × 0.83) + (6080 × 5.5) N
		= 415 + 30.4 = 445 kN
		(*Note:* The bond considerably contributes to the axial strength.)
Eq. (21.10)		[*Note:* For hollow sections, we use the formula $d = [6p_s Z_{xx}/p_c b]^{1/2}$ for depth to develop the bending capacity of the section.]

REFERENCES

[1] IS 800 (1984), *Code of Practice for General Construction in Steel* (Second Revision), BIS, New Delhi.

[2] *Handbook for Structural Engineers, 1, Structural Steel Sections*, BIS, New Delhi, 1974.

Analysis of Flexible Beams on Elastic Foundation

22.1 INTRODUCTION

A *rigid foundation* such as a footing constructed on soil can be assumed to settle uniformly under a concentrated load [1]. As we have the same settlement at all points under the footing, we can also assume that the ground pressure from below is uniform. On the other hand, if we place a concentrated load on a long flexible strip on elastic soil, the base pressure cannot be assumed to be uniform. As shown in Figure 22.1, the base pressure will be varying along its length, depending on the settlement of the strip along its length. Thus, the ground pressure under *flexible structure* on elastic foundation cannot be determined by the empirical method. In this chapter, we briefly examine the general theory of the behaviour of beams on elastic foundation.

22.2 METHODS OF ANALYSIS OF BEAMS ON ELASTIC FOUNDATION

The following are the commonly used methods for finding the pressures exerted by a concentrated load on a beam supported by an elastic foundation:

Method 1: The conventional method assuming that the beam is rigid, the settlement is uniform and the ground pressures are uniform.

Method 2: The Winkler method, using the modulus (coefficient) of subgrade reaction property of the elastic foundation.

Method 3: The elastic half space method, which is an elastic method, using the modulus of compressibility or stiffness coefficient.

Method 4: Simplified ACI method using Winkler theory.

Method 5: The method based on available published formulae for various specific cases.

Of these above methods, the Winkler and elastic half space models are more commonly used. They are also recommended in IS 2950, Part I for the design of raft foundations [2]. Any other method can be only an approximation, depending on the nature of the problem. Also the Winkler model has been found in practice to be an easy method giving good results. We will examine it in more detail. Methods 2–5 are briefly explained below. A detailed treatment of method 4 is given in Chapter 23.

22.3 COEFFICIENT OF SUBGRADE REACTION AND WINKLER MODEL

Winkler published his paper on 'Laws of Elasticity and Strength' in 1867, where he introduced the concept of elastic springs in which the reaction from the spring is taken as being proportional to the deflection of the spring. This is known as the *Winkler model.* The term *coefficient of subgrade reaction* was, however, first proposed by Zimmermann later in 1888 to represent the proportionality between the load on a railway sleeper and its settlement into the subgrade due to a load. We should know that this law of proportionality refers only *for a small region of loading*, the rail being considered as a deformable beam of infinite length resting on elastic supports. In later years, structural engineers took this concept to *beams supported continuously* on an elastic foundation also by assuming that the contact pressure will be proportional to the deflection of the beam all along its length. Thus, the subsoil has been replaced by a system of springs independent of each other. The stiffness of the spring is assumed as constant throughout its length. If necessary, it can be also varied to represent weaker spots in the foundation. (It is noteworthy that in Westergard's method of *design of rigid pavements*, he has taken the pavement to be resting on a liquid subgrade instead of springs.)

22.4 WINKLER SOLUTION FOR A CONTINUOUS BEAM ON ELASTIC FOUNDATION

A continuous beam means a beam of infinite length such as the tracks of a railway line. Hetenyi proposed the use of Winkler model for such a beam in 1945 [3], [4] and [5]. The differential equation for an ordinary beam of width B with a downward load q/m^2, can be written as

$$EId^4y/dx^4 = qB$$

where

$q \times B$ = Load per unit length of the beam (kg/m)

y = Deflection

For a beam resting on springs with an upward reaction proportional to the deflection, this equation reduces to

$$EId^4y/dx^4 = qB = (KB)y \tag{22.1}$$

where

K = Modulus of subgrade reaction in (kg/m³) (This is pressure for unit deflection.)
B = Breadth of beam
KB = Load per unit length for unit deflection (kN/m²)
KBy = Load per unit length kN/m.

22.4.1 Solution for a Column Load at *P* on a Beam of Infinite Length

For a beam of width B, as in Figure 22.1, of infinite length, if we consider a point P as the origin and for the positive values of x away from P, then

$$EId^4y/dx^4 = -KBy$$

or

$$\frac{d^4y}{dx^4} + \left(\frac{KB}{EI}\right)y = 0 \tag{22.1a}$$

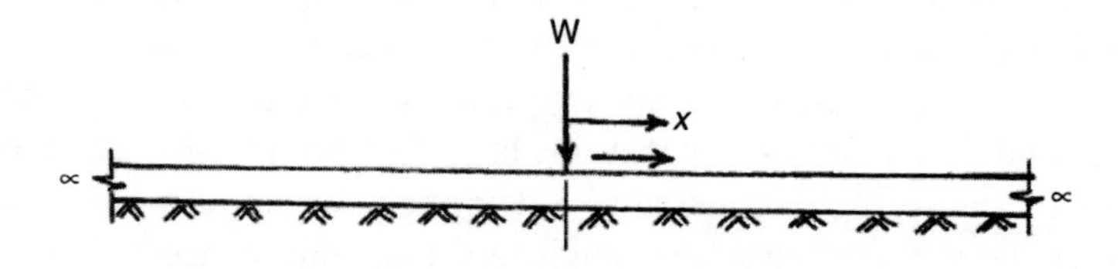

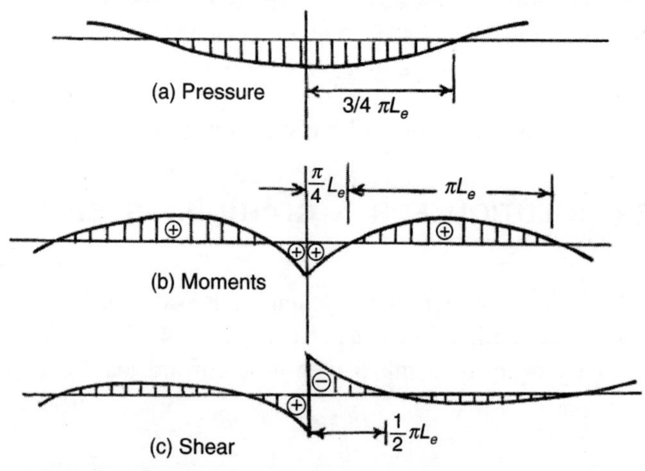

Figure 22.1 Distribution of pressure, moment and shear in a beam of infinite length on an elastic foundation ($L_e = 1/\lambda$) (For λ see Eq. 22.3).

We can see that in the above equation, if EI is very large (i.e. for a rigid beam), the pressure distribution can be taken as uniform as deflection is constant. This can be considered as a particular case of the general equation. Equation (22.1a) is a homogeneous linear differential equation and the general solution will give four constants. However, when we consider a very long beam, two of its terms will become zeros and we get the final solution for positive values of x with only two constants c_1 and c_2 as

$$y = e^{-\lambda x} (c_1 \cos \lambda x + c_2 \sin \lambda x) \tag{22.2a}$$

[*Notes:* 1. The value of λ is explained in Sec. 22.4.3.
2. Because of symmetry, the same solution can be used in the negative side also in Eq. (22.2).]

From the theory of structures (with $KB = k$), we get

$$y = +\frac{PL}{2k} = A_{\lambda x} \tag{22.2b}$$

$$\theta = \frac{dy}{dx} = -\frac{P\lambda^2}{k} B_{\lambda s} \tag{22.2c}$$

$$M = -EI \frac{d^2 y}{dx^2} = \frac{P}{4\lambda} C_{\lambda x} \tag{22.2d}$$

$$Q = -EI \frac{d^3 y}{dx^3} = -\frac{P}{2} D_{\lambda x} \tag{22.2e}$$

22.4.2 Moments and Shears in Long Beams due to Loads

The bending moment shear force and deflection at various points due to (a) a concentrated load at the centre of the beam, (b) a moment applied at the centre, (c) concentrated load at the free end, and (d) moment at the free end on a long beam of infinite length on elastic foundation has been worked out and tabulated by Hetenyi [3]. Of these, the effect of concentrated load and moment applied at the free ends and centre are of considerable importance. The results for these two cases can be summarized as in Tables 22.1 and 22.2. (see Sec. 24.2, Steps 8–11 for application.)

TABLE 22.1 Effect of Moment and Load Applied on Infinite (Long) Beams on Elastic Foundation

S. No.	*Load application*	*Formula for*		
		Moment	*Shear*	*Deflection*
1	Moment M applied at free end	$MA_{\lambda x}$	$-2MB_{\lambda x}$	$\left(\frac{-2M\lambda^2}{KB}\right)(C_{\lambda x})$
2	Load V applied at free end	$-\frac{V}{2} B_{\lambda x}$	$-VC_{\lambda x}$	$\frac{2V\lambda}{KB}(D_{\lambda x})$
3	Moment M at centre	$\frac{M}{2} D_{\lambda x}$	$-\frac{M}{2} A_{\lambda x}$	$\frac{M\lambda^2}{KB} B_{\lambda x}$
4	Load V at centre	$\frac{V}{4\lambda} C_{\lambda x}$	$-\frac{V}{2} D_{\lambda x}$	$\frac{V\lambda}{2KB} A_{\lambda x}$

TABLE 22.2 Hetenyi's Values of $A_{\lambda x}$, $B_{\lambda x}$, $C_{\lambda x}$, $D_{\lambda x}$ for Effects of Moments and Loads Applied on Infinitely Long Beams

λx	$A_{\lambda x}$	$B_{\lambda x}$	$C_{\lambda x}$	$D_{\lambda x}$
0.0	1.0000	0.0000	1.0000	1.0000
0.1	0.9907	0.0903	0.8100	0.9003
0.5	0.8231	0.2908	0.2415	0.5323
1.0	0.5083	0.3096	–0.1108	0.1988
1.5	0.2384	0.2226	–0.2068	0.0158
2.0	0.0667	0.1231	–0.1794	–0.0563
3.0	–0.0423	0.0070	–0.0563	–0.0493
4.0	–0.0258	–0.0139	0.0019	–0.0120
5.0	–0.0045	–0.0065	0.0084	0.0019
6.0	–0.0017	–0.0007	0.0031	0.0021
7.0	0.0013	0.0006	0.0001	0.0007
8.0	0.0003	0.0003	–0.0004	0.0000
9.0	0.0000	0.0000	–0.0001	–0.0001

22.4.3 Classification of Beams as Rigid and Flexible

The following are the factors that influence the results of the ground reaction of a beam on elastic foundation:

K = Modulus of subgrade reaction
B = Breadth of the beam
EI = Rigidity of the beam
λ = 'Characteristic coefficient' of the beam equal to

$$\lambda = \sqrt[4]{\frac{KB}{4EI}} \tag{22.3}$$

which has a dimension L^{-1}.

We can notice that λ multiplied by a distance x or λx is a dimensionless number. Also, the term $1/\lambda = L_e$ is called the *elastic length* of the system. If we designate L_B as the length of the real beams, then $\lambda L_B = L_B/L_e$ is a number. [Usually, the expression is derived for a breadth B equal to unity.] We may also write,

$$L_e = \frac{1}{\lambda} = \sqrt[4]{\frac{4EI}{KB}} \tag{22.3a}$$

which will be a length.

Vesic [6] has suggested that this quantity λL_B can be taken as an indication of the relative rigidity of the beam foundation system. Hence, if we consider $\lambda = 1/L_e$ (reciprocal of elastic length) and relate it to the actual length of the beam L_B, then we get the following relations. These are also shown by Figure 22.2.

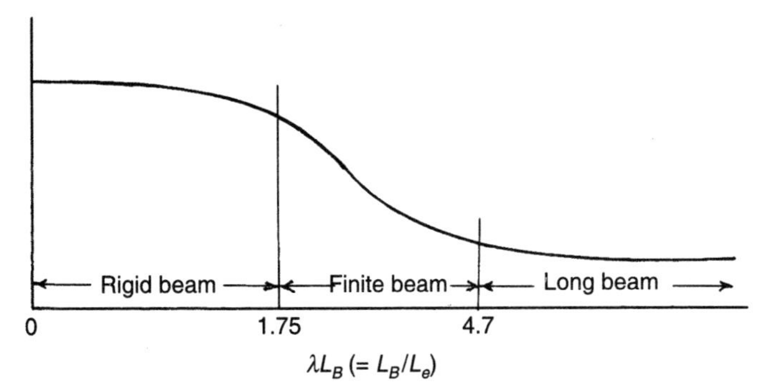

Figure 22.2 Method of classification of beams on elastic foundation (L_B is length of the beam and L_e is the characteristic or elastic length).

Case 1: λL_B is equal to or less than 1.75 (or $\pi/4$). (The beam acts as a rigid beam.)

Case 2: $\lambda L_B > 1.75 \leq \dfrac{3\pi}{2}$ or 4.7 (22.3b)

[The beam is a *flexible beam of finite length* or a short (but not rigid) beam.]

Case 3: $\lambda L_B > \dfrac{3\pi}{2}$ or 4.7

(The beam is a flexible beam of infinite length or a long beam.)

(*Note:* $1/\lambda$ is also designated as L_e = elastic or characteristic length. Case 3 can be computed by Tables 22.1 and 22.2. In Case 2, we have to use tables like items 6 and 7 in Sec. 22.4.4. Case 1 can be treated as a rigid beam.)

The above results can be further explained by stating that if the beam can be considered rigid, then the contact pressure can be estimated by simple statics. Similarly, when the beam is very long, the effect of a load at a point will not be felt at distances farther than $4L_e$ (or at the most $4.5L_e$) from the point of application as shown in Figure 22.1.

22.4.4 Winkler Solution for Short Beam on Elastic Foundation

Figure 22.3 shows *a short beam* subjected to a concentrated load. As distinct from a long beam, the solution for a short beam will have all the four constants. A closed form solution is rather complex and it is easier to handle the problem by influence coefficients. Different investigators have used different methods to solve the problem and some of these are briefly described below. *Of all these methods, the use of published tables such as item 6 or 7 below is the most practical method for beginners. Nowadays, computers using appropriate software are also available.*

1. **Successive approximation method.** In this method, an arbitrary pressure distribution is assumed and the deflection is determined by the moment area method. From the deflections, the pressures are determined using the Winkler model. These are corrected so that the CG of loads and reactions coincide. Based on the adjusted reactions, the deflection line is recalculated till convergence takes place. Such a method was first proposed by Ohde in 1942 [7].

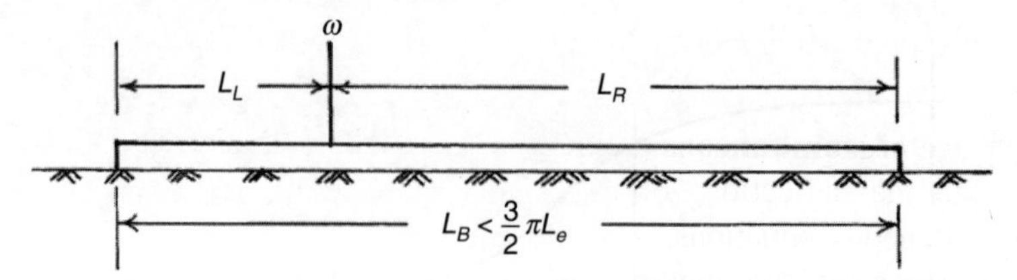

Figure 22.3 Finite beam (beam of definite length) on an elastic foundation.

2. **Finite difference method.** In 1957, K.C. Ray used the finite difference method to solve this problem [8], [9]. The equation to the beam with load can be written as

$$EI\,(d^4y/dx^4) + KBy = qB$$

$$(d^4y/dx^4) + \frac{KBy}{EI} = \frac{qB}{EI}$$

The beam is cut into 20 intervals of length δ and the above differential equation is reduced to the finite differential form

$$\frac{1}{\delta^4}(y_{n+2} + 4y_{n+1} + 6y_n - 4y_{n-1} + y_{n-2}) + \frac{KB}{EI}y_n = \frac{q_nB}{EI}$$

or

$$y_{n+2} - 4y_{n+1} + y_n\left(6 + \frac{KB\delta^4}{EI}\right) = \frac{(q_nB)\delta^4}{EI} = \frac{S\delta^4}{EI}$$

Assuming load P is distributed over length δ, we get $P = S\delta$. Hence,

$$y_{n+2} - 4y_{n+1} + y_n\left(6 + \frac{KB\delta^4}{EI}\right) = \frac{(q_nB)\,\delta^4}{EI} = \frac{P\delta^3}{EI} \tag{22.4}$$

In order to find the influence line for a load $W = 1$, the position of point $W = 1$ is varied through the length of the beam and each case is solved as an independent problem. The term

$$\frac{KB\delta^4}{EI} = \frac{KB}{EI}\left(\frac{L_B}{20}\right)^4 \tag{22.5}$$

is designated as m. The value of m can be varied from 0.0001–0.50 to represent the various cases of relative rigidity of the foundation with respect to the beam. Influence coefficients for bending moments shear and pressures have beem given by Ray in his publications.

3. **Finite element method.** In this method also, the beam is divided into a number of elements and a contact pressure is expressed in step loads. The flexibility matrix of the foundation due to this step loads is worked out for the appropriate nodal points, which is then inverted to get the stiffness matrix of the total system (foundation and structure). From this, the deflections of the nodal points can be worked out and hence the pressures. The finite element method and a similar finite grid method are illustrated

in the book *Analytical and Computer Methods in Foundation Engineering* by J.E. Bowles[20].

4. **By use of simultaneous equations.** Another method used to solve the problem is to match the deflections of the foundation and the beam at the nodal points using simultaneous equations. The pressures at the nodal points can be determined. The advantage of the method over finite element method is that it requires less computation.

5. **Iyengar's tables using series method.** Iyengar and Raman [10] have published influence coefficients for beams on elastic foundations by developing a solution to the differential equation in the form of a series. They assumed the solution for a short beam as containing the following four terms:

$$y = \begin{bmatrix} c_1 \sin \lambda x \sinh \lambda x + c_2 \sin \lambda x \cosh \lambda x \\ + c_3 \cos \lambda x \sinh \lambda x + c_4 \cos \lambda x \cosh \lambda x \end{bmatrix}$$

The solution is put in a series form as,

$$y(x) = \sum_{m=1}^{4} A_m \phi_m (x) \tag{22.6}$$

where A_m is a constant and is $\phi_m(x)$ a characteristic function representing the normal mode shapes of a freely vibrating beam, which represents the real beam. Influence coefficients are available for beams with λL = 0.10 to 9.0 in increments of 0.20 and these can be directly used for design calculations. In Iyengar's table, X denotes the position of load and Y denotes the distance from the left end of the point where the BM, SF and pressure are required.

6. **Wolfer tables using series method [11].** In this method, the problem of a beam on elastic foundation is represented in terms of the bending moment M.

$$EI(d^2y/dx^2) = -M \tag{22.7}$$

Also,

$$(d^2M/dx^2) = q = BKy$$

$$y = \left(\frac{1}{BK}\right)\left(d^2 M / dx^2\right) \tag{22.8}$$

We have,

$$\frac{1}{\lambda} = L_e = \left[\frac{4EI}{BK}\right]^{1/4} \tag{22.8A}$$

which is called *the elastic* or *characteristic length.* Therefore,

$$(1/BK) = (L_e^4/4EI)$$

From Eq. 22.8, we have

$$y = (L_e^4 / 4)(d^2 M / dx^2)$$

Differentiating twice, we get

$$EI\,(d^2 y / dx^2) = (L_e^4 / 4)(d^4 M / dx^4)$$

or

$$\left(\frac{L_e^4}{4}\right)\left(\frac{d^4 M}{dx^4}\right) + M = 0 \tag{22.9}$$

Putting back $1/L_e = \lambda$, we get the solution for the unit load as:

$$M = \begin{bmatrix} c_1 \cos \lambda x \cosh \lambda x + c_2 \sin \lambda x \sinh \lambda x \\ + c_3 \cos \lambda x \sinh \lambda x + c_4 \sin \lambda x \cosh \lambda x \end{bmatrix} \tag{22.10}$$

The equations are solved for various positions of load indicated as λ_L = 0–0.5 and λ_R = 0–0.5 in the Wolfer tables. In these tables,

$$\lambda_L = \frac{\text{Length of the beam to the left of load}}{\text{Elastic length}} \tag{22.11}$$

$$\lambda_R = \frac{\text{Length of the beam to the right of load}}{\text{Elastic length}}$$

Wolfer tables give the coefficients for bending moment, shear and contact pressures for various cases. With these publications, very detailed results can be obtained. This table is extensively used in Germany.

7. **Tables by Timoshenko and Gere [12].** Similar tables have been published by Timoshenko and Gere, which can be used for designing beams on elastic foundation using the Winkler model.
8. **By computer software.** The problem of beams on elastic foundation can be solved by the use of above mentioned tables (items 5, 6, 7 above) very easily. Otherwise, if a computer and the necessary software are available, the solution can be easily obtained. In any case it is advisable to have a background of the theory if the results are to be applied intelligently.

22.4.5 Limitations of Winkler Model and its Improvement

As we have already seen, the springs in the Winkler model act independent of each other so that there is no continuity at the ends of the foundation. As shown in Figure 22.4, in the real settlement of a foundation, there is always continuity and the *settlement extends beyond the ends of the beam.* This is a limitation of the Winkler model.

A number of other spring models, called *coupled models*, have been suggested to improve the simple model. But these only increase the complexity of the problem. It is generally felt that such improvements in the methods of computation are not warranted when our knowledge of the homogeneity of the soil as well as the modulus of subgrade reaction cannot be accurate.

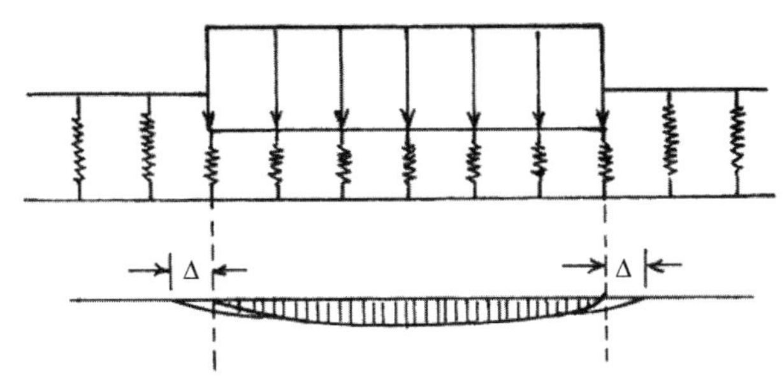

Figure 22.4 Winkler model for analysis of beams on elastic foundation. (*Note:* discrepency of deflection at the ends.)

What we can arrive at in any case is only a fair estimate, and never the exact values. Furthermore, the effect of settlements of the beam due to causes such as consolidation of soil is very large and it is to be further estimated for the final solution of the problem. Thus, if we want a more accurate solution, it is wiser to solve the problem by the recently developed and more realistic elastic half space model than to use improved spring models.

22.4.6 Approximate Values of Modulus of Subgrade Reaction (also called Subgrade Coefficients)

Textbooks dealing with soil mechanics should be referred for a full understanding of the difference between E_s, modulus of elasticity (deformation modulus) expressed in kg/cm^3 and K, modulus (coefficient) of subgrade reaction expressed in kg/cm^3. IS 2950, Part 1 (second revision) on the design of raft foundation [2] recommends the following values. (From Eq. (22.3), we can see that the higher the values of K, the larger will be λ and λL_B of Figure 22.2.)

TABLE 22.3 Approximate Values of Modulus of Subgrade Reaction [IS 2950 (1981, Second Revision) Table 1]
(a) Cohesionless Soils (values for 30 × 30 cm plate)

SPT value	*Relative density*	*K value* (kg/cm^3)	
		Dry on moist	*Submerged*
<10	Loose	1.5	0.9
10–30	Medium	1.5 to 4.7	0.9 to 2.9
>30	Dense	4.7 to 18.0	2.9 to 10.8

(b) Cohesive Soils (values for 30 × 30 cm plate)

q_u = *Unconfined camp: strength* (kg/cm^2)	*Consistency*	*K value* (kg/cm^3)	*SPT*
1–2	Stiff	1.2 to 2.7	<10
2–4	Very stiff	2.7 to 5.4	10 to 30
> 4	Hard	5.4 to 10.8	> 30

Notes: (a) 1 kg/cm^2 = 100 kN/m^2.

(b) For clays, a value of $K = 120q_u$ in kN/m^2 (where q_u is unconfined strength in kN/m^2) is commonly used for preliminary estimate of K.

(c) Estimate q_u from $q_u = 10N$ kN/m^2.

(d) We correct the value of K for a 30 × 30 cm plate for a plate $B \times B$ expressed in metres as follows:

$$K_{\text{corrected}} = \frac{K}{B} \text{ for clay} \tag{22.12a}$$

$$K_{\text{corrected}} = K\left(\frac{B+0.3}{2B}\right)^3 \text{ for sand} \tag{22.12b}$$

For a foundation of $B \times L$, the correction will be

$$K_{\text{corrected}} = K_{B\times B}\left(\frac{1+0.5B/L}{1.5}\right) \tag{22.12c}$$

(e) If the beam is a very long foundation with width B, this reduces to

$$K_{\text{corrected}} = 0.67K_{B\times B} \tag{22.12d}$$

(f) In a mat foundation, we take B as the average spacing of columns.

Example: Estimate the modulus of the subgrade value of a 30 cm square of a stiff clay of SPT value 10.

Unconfined strength $q_u = 10N$ kN/m^2 (or N/10 in kg/cm^2)

$$q_u = 10 \times 10 = 100 \text{ kN/m}^2$$

$$K = 120 \times q_u = 120 \times 100 = 1.2 \times 10^4 \text{ kN/m}^3 = 12 \text{ MN/m}^3$$

22.5 ELASTIC HALF-SPACE OR MODULUS OF COMPRESSIBILITY METHOD FOR ANALYSIS OF BEAMS ON ELASTIC FOUNDATION

The third method (as mentioned in Sec. 22.2) that is used to determine the contact pressure on a foundation is the elastic half space method, which is also known as the *stiffness coefficient method*. In the theory of consolidation, we have the equations

$$a_v = \Delta e/\Delta\sigma$$

$$E_{st} = \text{stress/strain} = \frac{\Delta\sigma}{(\Delta e/1+e)} = \frac{1+e}{a_v} = \frac{1}{m_v} \tag{22.13}$$

The quantity m_v is called the *modulus of compressibility* (or *coefficient of volume compressibility*). The reciprocal of the above modulus is called the *stiffness coefficient* E_{st}. In our usual analysis, E_{st} is considered as a constant along the depth of the soil. However, if necessary, improvements can be also made by assuming it as varying with depth.

We can imagine a three-dimensional space divided into two by an imaginary horizontal infinite plate (out of which the upper half is empty and the lower half is filled with an elastic homogeneous material which behaves elastic in all directions). Such a model is the elastic half space model. Fundamental deformation equations have been derived by Boussinesq and Schleicher for deformation of such a space in terms of the stiffness coefficient E_{st}.

For a beam of reinforced concrete, the problem reduces to a *beam of length* L_1, width B and thickness t, flexural stiffness D, Young's modulus E_1 and Poisson's ratio μ_1. For the *soil, the properties* are E_{st} and μ_2. Many cases of loading can be assumed, and for closed form solution, the easiest will be the case of a uniformly loaded strip load. The basis of the analysis is to match the displaced shape of the beam and the surface of the half space under the action of the applied loading.

Though closed form solutions for beams on elastic half space are available, the use of computers has made solution of these problems by numerical methods easier and more popular. The method frequently used is to cut the beam into a number of elements (say, 10) as shown in Figure 22.5. Each part is connected to the foundation below by a rigid bar through which the unknown contact pressure acts. The vertical deflection of each segment is affected by the contact pressure of the adjacent element also, and its magnitude can be worked out by the influence coefficient method.

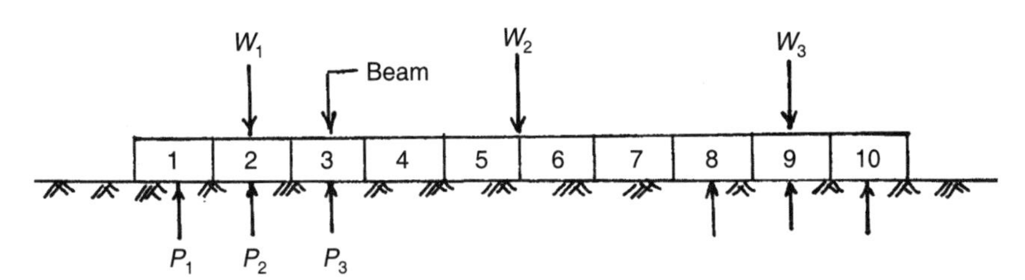

Figure 22.5 Determination of contact pressure under beams on elastic foundation.

We establish the compatibility conditions of deflection of external load and contact pressure as well as equilibrium condition between external loads and contact pressures by equations of the load. From a series of equations, the contact pressures can be determined. The book *Numerical Methods in Geotechnical Engineering* by Desai and Christin [13] covers this subject with examples and can be used as a reference.

In Europe, the work along these lines was started by Ohde, Kany and others around 1955. The publication by Kany [14] gives a clear account of the method used. The procedure can be also extended to rafts as a plate as explained in one of the recent books by Hemsley [15]. Another recent method for beams and plates on elastic foundation is the stiffness method of analysis by Cheung [16], [17], using the methods of matrix analysis of structures.

22.6 SIMPLIFIED ACI METHOD

The fourth method is the simplified ACI methods. Fritz Kramrisch and Paul Rogers described in 1961 [18] a procedure for the design of a certain category of beams (that we normally meet in practice) that are supported on Winkler foundation. This method has been recommended by ACI [19] and also included in IS 2950 (Part I), 1981 (Second Revision) [2]. It is to be noted that in this method, the total length of the beam between exterior columns should be at least $(3 \times 1.75)/\lambda = 5.25/\lambda$. Such beams are long and the end conditions do not influence each other. The simplified ACI method is described in detail in Chapter 23.

22.7 FORMULAE FOR CONTACT PRESSURES UNDER PERFECT RIGID STRUCTURES

The fifth method is to use readymade formulae. IS 2950 [(Part I), 1973, First revision] [2], [Appendix E] gives formulae derived from Boussinesq's equation for contact pressure distribution below rigid foundations, on elastic isotropic half space of depth not less than the width of the *rigid structure*. If the stresses are large, the calculated maximum values should be rounded off to the bearing capacity values and redistributed to maintain equilibrium conditions. Formulae are available for rigid circle, rectangle or strip foundations. As it uses Boussinesq's theory, the expressions are independent of the soil properties.

22.8 SELECTION OF SUITABLE MODEL FOR BEAMS ON ELASTIC FOUNDATIONS [K from E_s]

Vesic [6] has shown that for a long beam placed on an infinite depth of soil with a constant modulus of elasticity of soil, E_s and apparent modulus of subgrade reaction K, the value of KB is given by the following formula [see also Eq. (22.12)]. The value recommended by IS 2950 (Part 1), 1981, Cl. B-3.3 is,

$$KB = 0.65\left(\frac{E_s B^4}{E_B I}\right)^{1/12}\left(\frac{E_s}{1-\mu^2}\right) \tag{22.14}$$

where

K = Modulus of subgrade reaction in kN/m^3
E_s = Young's modulus of the soil
E_B = Young's modulus of beam [$E_B I$ is the rigidity of the beam]
B = Width of the beam
I = MI of beam cross-section
μ = Poisson's ratio of soil

The theoretical works by Gibson [21] as well as by Carrier and Christian [22] have shown that, when the E_s values of the soil are assumed to increase with depth, the settlement of the beam behaves according to the Winkler theory. Massalskii [23] and Ward et al. [24] give experimental evidence for such behaviour. These investigations show that improvements in the mathematical model are not as important as choosing the value of the soil parameters for the modulus and also the procedure to include the right rigidity of the structure in the analysis. A more recent approach to the problem is to use the finite element method to model the soil to obtain an apparent spring constant, which can then be used for easy structural analysis. Actual field measurements and comparison with analysis by the two methods (Winkler model and Elastic half space model) have revealed the following:

1. Both models give fairly good results, which are conservative.
2. The superstructure in many cases gives more stiffness to the foundation than usually assumed in the analysis.
3. The highly indeterminate foundation is not as much stressed as indicated by both types of analysis.

These results confirm that regardless of how the foundation is modelled, the important and difficult problem is the selection of the soil parameters.

22.9 ANALYSIS OF WINKLER AND ELASTIC HALF SPACE MODEL BY COMPUTERS

The modulus of subgrade reaction method was popular till recently because of its simplicity and availability of published values of influence coefficients. However, with the availability of computers, the elastic half space method has also now become popular. Many large organizations use computer-aided design programmes based on this elastic theory. Any complex foundation system can now be modelled by the finite element method to give the resultant contact pressures and other design parameters. Such an analysis is needed for complex cases but for ordinary cases and regular layouts, approximate solutions give satisfactory results. However, the determination of exact soil parameters needed for a correct solution is a serious problem yet to be solved.

22.10 EFFECT OF CONSOLIDATION SETTLEMENT

When the foundation is on clayey soils, the effect of consolidation settlements, which is non-linear as well as time-dependant character, should also be taken into account in designs. Terzaghi's theory, or a modified form of his theory, as given in Appendix A, is used for the calculation of the consolidation effects. We should remember that it is most important that we must represent the ultimated deformability of the soil somehow in calculations. All the methods of soil structure interaction finally reduce to the problem as to how to represent the soil deformation and how to represent the structure. Because of the consolidation effects in clayey soils, all the present representation of the stiffness of the soil is only very approximate imitations of the reality. Similarly, the superstructure has rigidity and its rigidity has also to be represented. Thus, we should be aware that very sophisticated improvements in the analysis of a raft foundation as a structure are not warranted unless equal improvements are made in modelling the soil also. One of the methods is to use E_{st} values, shown in Sec. 22.5, which will include consolidation settlement also.

22.11 LIMITATIONS OF THE THEORY

Footings, which are planned as rigid members, can be solved as rigid beams on elastic foundations. However rafts are treated as plates on elastic foundations. Cutting rafts into beams and solving them as beams on elastic foundation is an approximation that is applicable only to the design of rafts with a regular layout of columns. For irregular layout of rafts, computer methods have to be resorted to [15].

22.12 SUMMARY

This chapter provided a mathematical definition of elastic length L_e (or characteristic coefficient λ) of a beam on elastic foundation and showed how this parameter could be used to find whether the beam would behave as a rigid, semi-rigid or flexible foundation.

EXAMPLE 22.1 (Analysis of a beam on elastic foundation)

A combined rectangular footing in medium dense sand is 7.5 m in length, 1.3 m in width and 450 mm in depth. Loads of 750 kN are placed at 2.5 m at LHS and a load of 300 kN at 0.6 m from the RHS. Find the BM and SF in the beam.

Reference	*Step*	*Calculation*
	1	*Estimate subgrade modulus* From Table 22.3, $K = 5 \times 10^4$ kN/m^3 (assumed) $K_{\text{corrected}}$ for 1.3 × 7.5 m size [Eq. (22.12)] $= K\left(\dfrac{B+0.3}{2B}\right)^2\left(\dfrac{1+0.5B/L}{1.5}\right)$ $= 5\times10^4\left(\dfrac{1.6}{2.6}\right)^2\left(\dfrac{1+(0.5\times1.3)/7.5}{1.5}\right)$ $= 5 \times 10^4[0.379][0.72] = 1.36 \times 10^4$ kN/m^3 [We can also estimate *KB* from Eq. (22.14)]
	2	*Find λ and λL* Assume $E_c = 20 \times 10^6$ kN/m^2 $I = 1.3(0.45)^3/12 = 9.9 \times 10^{-3}$ m^4 $\lambda = \sqrt[4]{KB/4EI} = \sqrt[4]{\dfrac{1.36\times10^4\times1.3}{4\times20\times10^6\times9.9\times10^{-3}}} = 0.386$
Eq. (22.3)		$\lambda L = 0.386 \times 7.5 = 2.86 > 1.75$ and < 4.7 The beam is a flexible beam of finite length (not a flexible infinitely long beam or a rigid beam)
Sec. 22.4		(Any of the methods given in Sec. 22.4 are applicable, but the problem can be solved *easily by* values of coefficients for BM from Tables referred in Sec. 22.4.4, items 5, 6, and 7 or by computers with software.)

REFERENCES

[1] Schultz, E., *Distribution of Stress Beneath a Rigid Foundation*, Technische Hochschule Achen, Germany, 1971.

[2] IS 2950, *Part I–Code of Practice for Design and Construction of Raft Foundation* (First Revision, 1973 and Second Revision, 1981), BIS, New Delhi.

[3] Hetenyi, M., *Beams on Elastic Foundation*, University of Michigan Press, Michigan, U.S.A., 1945.

[4] Hetenyi, M.A., 'General Solution for the Bending of Beams on Elastic Foundations by Arbitrary Continuity', *Journal of Applied Physics*, Vol. 21, 1950.

[5] Seeley, F.B. and J.O. Smith, *Advanced Mechanics of Materials*, John Wiley and Sons, London, 1952.

[6] Vesic, Alexander, Beams on Elastic Sub-grade and Winkler Hypothesis, Proceedings of the 5th International Conference on Soil Mechanics and Foundation Engineering, Paris, 1961.

[7] Ohde, *Calculation of Soil Pressure Distribution under Foundations* (in German), Der Bawingenicur, Bawng, Berlin, 1942.

[8] Ray, K.C., Finite Beams on Elastic Soils Influence Lines, *Journal of Institution of Engineers* (India), Vol. 38, Oct. 1957.

[9] Ray, K.C., 'Influence Lines for Pressure Distribution under a Finite Beam on Elastic Foundation', *Journal of the American Concrete Institute*, Detroit, Dec. 1958.

[10] Iyengar, K.T.S. and Raman S. Anantha, *Design Tables for Beams on Elastic Foundations*, Applied Sciences Publishers, London, 1979.

[11] Wolfer, K.H., *Beams on Elastic Foundation, Influence Tables* (English Edition), Bauverlag, GMBH Wiesbaden and Berlin, 1978.

[12] Timoshenko and M. Gere, *Beams on Elastic Foundation, Charts and Tables of Coefficients*, McGraw Hill, New York, 1961.

[13] Desai, C.S. and T.T. Christin, *Numerical Methods in Geotechnical Engineering*, McGraw-Hill, New York, 1977.

[14] Kany, *Calculation of Raft Foundation* (in German), Verlag W. Ernst and Sohn, Berlin, 1959.

[15] Hemsley, J.A., *Elastic Analysis of Raft Foundations*, Thomson Teltord, London, 1998.

[16] Cheung, Y.K. and O.C. Zienkiewiez, 'Plates and Tanks on Elastic Foundation: An Application of Finite Element Method', *International Journal of Solids and Structures*, 1965.

[17] Cheung, Y.K. and D.K. Nag, Plates and Beams on Elastic Foundations: Linear and Non-Linear Behaviour, *Geotechnique*, 1968.

[18] Kramrisch, F. and Paul Rogers, Simplified Design of Combined Footings, *Journal of the Soil Mechanics and Foundation Division*, SM 5, ASCE, Oct. 1961.

[19] ACI Committee 436, Suggested Design Procedures for Combined Footings and Mats, *Journal ACI*, Oct. 1966.

[20] Bowles, J.E., *Analytical and Computer Methods in Foundation Engineering*, McGraw-Hill, New York, 1974.

[21] Gibson, R.E., Some Results Governing Displacements on a Non-homogeneous Elastic Half Space, *Geotechnique*, Vol. 17, 1967.

[22] Carrier, W.D. and J.T. Christian, 'Analysis of an Inhomogeneous Elastic Half Space', *Journal of Soil Mechanics and Foundation Division*, ASCE, March 1973.

[23] Massalskii, E.K., Experimental Study of Flexible Beam on Sand Foundation, *Soil Mechanics and Foundation Engineering*, No. 6, ASCE, 1964.

[24] Ward et al., Geotechnical Assessment of a Site at Mudford Nortolk for a Large Proton Accelerator, *Geotechnique*, Vol. 18, 1968.

ACI Method for Analysis of Beams and Grids on Elastic Foundations

23.1 INTRODUCTION

In this chapter, we deal with the ACI simplified method of analysis of beams on elastic foundation. IS 2950 [Part 1 (Second Revision), 1981 [1]] on design of raft foundations has also adopted the ACI Committee 436 [2] recommendations of the use of a simplified method for solving beams with column loads on the Winkler foundation (as in grid foundation) obeying the following four conditions:

1. There should be a minimum of three bays [four column loads on the beam].
2. Variation of adjacent span lengths should not be more than 20%.
3. The average column spacing should be such that the value of λL_B *should not be less than* 1.75 *and should not be more than* 3.50 (see Figure A.4). (L_B = average span length)
4. The column loads should not vary by more than 20%.

Rafts with regular column layouts can be cut into beams on elastic foundation and analysed by this method.

23.2 DERIVATION OF THE METHOD

We briefly derive the formulae and study an example to see how it is applied to determine the pressures and moments in a beam on elastic foundation. The full derivation is available in Ref. [3].

1. **Formula for value of negative moment below column load point.** Let $1/\lambda = L_e$. We found in Chapter 22 [Eq. (22.2d)] that the bending moment at x of an infinitely long beam on elastic foundation due to a column load W_1 at point C can be written in the following form (see Figure 23.1):

$$M_x = \frac{W_1}{4\lambda} C_{\lambda x}$$

where

$$C_{\lambda x} = e^{-\lambda x} (\cos \lambda x - \sin \lambda x)$$

Thus, if there are several loads W_1, W_2, ..., the influence of the various loads for a point C can be represented as follows:

$$M_e = \frac{W_1}{4\lambda} C_{\lambda x1} + \frac{W_2}{4\lambda} C_{\lambda x2} + \frac{W_3}{4\lambda} C_{\lambda x3} \tag{23.1}$$

where x_1, x_2, x_3 are the distances of loads W_1, W_2, W_3 from point C.

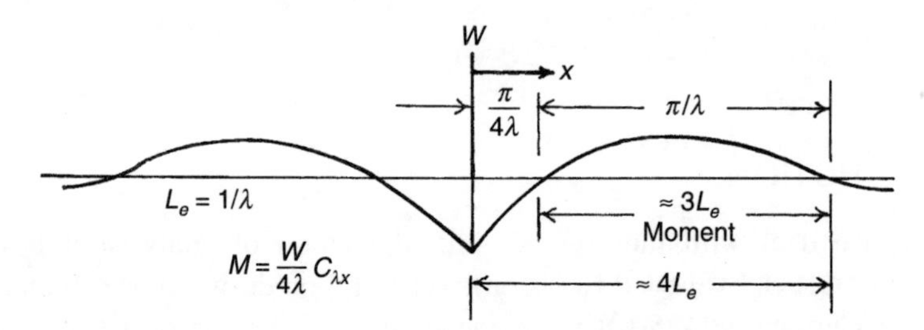

Figure 23.1 Length of influence on bending moment due to a load on a beam on an elastic foundation. (It extends to only for a length approximately equal to $4L_e$.)

Now, if we restrict the spacing of the loads L_A *to be between* $1.75/\lambda$ *and* $3.50/\lambda$, the trigonometric function for $C_{\lambda x}$ can be approximated to a straight line. Again, if the loads are more or less equal, then the negative moment under an internal column M_i, as shown in Figure 23.2, can be written down as follows [3]:

$$M_i = \frac{W_1}{4\lambda}(0.24\lambda L_A + 0.16) \not> -\frac{W_1 L_A}{12} \tag{23.2}$$

where

W_1 = Column load at point
L_A = Average length of the adjacent spans or average column spacing
λ = Characteristic length

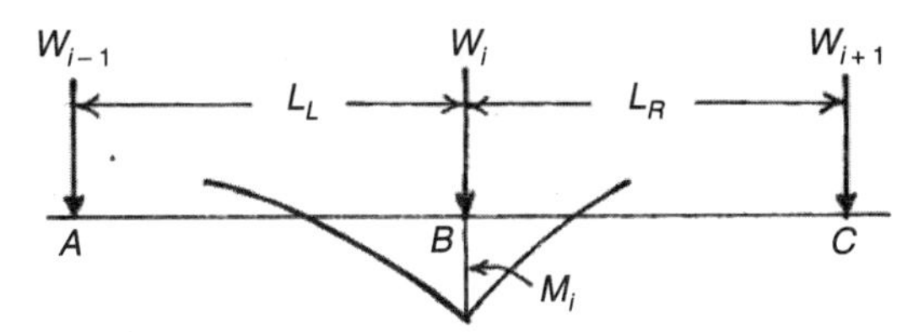

Figure 23.2 Negative moment under an interval column on a beam on elastic foundation.

The right-hand side of the equation shows that the *negative bending moment* at the support cannot be greater than the value for a continuous beam with uniform reaction as in a rigid beam.

The lower limit for the application of the above formula can be evaluated as $1.78/\lambda$, which can be taken for simplicity as $1.75/\lambda$. For column spacings that are closer, moments will be greater than those in Eq. (23.2) as it will behave as a rigid beam. In any case (we know from the case of a fixed beam), the maximum value at supports can be only less than $WL_\lambda/12$ as for a rigid beam. On the other hand, if spacing is greater than $3.5/\lambda$, then the influence of one load at the column has no influence at the adjacent column point as the beam is very flexible, as shown in Figure 23.1.

In his original paper, Kramrisch [3] showed by worked out examples that for values between $1.75/\lambda$ and $3.50/\lambda$ with equal loads and equal spacing and with varying ratios of stiffness of foundation and stiffness of structure, the greatest deviation from exact value of M_i from Eq. (23.2) is only about 3.5%. Even with a deviation of 20% in the column spacing as well as column loads, we obtain only a variation within 16% with considerably smaller average deviation. These variations are well within the range of redistribution on moments possible in reinforced concrete structures and are acceptable for design.

2. **Contact pressures in interior spans.** Our next aim is to determine the contact pressures in the interior span so that we can calculate the total span moment M_0 due to contact pressures. For this purpose, we assume (for the limitations specified by item 3 in Sec. 23.1) a linear distribution of the soil pressure with a maximum value below the load point (columns) and a minimum at the centre of adjacent spans, as shown in Figure 23.3.

   We also assume that the negative moment M_i is not affected much by the additional loads beyond the adjacent columns and remains as given by Eq. (23.2). Let

   p_i = Contact pressure below column i (Maximum value)

   p_l = Contact pressure at the centre of span left of i where we consider load W_i

   p_r = Contact pressure at the centre of span right of $i - 1$ when we consider load W_{i-1}

   $p_m = (p_l + p_r)/2$

   L_L = Length of left span from column i

   L_R = Length of right span from column i

   $L_A = (L_L + L_R)/2$

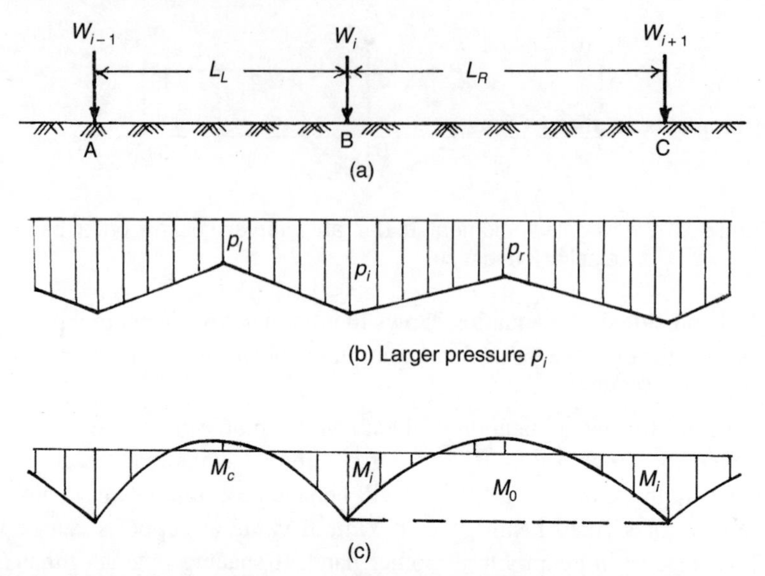

Figure 23.3 Beam resting on an elastic foundation and subjected to column (concentrated loads). (a) Beam, (b) contact pressure, (c) bending moment.

Case 1: Let us first take the case of equal spans equal to L_A and equal loads. Let low pressure values $p_l = p_r = p_m$. We can obtain the magnitude of the negative moment by considering *AB* and *BC* in Figure 23.3 as fixed beams.

$$M_i = \frac{p_m L_A^2}{12} + \frac{(p_i - p_m)L_A^2}{32} = \frac{L_A^2}{96}(3p_i + 5p_m) \tag{23.3}$$

We can see from Figure 23.3 and Figure 23.4 (page 340) that

$$W_i = L_A(p_m + p_i)/2 \quad \text{or} \quad p_m = \frac{2W_i}{L_A} - p_i \tag{23.4}$$

Using Eqs. (23.4) and (23.3), we get the value of p_i, the pressure under W_i as

$$p_i = \frac{5W_i}{L_A} + \frac{48M_i}{L_A^2} \quad (M_i \text{ is to be taken as negative}) \tag{23.5}$$

From Eqs. (23.4) and (23.5), we obtain

$$p_m = \frac{-48M_i}{L_A^2} - \frac{3W_i}{L_A} \tag{23.5a}$$

Case 2: Now, let us take the case where L_L and L_R are different.

Again, let us find W_i in relation to the values of p_l and p_r, on either side of p_i at the mid-span points. When spans are of different L_L and L_R, by equating reactions to W_i, we get

$$W_i = \left(\frac{p_l + p_i}{2}\right)\frac{L_L}{2} + \left(\frac{p_i + p_r}{2}\right)\frac{L_R}{2} \quad (23.6)$$

By taking moments so that CG of the area fall along W_i, we obtain

$$p_l = \frac{2W_i}{L_A}\left(\frac{L_R}{L_L}\right) - p_i\left(\frac{L_A}{L_L}\right) \quad (23.6a)$$

$$p_r = \frac{2W_i}{L_A}\left(\frac{L_L}{L_R}\right) - p_i\left(\frac{L_A}{L_R}\right) \quad (23.6b)$$

When the spans are equal,

$$p_l = p_r = p_m = \frac{p_l + p_r}{2} = (2W_i / L_A) - p_i \quad (23.6c)$$

$$p_m = -\frac{3W_i}{L_A} - \left(\frac{48M_i}{L_A^2}\right) = \frac{-48M_i}{L_A^2} - \frac{3W_i}{L_A} \quad (23.6d)$$

Equations (23.6a) and (23.6b) give the required pressure distribution when L_L and L_R are different.

In actual practice, if there is a variation in either the load or the span length, we first find the mid-point pressure in a span first as caused by the load on the right, and then as caused by the load on the left. The mean of the two values is then assigned to the mid-point for the calculation of M_0 in the span.

3. **Total positive moment in interior spans.** When we have the ordinates of the pressure diagram in a span L_1 (as shown in Figure 23.4 with ordinates p_1, p_c and p_2), the value of the total span moment can be calculated as M_0 and will reduce to the following:

$$M_0 = \frac{L_1^2}{48}(p_1 + 4p_c + p_2) \quad (23.7)$$

where

L_1 = Span
p_1 = Pressure ordinate on the left of span
p_2 = Pressure ordinate on the right of span
p_c = Pressure ordinate at the centre of span

The positive moment M_c at the centre of the beam, as shown in Figure 23.3, is obtained from the following relation:

$$M_0 = M_c + M_i \quad (23.8)$$

where

M_0 = Simply supported moment at the centre
M_1 = Average negative moment under the columns at each end of the bay shown in the Figure 23.3
$M_i = (M_{il} + M_{ir})/2$

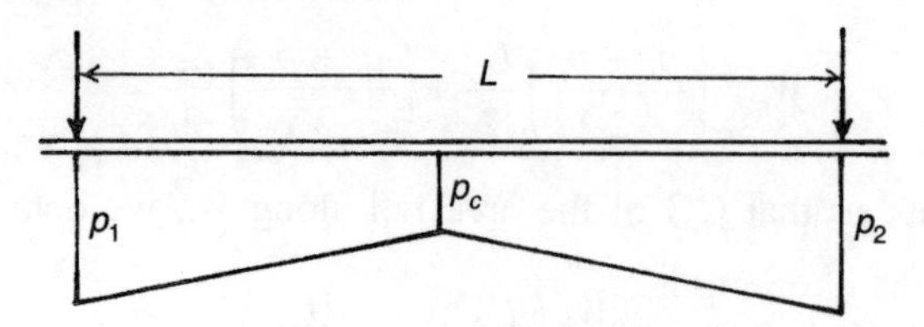

Figure 23.4 Calculation of total span moment M_0 from contact pressure in the span.

4. **Moments in exterior span of beam.** For determining the distribution of contact pressure under the external column, we make the following two assumptions:

   (a) The load of the exterior column is equal to the resultant of the subgrade reaction starting from the centre of the first bay to the end of the beam, i.e. to the tip of the projection as in Figure 23.5.

   (b) The contact pressure from the edge to the centre of the adjacent span will have a maximum value under the external column load and a minimum at the tip of the projections, and also another minimum at the centre of the first bay as shown in Figure 23.5. The distribution is assumed linear. The contact pressure at the tip of the projection cannot exceed this value under the exterior column.

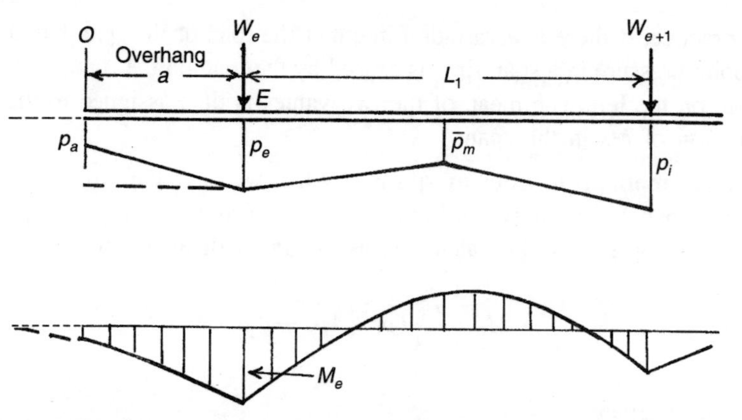

Figure 23.5 Behaviour of end span of a beam or elastic foundation loaded with concentrated loads.

Let us now take the exterior span, which is different from the interior spans. We also assume the projection is less than the elastic length L_e. Using a procedure similar to that explained above, Kramrisch has shown that the moment under the external column (E in Figure 23.5) which has a projection can be found as explained below [3].

The moment under the external column is due to three causes: (i) the load itself, (ii) the influence of the next interior column, and (iii) the influence of the free end. If we take M_e as the moment under the exterior column, the effect of the first two causes above is to produce a moment as given below (for derivation, see Refs. [2] and [3]):

$$M_e = -\frac{W_e}{4\lambda}(0.13\lambda L_1 + 1.06\lambda a - 0.50) \tag{23.9}$$

where a is the length of the projection and L_1 is the length of the exterior span. If we again assume that the length of projection is very small, it may require high subgrade reaction to produce the above value of moment. However, the maximum value of p_a can be only equal to the pressure under the external column. By equating the load and pressure, we get the maximum value of p_e as follows (using symbols shown in Figure 23.5):

$$W_e = p_e a + \frac{1}{2}(pm_e + \bar{p}_m)(L_1/2)$$

Hence,

$$p_e = (4W_e - \bar{p}_m L_1)/(4a + L_1) \tag{23.10}$$

where $\bar{p}_m$, as shown in Figure 23.5, is the pressure at the mid-point of the first interior span. Using the above value of p_e, the maximum possible value of M_e will be given by the following equation:

$$M_e = -p_e(a^2/2) = \frac{(4W_e - \bar{p}_m L_1)}{(4a + L_1)}\left(\frac{a^2}{2}\right) \tag{23.11}$$

The lesser of the two values obtained from Eqs. (23.9) and (23.11) is to be taken as the design value for M_e.

5. **Contact pressure under exterior columns.** The contact pressure in *external span* works out as follows. It can be shown that if Eq. (23.9) governs, referring to Figure 23.5, we get the following values for p_e and p_a, where $p_e > p_a$:

$$p_e = \left(4W_e + \frac{6M_e}{a} - p_m L_1\right)\Big/(a + L_1) \tag{23.12}$$

$$p_a = -\frac{(3M_e)}{a^2} - \frac{p_e}{2} \ (M_e \text{ is negative}) \tag{23.12a}$$

If Eq. (23.11) governs, then Eq. (23.1) gives the value of the end pressure as follows when ($p_e = p_a$):

$$p_e = (4W_e - p_m L_1)/(4a + L_1) \tag{23.12b}$$

The procedure using the above method is illustrated by Example 23.1.

23.3 DESIGN PROCEDURE

The following steps, as given in IS 2950 (1981), Appendix E, are recommended in the design of beams on elastic foundation by the above ACI method.

1. Consider interior spans

Step 1: Calculate λ and the elastic length L_e and *check whether the four conditions for ACI method*, given in Sec. 23.1, are applicable.

Let

$$\lambda = \left(\frac{KB}{4EI}\right)^{1/4} = \frac{1}{L_e} \quad \text{(From Eq. (22.3))}$$

Step 2: Calculate the negative moment of the interior span using Eq. (23.2),

$$M_i = -\frac{W_i}{4\lambda}[0.24\lambda L_A + 0.16] \ngtr -\frac{W_i L_A}{12}$$

Step 3: Determine the *maximum contact pressure* p_i under load using Eq. (23.5), taking M_i as negative.

$$p_i = \frac{5W_i}{L_A} + \frac{48M_i}{L_A^2}$$

Step 4: Determine the *minimum contact pressures* at the centres of the left and right spans (interior spans) from Eqs. (23.6a) and (23.6b) respectively.

$$p_l = \frac{2W_i}{L_A}\left(\frac{L_R}{L_L}\right) - p_i\left(\frac{L_A}{L_L}\right)$$

$$p_r = \frac{2W_i}{L_A}\left(\frac{L_L}{L_R}\right) - p_i\left(\frac{L_A}{L_R}\right)$$

When spans are equal, using Eq. (23.5a), we get

$$p_m = \frac{2W_i}{L_A} - p_i = \frac{-48M_i}{L_A^2} - \frac{3W_i}{L_A}$$

In practice, if the column loads are different, we first find the values of the pressure for the point as obtained from the load situated on the right of the point and then find the value of the pressure for the same point from the load situated on the left of the point and take the mean value as the real value.

$$p_m = (p_{mr} + p_{ml})/2$$

Step 5: Calculate M_0 using Eq. (23.7).

$$M_0 = (L_A^2/48)(p_l + 4p_c + p_r)$$

Step 6: Determine the central moment M_c using Eq. (23.8).

$$M_c = M_0 - M_i$$

2. Consider exterior span

Step 7: Find moment under exterior column using Eq. (23.9).

$$M_e = -\frac{W_e}{4\lambda}(0.13\lambda L_1 + 1.06\lambda a - 0.50)$$

Step 8: Find pressure p_e under exterior load using Eq. (23.12) (M_e to be taken as –ve).

$$p_e = \left(4W_e + \frac{6M_e}{a} - p_m L_1\right) \Big/ (a + L_1)$$

Find also pressure p_a at exterior load using Eq. (23.12a) (M_e to be taken as –ve).

$$p_a = (3M_e/a^2) - (p_e/2)$$

Step 9: Find M_e assuming $p_a = p_e$ in Eq. (23.12b) to get

$$p_a = p_e = (4W_e - p_m L_1)/(4a + L_1)$$

$$M_e = -\frac{p_e a^2}{2} \quad \text{from Eq. (23.11)}$$

Take the lesser value from Steps 7 and 9 for the design.

Step 10: Determine the positive moment in exterior spans.
M_e and M_i are known. From Eq. (23.7), we get the following:

$$M_0 = \frac{L^2}{48}(p_e + 4p_c + p_i)$$

$$M_{ce} \text{ +ve} = M_0 - \text{(average moment below of 2 adjacent supports)}$$

Step 11: Sketch the pressure and moment diagrams.

These steps are illustrated in Example 23.1.

23.4 ANALYSIS OF GRID FOUNDATIONS

ACI method is very suitable for the analysis of grid foundations. The grid foundation consists of a system of intersecting strip foundations as shown in Figure 23.6. In most such cases, an inverted *T* beam type of foundation with slab below or U beam with slab above will be found to be more rigid and easy to construct.

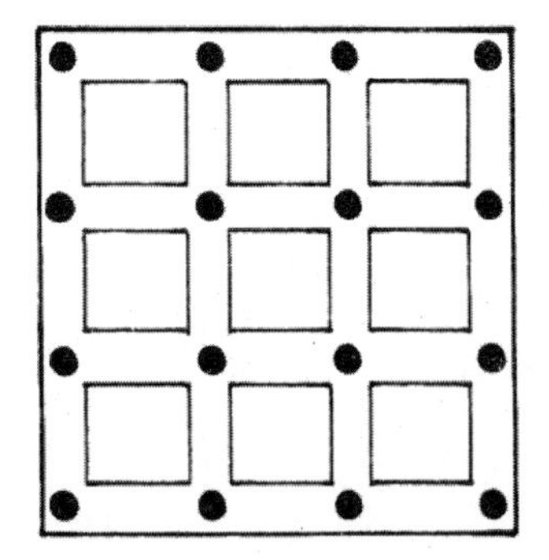

Figure 23.6 Layout of grid foundation.

The Report of ACI Committee 436 recommends '*grid foundations to be analysed as independent strips using column loads proportional in direct ratio to the stiffness of the strips acting in each direction*' [2]. In this method, each column load will be *proportioned in each direction in the inverse proportion to the distance to the next column point*. The strips need to be designed only with these assigned column loads. It is claimed that this grid action occurs in grid arrangement occupying even up to 70% of the plan area.

However, in the example worked out by Fritz Kramrisch and Rogers in their paper [3] from which the above simplified ACI method was evolved, the authors have analysed a grid foundation using the full load in each column taken in both directions for illustration of this method. This is an extension of the method of the raft design as an inverted floor in which the raft is divided into strips in the *X*- and *Y*-directions and each strip designed for full column loads in both directions. The second method gives a good conservative design of the grid system.

23.5 SUMMARY

Beams on elastic foundation which obey certain limitations can be analysed by a simple procedure proposed by ACI. This chapter explains this method in detail.

EXAMPLE 23.1 (ACI method of analysis of beams on elastic foundation)

A slab 450 mm thick and 5.4 m breadth supports four column loads at a spacing of 5.4 m and overlong of 1.5 m as shown in Figure E23.1. Assuming a coefficient of subgrade reaction of 3.7 kg/cm^3 (3700 t/m^3) and E of concrete as 20 × 10^4 kg/cm^2, find the BM and pressure distribution using the ACI method of analysis of beams on an elastic foundation.

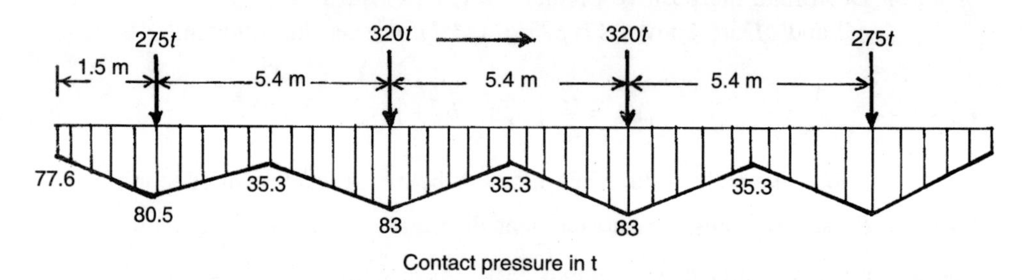

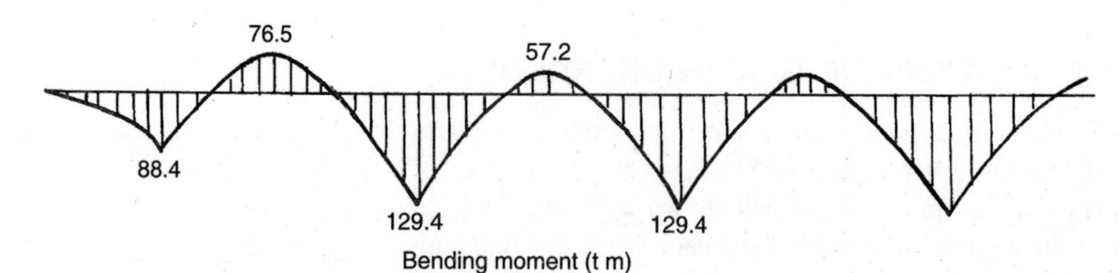

Figure E23.1

Reference	*Step*	*Calculation*
Eq. (22.8A)	1	*Calculate elastic length and check the four conditions*
		$L_e = \text{elastic length} = \left(\frac{4EI}{KB}\right)^{1/4} = 1/\lambda$

Reference	*Step*	*Calculation*
Sec. 23.3		Span length L_A = 5.4 m and b = 5.4 m; d = 0.45 m
		Value of $I = \frac{5.4(0.45)^3}{12} = 0.041\ \text{m}^4$
Given		K = 3700 t/m³ (3.7 kg/cm³); E_c = 20 × 10⁴ kg/cm² (20 × 10⁵ t/m²)
Eq. (22.8A)		$\frac{1}{\lambda} = L_e = \left(\frac{4 \times 20 \times 10^5 \times 0.041}{5.4 \times 3700}\right)^{1/4} = 2.01\ \text{m};\ \lambda = \frac{1}{2.01}$
		$\lambda L_A = \frac{L_A}{L_e} = 5.4/2.01 = 2.68$
Sec. 23.1		2.68 > 1.75 and < 3.5
		ACI method is valid as all the necessary four conditions given in Sec. 23.1 are satisfied.
		(a) Consider interior spans
	2	*Calculate M_i negative moment for interior span (load* 320 t)
Eq. (23.2)		$M_i = -\frac{W_i}{4\lambda}(0.24\lambda L_A + 0.16)$
		$= \frac{320 \times (2.01)}{4}\left(\frac{0.24 \times 5.4}{2.0} + 0.16\right) = -129.4\ \text{tm}$
		Largest value ($W_iL/12$) = –144 tm
		129.4 being lesser than 144 is admissible.
	3	*Determine maximum pressure p_i (per metre length)*
Eq. (23.5)		$p_i = \frac{5W_i}{L_A} + \frac{48M_i}{L_A^2}$ (M_i is negative)
		$= \frac{5 \times 320}{5.4} - \frac{48 \times 129.4}{(5.4)^2} = 83$ t/m for width B
		Pressure $= \frac{83}{B \times 1}\ \text{t/m}^2 = \frac{83}{5.4} = 15.37\ \text{t/m}^2$
		(Add to this the self-weight of slab as UDL)
		Self-weight = 0.450 × 2.4 = 1.08 t/m²
		Total pressure p_i = 16.45 t/m²
	4	*Determine the minimum pressure at centre span*
Eq. (23.6d)		$p_l = p_r = p_m = -\frac{48M_i}{L_A^2} - \frac{3W}{L_A}$ (as spans are equal)
		(As M_i is negative) $= \frac{48M_i}{L_A^2} - \frac{3W}{L_A}$
Step 2		$= \frac{48 \times 129.4}{(5.4)^2} - \frac{3 \times 320}{5.4}$ (Interior spans load 320 t)

Reference	*Step*	*Calculation*
		= 35.3 t/m for width
		With self wt. $= \dfrac{35.3}{5.4} + 1.08 = 7.62\ \text{t/m}^2$
	5	*Calculate M_0*
Eq. (23.7)		$M_0 = (L^2/48)\ (p_l + 4p_c + p_R)$ (where p is to be in t/m)
		$= \dfrac{(5.4)^2}{48}[83 + (4 \times 35.3) + 83]$
		= 186.6 tm
	6	*Determine M_c (positive moment)*
Eq. (23.8)		M_c = 186.6 – 129.4 = +57.2 t/m
		(Step 2. The largest value of $M_1 = M_2$ = 144 tm)
		(b) Consider exterior spans
	7	*Find M_e under exterior column by Eq.* (23.9)
		Load = 275 t; over hand = a = 1.5 ms; interior span = L_1 = 5.4 m
Eq. (23.9)		$M_e = -\dfrac{W_e}{4}(L_e)[0.13(L_1/L_e) + 1.06(a/L_e) - 0.50]$
		where $L_e = 1/\lambda$ characteristic length = 2.01 m
		$= -\left(\dfrac{275 \times 2.01}{4}\right)\left(\dfrac{0.13 \times 5.4}{2.01} + \dfrac{1.06 \times 1.5}{2.01} - 0.50\right)$
		= –88.44 t/m (first value)
	8	*Find pressure at exterior load using Eq.* (23.12)
Eq. (23.12)		$p_e = \left(4W_e + \dfrac{6M_c}{a} - p_m L_1\right) \Big/ (a + L_1)$ (with sign as M_c negative)
		$= \left(4 \times 225 - \dfrac{6 \times 88.44}{15} - 35.3 \times 5.4\right) \Big/ (0.5 + 5.4)$
		p_e = 80.55 t/m (for width B)
		Pressure $= \dfrac{80.55}{5.4} + 1.08 = 15.99\ \text{t/m}^2$
		Find also pressure at the exterior end (with M_e as negative)
Eq. (23.12a)		$p_a = -(3M_e/a^2) - (p_e/2)$
		$= \dfrac{3 \times 88.44}{(1.5)^2} - \dfrac{80.55}{2} = 77.66\ \text{t/m}$
		Pressure $= \dfrac{77.66}{5.4} - 1.08 = 15.46\ \text{t/m}^2$
	9	*Again find M_e by Eq. (23.12b) assuming $p_a = p_c$*

Reference	Step	Calculation
		$p_a = p_e = (4W_e - p_m L_1)/(4a + L_1)$
		(p_m in Step 4 = 35.3 t/m)
Eq. (23.12b)		$p_e = \dfrac{(4 \times 275 - 35.3 \times 5.4)}{(4 \times 1.5 + 5.4)} = 79.79$ t/m
		$M_e = -\dfrac{p_e a^2}{2} = -\dfrac{79.79 \times (1.5)^2}{2} = -89.76$ tm (second value)
		M_e is lesser of Eqs. (23.9) and (23.11) L_e Steps 7 and 9
		= –88.4 tm
Step 2	10	*Determine positive moments in the exterior spans*
		$M_e = -88.44$ and $M_i = -129.4$ (for exterior span)
Eq. (23.7)		$M_0 = \dfrac{L^2}{48}(p_e + 4p_c + p_i)$
		$= \dfrac{(5.4)^2 \times [80.55 + (4 \times 35.3) + 83]}{48}$
		= 185.4 tm
As in Eq. (23.8)		$M_c = 185.4 - \left(\dfrac{88.44 + 129.4}{2}\right) = 76.5$ tm
		(+ve moment external span)
	11	*Sketch the contact pressure BM and SF diagrams*
		These are drawn as Figure E23.1.
		(*Note*: Design of footing to be based on moment at the face of the column only and not at the centre of the column.)

Notes:

1. This problem can be worked by using Tables giving coefficients for beams on elastic foundation available in Refs. 8 to 12 of Chapter 22 with much more ease and accuracy. No lengthy calculations are necessary and restrictions of Sec. 23.1 do not apply to these tables.
2. Even though we carry out the analysis using centre-to-centre distances for detailing of steel we need to consider only the moments at the face of the columns.

REFERENCES

[1] IS 2950, Part I (Second Revision) 1981, *Code of Practice for Design and Construction of Raft Foundation*, Bureau of Indian Standards, New Delhi.

[2] American Concrete Institute, *Suggested Design Procedure for Combined Footings and Mats*, Report of ACI Committee 436, Detroit, Michigan, 1966.

[3] Kramrisch, F. and P. Rogers, Simplified Design of Combined Footings, *Journal of the Soil Mechanics and Foundation Division, Proc. ASCE*, Oct. 1961.

Analysis of Flexible Plates on Elastic Foundations

24.1 INTRODUCTION

We have found that an approximate analysis of beams and plates on elastic foundation can be made by one of the following methods, depending on its relative rigidity and column spacing L_B:

1. As a rigid beam when column spacing L_B and dimensions of beam are such that λL_B is less than 1.75.
2. Using influence tables if $\lambda L_B > 1.75 < 4.7$.
3. By ACI approximate method if λL_B is between 1.75 and 3.5 (Chapter 23).
4. In this chapter, we deal *with the fourth approximate method* used for analysing a very flexible elastic plate on elastic foundation using Hetenyi's solutions as given by ACI Committee 436 [1], [2] and also IS 2950 (Part I), [3] on raft foundations. In this method, we find the radial and tangential moments (M_r and M_t) and the shear produced by a concentrated load at a radial distance r and then derive M_x and M_y.

24.2 DESCRIPTION OF ACI PROCEDURE—ELASTIC PLATE METHOD

The following is the procedure usually followed (the method is based on the theory of plates on elastic foundation by Hetenyi [1]):

Step 1: Determine the minimum raft depth for maximum punching shear. Let the raft depth adopted be t.

Step 2: Establish the corrected value of the coefficient of subgrade reaction.

Step 3: Determine the flexural rigidity D of the foundation *per unit width.*

$$D = \frac{E_c t^3}{12(1-\mu^2)} \text{ (in kN-m)} = EI \tag{24.1}$$

where

E_c = Modulus of elasticity of concrete ($20\text{–}25 \times 10^6$ kN/m^2)

μ = Poisson's ratio of concrete ≈ 0.15

Step 4: Determine the *radius of effective stiffness* L_r, which gives us a measure of the flexibility of the raft by the following formula. (This is different from L_c, the elastic length L_e of a beam of Sec. 24.1.)

$$\frac{1}{\lambda} = L_r = \left(\frac{D}{K_{\text{soil}}}\right)^{1/4} \text{ (Dimension of } L_x \text{ will be in metres)} \tag{24.2}$$

where

K_{soil} = Coefficient of subgrade reaction adjusted for column spacing in kN/m^3

The *radius of influence of the column* load is approximately $4L_r$ as shown in Figure 24.1.

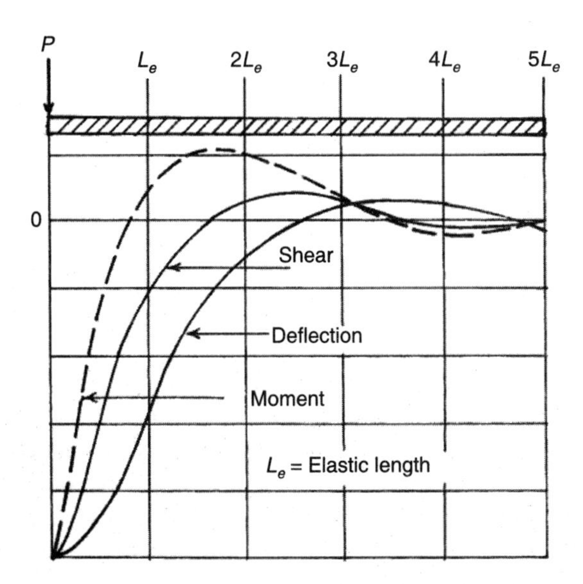

Figure 24.1 Distance from load affected by bending moment, deflection and shear due to a concentrated load on a beam on elastic foundation.

Step 5: From Hetenyi's solution, M_r and M_t, in kNm per metre width due to a load P (in kN) can be determined for any point distant $x = r/L_r$, by using Z factors from the following equations [1], [2]:

Radial moment,
$$M_r = -\frac{P}{4}\left[Z_4 - \frac{(1-\mu)}{x} Z_3'\right] = PC_1 \qquad (24.3)$$

Tangential moment,
$$M_t = -\frac{P}{4}\left[\mu Z_4 + \frac{(1-\mu)}{x} Z_3'\right] = PC_2 \qquad (24.4)$$

Also, shear,
$$V = -\frac{P}{4L} Z_4' = (P/L)C_3 \qquad (24.5)$$

Deflection,
$$\Delta = \frac{PL^2}{4D} Z_3 = (PL^2/D)C_4 \qquad (24.6)$$

where the Z values are obtained from Figure 24.2 for the values corresponding to r/L.

Shukla's chart[4]: We can further simplify Figure 24.2 to Figure 24.3, proposed by Shukla, To find C_1 to C_4 of Eq. 24.3 [4]. Figure 24.3 is easier to use than Figure 24.2. (It is plotted by taking $\mu = 0.25$.)

Step 6: Convert radial M_r and M_t obtained into rectangular M_X and M_Y by the following transformation (see Figure 24.2):

$$M_X = M_r \cos^2 \theta + M_t \sin^2 \theta \qquad (24.7)$$

$$M_Y = M_r \sin^2 \theta + M_t \cos^2 \theta \qquad (24.8)$$

where θ is the angle as defined in Figure 24.2. The value of $x = r/L$ and L is defined in Eq. (24.2).

[*Notes:* 1. It is important to note the direction of M_x and M_y. They are moments as shown in Figure 24.2. M_y is the moment about the X-axis in the Y-direction. M_x the moment about the Y-axis and M_y the moment about in the X-axis.

2. As at $x = 0$ values of moments and shears will be theoretically infinite, we calculate these values at column faces and assume they are uniform under the column.]

At points where more than one load is effective, find the effects individually and add their effects. When the edge of the mat is within the radius of influence, we first calculate the moments and shear perpendicular to the edge of the mat assuming the mat is infinitely large. Then, we apply these as edge loads with opposite sign to satisfy the edge condition. For this purpose, we use the method as applied for beams on elastic foundation.

Step 7: The shear V *per unit width* of the mat is obtained from the following equation:

$$V = -\frac{P}{4L} Z_4' = C_3(P/L) \qquad (24.9)$$

where C_3 is obtained from Figure 24.3.

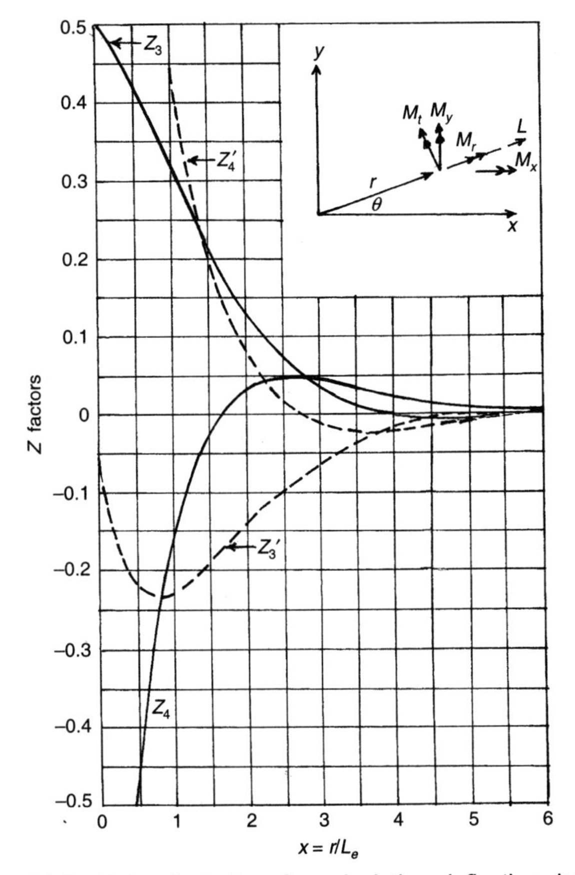

Figure 24.2 Hetenyi's factors for calculating deflection, bending moment and shears in a flexible plate [1][5].

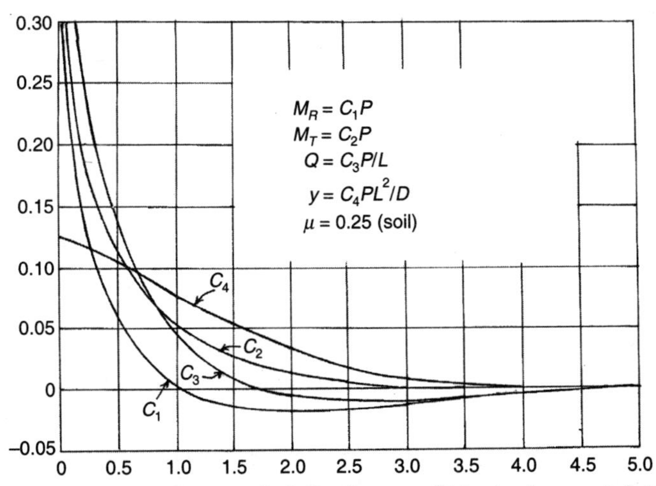

Figure 24.3 Moment, shear and deflection coefficients for an infinite mat on elastic foundation [4].

Step 8: A stiff deep wall is treated as a line load distributed through the wall to the mat. The mat may be divided into strips of unit width perpendicular to the wall, using the method of beams on elastic foundations.

Step 9: All moments and shears for each column and wall loads are superimposed to get the final moments and shears.

Step 10: Correction for edge effects. If the edge of the mat is at a distance less than $4L_e$ as described from an individual columns load, a correction has to be applied for the edge effect as follows (as in reality the moment and shear at the edge can be assumed as zero).

(a) Calculate the moments, shears and deflection at the edge of the mat due to column loads within the radius of influence $4L_e$ from the edge.

(b) Divide the mat into strips in the *X*- and *Y*-directions. Assuming the strips are semi-infinite beams, apply moments and shear equal and opposite to those obtained in (a) above.

(c) Calculate the moment, shear and deflection by Hetenyi's table assuming the mat is infinitely long as explained in Sec. 22.4.2.

Taking

$$\lambda = \left(\frac{KB}{4EI}\right)^{1/4}$$

We get these as

$$M = M_1 A_{\lambda x} - (P_1/\lambda) B_{\lambda x}$$

$$Q = -2M_1 \lambda B_{\lambda x} - P_1 C_{\lambda x}$$

$$w = -\frac{2M_1\lambda^2}{K} C_{\lambda x} + \frac{2P_1\lambda}{K} D_{\lambda x}$$

where b is the width of the mat strip (1 m) and
I_b is the MI of the mat strip

24.3 SUMMARY

This chapter explains the design of a flexible plate on an elastic foundation with column loads, by using the chart given by ACI Committee (Figure 24.2) (or Sukula's chart (Figure 24.3) based on that chart) together with Hetenyi's solution for beams on elastic foundation described in Sec. 22.4.2. Finite difference method and finite element method can be also used for analysis of mat foundations. These are explained in reference 4 and 5.

EXAMPLE 24.1 (Analysis of a flexible mat foundation)

A slab raft supporting 16 columns, is 460 × 460 mm in size and spaced at 4.5 m in both directions and loaded with column loads, as shown in Figure E24.1. The depth of the slab, as required from punching shear, is 600 mm. M-20 grade concrete is used for the construction. Assume the effective modulus of subgrade of soil is 5.55 kg/cm^3. Indicate how to analyse the slab for BM and SF along the centre line of the columns and the centre line of the mid-spans between columns.

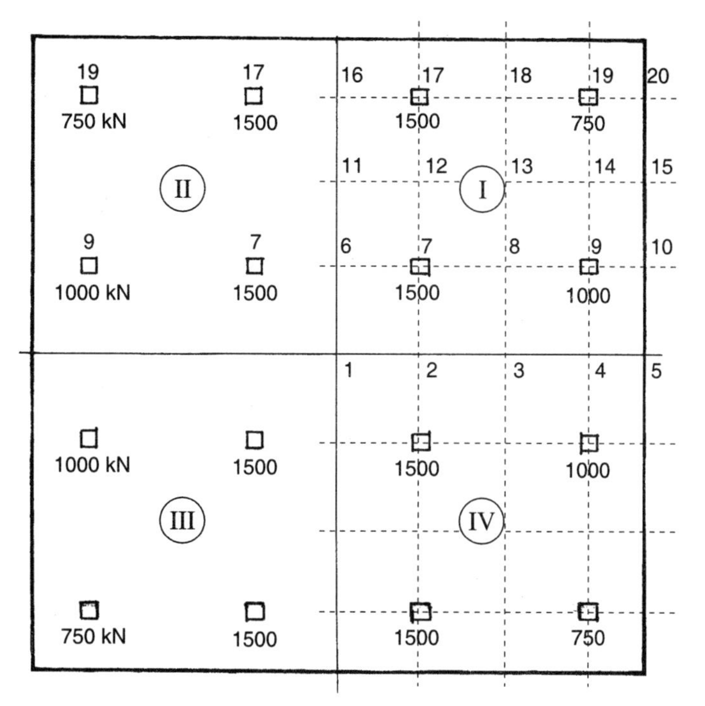

Figure E24.1

Reference	*Step*	*Calculation*
	1	*Calculate E_c and flexural rigidity D*
		$E_c = 5000\sqrt{f_{ck}} = 5000\sqrt{20} = 22.4$ kN/mm^2
		$= 22.4 \times 10^6$ kN/m^2
IS 456		$D = \dfrac{E_c t^4}{12(1-\mu^2)}$; $\mu = 0.15$
		$= \dfrac{22.4 \times 10^6 \times (0.6)^3}{11.73} = 412 \times 10^3$ kNm
	2	*Find effective subgrade reaction*
		Assume = k_c = 5.55 kg/cm^3 = 55.5 × 10^6 N/m^3
		= 55.5 × 10^3 kN/m^3
	3	*Calculate radius of effective stiffness L_r*
		$L_e = \left(\dfrac{D}{k}\right)^{1/4} = \left(\dfrac{412 \times 10^3}{55.5 \times 10^3}\right)^{1/4} = 1.65$ m
		Radius of influence = $4L$ = 4 × 1.65 = 6.6 m

Reference	Step	Calculation
		This indicates that any load within a distance of 6.6 m from that point will have effect at that point.
		(As the spacing of loads is only 4.5 m, adjacent loads will influence each other.)
	4	*Proceed to find M_X and M_Y at the cardinal points*
		(Let us divide the slab into four quadrants I–IV counter-clockwise and number the column line points and their centre line points as shown in Figure 24.1.)
Figure E24.1		First, mark off the cardinal points along the column lines and also the mid-span lines as shown in Figure E24.1. (Along column strips and middle strips.) We first find M_r and M_t and then find M_x and M_y by Eqs. 24.8 and 24.9.
		(a) *First, let us find the moments and shear at point* 1 *at centre of the mat. (There is no load at this point.)*
		Point 1 is under the influence of 4 equidistant loads, each of 1500 kN.
		Spacing of column = 4.5 m
		Polar coordinates to point = r (from point 1)
		$r = \sqrt{2} \times 2.25 = 3.18\ \text{m}$
		$\frac{r}{L_r} = \frac{3.18}{1.65} = 1.93$, $\phi = 45°$ and $\cos^2 \phi = \sin^2 \phi = 0.5$
		(b) *Find C values for r/L* = 1.93 *for calculating moments and deflection (we use Shukla's chart)*
		C_1 for M_r = – 0.0195 (in kNm/m)
		C_2 for M_t = + 0.013 (in kNm/m)
		C_4 for deflection = 0.035
		Load = 1500 kN
Figure 24.3		(c) *Calculate M_r, M_t, M_X and M_Y (Assume μ = 0)*
		$M_r = -\ 0.0195\ (4 \times 1500) = -117$ kNm/m width
		$M_t = +0.13\ (4 \times 1500) = 78$ kNm/m
		$M_X = M_r \cos^2 \theta + M_t \sin^2 \theta$
		$M_X = M_Y = -117(0.5) + 78(0.5) = -19.5$ kNm
		Deflection = $C_4 \frac{PL_r^2}{D} \times 1000$ mm
		$= \frac{0.035 \times (4 \times 1500) \times (1.65)^2 \times 10^3}{412 \times 10^3} = 1.42$ mm
Figure 24.3	5	*As a second example, let us consider effects at a load point, say, point* 1.7 *(first quadrant point* 7*)*

Reference	*Step*	*Calculation*

This is under the influence of I_7 (load no. 7 in 1st quarter) and other eight loads as shown in Table.

(a) *Effect of load at* 1.7

Under the load $r/L_r = 0$, the effect is infinitely large. Thus, take a point at the column face and assume it is the same in the column area.

Col. size. = 460 × 466 mm. Hence, r = 230 mm.

Find the effect of all the other eight loads also given below.
(Symbol I.7 represents load at first quadrant, II.7 corresponding load at second quadrant)

Table E24.1 for calculation of M_X, M_Y and Δ at point 7

Load point	Load (kN)	r/L_r	M_r	M_t	ϕ (degrees)	M_X*	M_Y*	Δ
I–7	1500							
I–9	1000							
I–17	1000							
I–19	750							
II–7	1500							
II–17	1000							
III–7	1500							
IV–7	1500							
IV–9	1000							

Notes: (a) I–IV denotes the quadrant.

(b) Calculate M_x and M_y using Eqs. 24.7 and 24.8.

Step 6 — *Apply edge correction. Calculate the moments and shears at free edge points and reduce it to zero*

We calculate the moments at points on the free edge due to loads within a distance $4L_r$ (where L_r = radius of effective stiffness).

As the moments and shears at these edges should be zero, apply equal and opposite forces and calculate its influence at the various points. For this purpose, the mat is divided into strips of unit width in both directions. Assume these strips are semi-infinite beams. Calculate the moments and shears at the required points using Tables 22.1 and 22.2.

Make the above correction for the moments and shears. The results obtained, as in Steps 4 and 5, with the corrections are the final values [4].

REFERENCES

[1] Hetenyi, M., *Beams on Elastic Foundation*, The University of Michigan Press, Ann Arbor, 1946.

[2] ACI Committee 336, *Suggested Design Procedure for Combined Footings and Mats*, American Concrete Institute, 1966.

[3] IS 2950–1981: Part I (Design), *Code of Practice for Design and Construction of Raft Foundations*, BIS, New Delhi.

[4] Sukula, S.N., 'A Simplified Method for Design of Mats on Elastic Foundations', *Journal of the American Concrete Institute, Proceedings*, Vol. 81, Sept.–Oct. 1984.

[5] Winterkon, H.F. and H.Y. Fang, *Handbook on Foundation Engineering*, Van nostrand and Reinhold, New York, 1975.

CHAPTER 25

Shells for Foundations

25.1 INTRODUCTION

Shell foundations are considered cost-effective when heavy loads are to be carried by weak foundation soils. Such situations require large-sized foundations because of the low bearing capacity. If we use bending members such as slabs and beams, the bending moments and shears in them will be large and the sections required will also be large. Shells which act mostly in tension or compression will be more efficient and economical in such situations. Even in smaller foundations, the amount of materials that is necessary for a shell to carry a load will be considerably less than that required for bending members such as beams and slabs. However, the labour involved in shell construction (in forming the shell surface, fabricating steel, supervision, etc.) will be more than that is necessary in conventional type of foundations. Thus, in such special situations, one can consider the use of shells as foundations.

In this chapter, we examine the general features of the shells commonly used in Civil Engineering (for roofs, water tanks, foundations, etc.) and then study the special features of the most commonly used shapes of shells for foundations, namely, the hyperbolic paraboloid and the cone.

However, we must also be aware that arches and many other forms of shells such as inverted barrel shells, folded plates, etc. can also be used as foundation structures. Compared to roofs, these shells when used as foundations will be smaller in spans and also in rise to thickness ratio. We must note that the intensity of loads the shells have to carry as foundation structures will be very much larger than in roofs.

25.2 CLASSIFICATION OF SHELLS

The shapes of shells commonly used in civil engineering are shown in Figure 25.1. They are generally classified, in structural engineering, into the following two groups:

1. Rotational shells
2. Translational shells

Rotational shells, also called *shells of revolution*, are formed by the rotation of a plane curve (called the meridian or the generating curve) about the axis. For example, a full vertical circle rotating about a vertical axis produces a sphere. A half circle will produce a dome.

Translational shells are formed by one curve (of curvature $1/r_1$, say, of positive curvature) moving over another curve (of curvature $1/r_2$). The moving curve is called the *generatrix* and the other *directrix*, the function of the two curves being interchangeable.

The Gaussian curvature K of the shell is defined as,

$$K = \left(\frac{1}{r_1}\right) \times \left(\frac{1}{r_2}\right)$$

It can be *positive* when both curvatures are in the same direction. It is *zero* when one has infinite radius of curvature (i.e. curvature is zero as in a straight line), and *negative* when the curvatures are in the opposite direction. Surfaces r_1 and r_2 in the same direction (K +ve) are called *clastic shells* and those with r_1 and r_2 in the opposite direction (K –ve) *synclastic shells*.

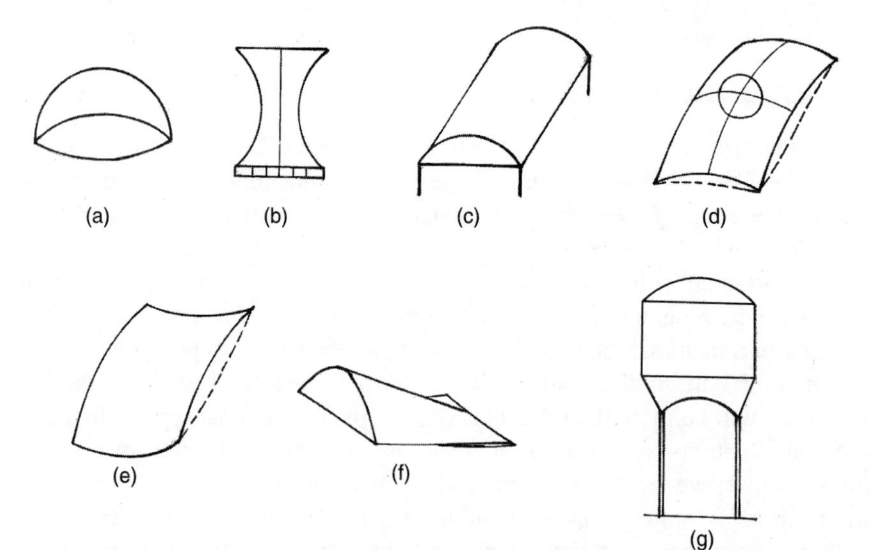

Figure 25.1 Commonly used shells and their classification: (a) Dome, (b) Hyperboloid, (c) Cylindrical shell, (d) Elliptic (circular) paraboloid, (d) Hyperbolic paraboloid, (f) Conoids, (g) Water tank made of a combination of shells.

25.3 COMMON TYPES OF SHELLS USED

A brief description of the common types of shells used in Civil Engineering practice [1], [2] is given as follows: (Figure 25.1)

1. **Domes.** Domes are surfaces of revolutions obtained by the rotation of a plane curve about an axis lying in the plane of the curve. The curve is called the *meridian* and its plane the *meridinial plane*. It is a shell of a positive Gaussian curvature and, as explained below, the membrane theory gives fairly good results for design throughout the system, except for a small distance near the boundaries. If the curve is a circle, we get a spherical dome. We can have a conical dome, elliptic dome etc.
2. **Hyberbolic shells.** A hyperbolic shell is produced by the rotation of a hyperbola about the y-axis. Such shells are used as walls of cooling towers.
3. **Cylindrical shells.** The middle surface of a cylindrical shell is formed by translation of a plane curve K (the generatrix) along a straight line P. Depending on the curve K, we can have "circular, cylindrical, elliptic cylindrical, paraboloid cylindrical" and other types of shells. Circular cylindrical shells can be long when their length is large compared to their radius, otherwise short—they are commonly used for roofs.
4. **Paraboloidal shells.** A paraboloid is produced when both the generatrix and the directrix are quadratic parabolas. They are divided into elliptic paraboloids, circular paraboloids, or hyperbolic paraboloids according to the intersections of the paraboloid with an arbitrary horizontal plane z = constant, which will result in an *ellipse, circle* or *hyperbola*.

   Thus, when two parabolas are unequal, both pointing downwards (positive Gaussian curvature), we get an *elliptical paraboloid*. When they are equal, we get a *circular paraboloid*. With two parabolas of opposite curvatures (negative Gaussian curvatures), we get a *hyperbolic paraboloid* (or hypar) shell.
5. **Conoids (skew shells).** The middle surface of a conoid is formed by the movement of a straight line P (generatrix) along a plane curve K at one end and another straight line P_0 at the other end, the straight line being parallel to the plane of the curve K. (*Note:* It can be shown that if curve K is replaced by a straight line raised at one end, the surface formed is a hyperbolic paraboloid. See Sec. 25.6.)
6. **Combination of shells.** Many structures such as Intze type water tanks are formed by a combination of different types of shells.

25.4 SIGNIFICANCE OF GAUSSIAN CURVATURE

When shells are discontinued at their edges by edge beams, bending occurs at these junctions. One important characteristic difference among the shells of different Gaussian curvature is the effect of edge disturbances as given below.

1. For the shells of positive Gaussian curvature, the edge effects tend to damp rapidly and are usually restricted to a small distance from the edge. In these cases, the membrane theory will give good results. This is the case with spherical domes.

2. For the shells of zero Gaussian curvature, the edge effects are also damped but less than those with positive curvature.
3. For the shells of negative curvature, the boundary effects are significant over a fairly large part of the shell. Thus, the edge effects of hyperbolic paraboloid shells are not insignificant, and a wide edge beam should be provided by gradual thickening of the edges.

25.5 TYPES OF SHELLS USED IN FOUNDATIONS [2]

Shell surfaces are not popular for use as foundations due to such reasons as the difficulty in exactly shaping the surface for the foundation, and casting the concrete in slopes. Domes (for circular spacing of columns), elliptic paraboloids (for unequal spacing in the *XY*-directions) circular paraboloids (for equal spacing of the column in the *XY*-directions) are all theoretically possible for foundations. But even though formation of these surfaces for roofs is easy, it is much more difficult for foundation than using conventional foundations such as rafts and piles.

However, because of the easiness in forming the casting surface of the cone and the hyperbolic paraboloids, these two shapes have been adopted, to a limited extent, in practical construction. The Bureau of Indian Standards has also published IS 9456 (1980) Code of Practice for the Design and Construction of Conical and Hyperbolic Paraboloidal Type of Shell Foundations [3][4]. We will also base our design on these codes. In this chapter we will examine the general features of these two types of shells. We will separately deal with their analysis and design in Chapters 26 and 27.

25.6 HYPERBOLIC PARABOLOIDS (HYPAR SHELLS)

We have seen that hyperbolic paraboloids (also known as *hypar* shells) are surfaces formed by two hyperbolas of opposite curvatures (concave and convex curvatures), one moving over the other forming translational shells of negative Gaussian curvature. The equation to the concave parabola is of the form $z = kx^2$.

It is a very important property of hypar surface that part of the surface, as shown in Figure 25.1, can be also formed as a *warped surface with straight line generetrix*. It becomes a ruled surface. This gives us immense advantage as the ground to be made ready to receive the concrete can be easily formed. In roofs, the form work for hyperbolic paraboloids can be easily assembled by straight planks placed with the proper slopes between the sides. It makes the construction of the formwork for concreting very easy. In foundation, the surface can be formed easily on the ground.

We can describe that part of the hypar that we generally use for roofs and foundation as follows. As shown in Figure 25.2(a), the rectangle *OXAY* is a horizontal plane of stretchable material. If we lift *A* to *A′* by a distance (h), (not less than 1/5 the longer side) point *P* on the surface will have the following coordinate in the z-direction:

$$z = \left(\frac{yh}{b}\right)\left(\frac{x}{a}\right) = \left(\frac{h}{ab}\right) yx = kxy$$

where *k is called the warp of the hyperbolic paraboloid.* As the above expression is the equation to a *hyperbola*, the surface is called a *hyperbolic paraboloid or hypar* (in contrast to the circular and elliptic paraboloids).

We can also look upon the hypar surface as formed by two curves, namely convex and concave parabolas (at right angles), which will act in tension and compression according to the loading and their position in the surface.

25.7 COMPONENTS OF A HYPAR FOOTING

A column footing can be made up of four hypar joined together, as shown in Figure 25.2. We will see in Chapter 26 that the hypar surface, when loaded with vertical loads, the surface will be in pure shear and consequently, subjected to compression and tension. These forces at the boundary have, however, to be resisted by the boundary elements, which will be in tension or compression. Thus, the hypar will have to be bounded by sloping *ridge beams* in compression and *ground edge beams* in tension.

It should be remembered that the hypar is a shell with negative Gaussian curvature and the edge disturbance will be felt by the shell adjacent to these beams to a fairly very large extent unlike that in a dome. Thus, it is usual to thicken the ridge beam, gradually starting from the shell to a fairly large width.

[According to the ASCE recommendations, for cylindrical roofs with zero Gaussian curvature, the thickening of such shells near the edges should extend to a width b

$$b = 0.38\sqrt{Rd}$$

where R = Radius of curvature
d = Thickness of the shell.]

25.8 USE OF HYPAR SHELLS IN FOUNDATION

As shown in Figure 25.2, for a simple footing under a column, four hypar elements have to be joined together by means of the ridge beams and edge beams. A combined footing and raft can also be formed as shown in the figure. In the latter case, we will have, in addition, the valley beams joining the system as shown in Figure 25.3. Thus, hypar foundation will have the following elements:

1. Shell proper
2. Ridge beam
3. Edge beam
4. Valley beams (in combined footings and rafts)

We will deal with the analysis of the forces in these members and their designs in Chapter 26.

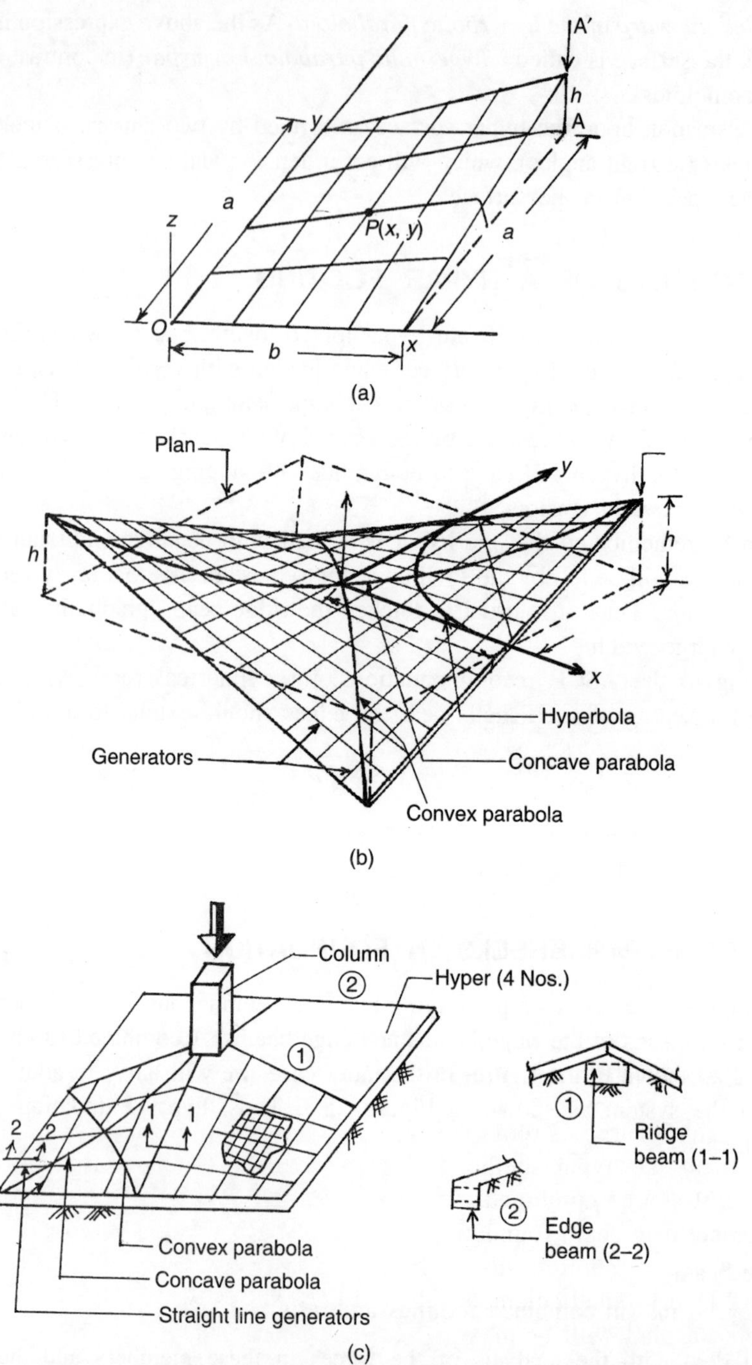

Figure 25.2 Formation of hyperbolic paraboloids (Hypar): (a) Hypar formed by lifting a corner of a rectangle, (b) Hypar formed by two parabolas, (c) Formation of hypar shell by four hypars, ① Ridge beam, ② Edge beam.

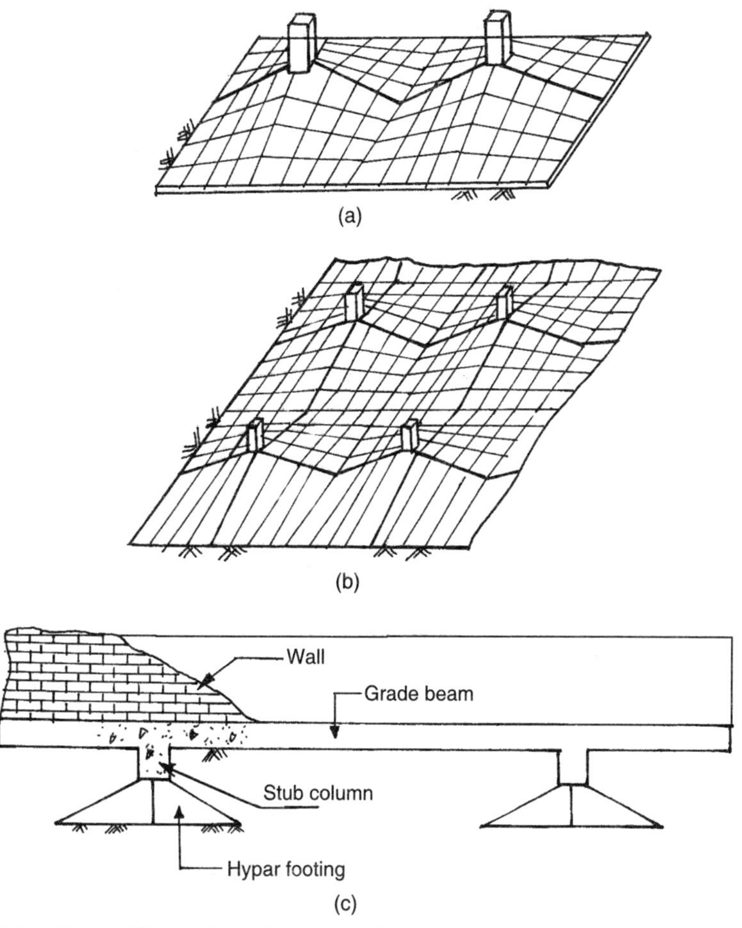

Figure 25.3 Use of hypar foundation: (a) As combined footing, (b) Raft foundation, (c) Under walls with grade beams and stub columns.

25.9 CONICAL SHELL AS FOOTING

A conical shell can be used as roofs and as a single foundation for columns (Figure 25.4). As foundations, unlike hypar shells, it cannot be combined into combined footings and rafts. However, it can be combined with other footings to form inverted dome combined with cone, conical ring beam foundation, etc. in special situations. With column load, the shell will be under compression along its length and ring tension along the circumference. Due to small diameter at the top and larger diameter at the bottom, the compression in the shell will be maximum at the top and minimum at the bottom. On the other hand, the ring tension will be minimum at the top and maximum at the bottom. It is usual to provide a *ring beam* at the bottom to restrict the cracking of the cone under the maximum tension. At the top, we cannot have a point loading and the application of load will be through a definite width below from the apex towards the base. Thus, the

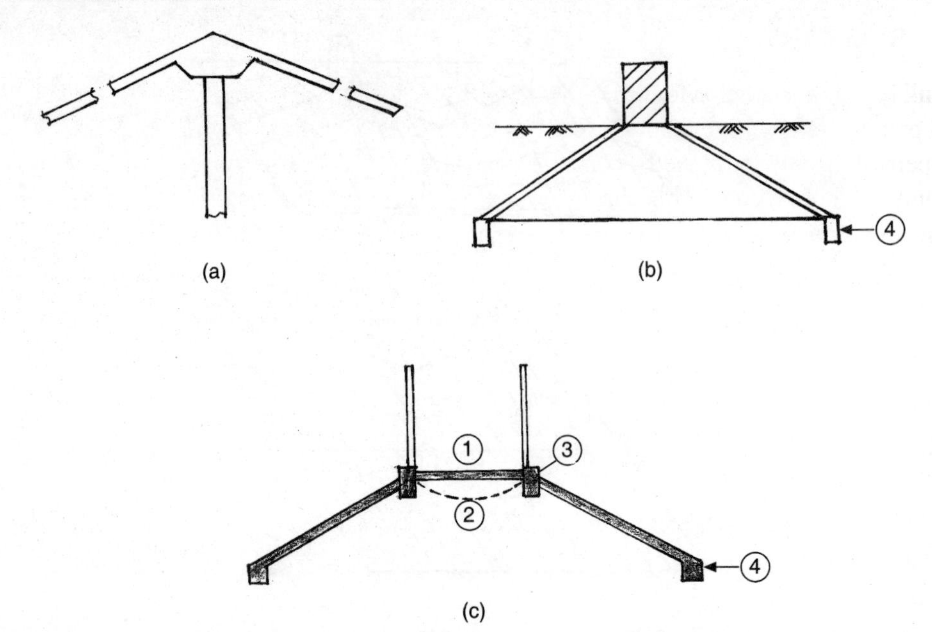

Figure 25.4 (a) Conical roof, (b) Conical foundation for a column, (c) Conical foundation for a circular shaft with a ① Circular raft on, ② Inverted dome segment and ③ ring beam, ④ Note thickening of edges at the bottom of the cone.

cone will act as the frustum of the cone. Thickening at the top by a ring beam is necessary to take care of discontinuity.

Even though high rise and steep slopes are commonly used when these shells are used for roofs, it is not economical or convenient for construction to have a steep and tall shell as foundation. Concrete cannot be deposited on steep slopes without form work. It is difficult to maintain concrete on slopes more than 1 vertical to 1.5 horizontal (i.e. 33.7°) without form-work. A rise of 1 in a total base width of 4 (or 1 in 2 of half base width) is usually adopted and this will be convenient for concreting.

The elements to be analysed and designed in a conical footing are:

1. Shell surface for tension and compression
2. Top column to cone junction
3. Ring beam at the bottom

The analysis and design of these will be covered in Chapter 27.

[*Note:* As shells are assumed to take most of the loading on them not by bending and shear but mostly by compression and tension, we do not check for punching shear at the place of transfer of column load to hypar or cone. However, we should always check whether the vertical upward component of the "*capacity of the shell in compression* at the junction of shell and column" is able to comfortably balance the vertical column *load acting downwards* so that the equilibrium of forces can be maintained.]

25.10 SUMMARY

We think of using shells for foundation when the foundation is of large span and the allowable ground pressure is low. Even though shells of various shapes are used as roofs, only a few types like hyperbolic paraboloids shells, conical shells and cylindrical shells in the form of arches are commonly used in practice. This chapter briefly describes the general aspects of shell structures and also the properties of the hyperbolic paraboloid and the cone as foundation structures.

REFERENCES

[1] Fischer, L., *Theory and Practice of Shell Structures*, Wilhelm Ernst and Sons, Berlin, 1968.

[2] Ramasamy, G.S., *Design and Construction of Concrete Shell Roofs*, CBS Publishers, 1986.

[3] IS 9456, 1980 (Amendment, No. 1, 1982), *Code of Practice for Design and Construction of Conical and Hyperbolic Paraboloid Shell*, BIS, Delhi.

[4] Ninan, P.K., *Shell Foundations*, Narosa Publishing House, Delhi, 2006.

CHAPTER 26

Hyperbolic Paraboloid (Hypar) Shell Foundation

26.1 INTRODUCTION

We have already seen how a hyperbolic paraboloid surface can be formed by lifting up or down of one corner of a rectangle of stretchable material with all the other three corners kept at the same level. The surface of these shells can be also formed by moving a straight edge between the opposing ridge and baselines as shown in Figure 26.1, which makes the layout of these shells on the ground very easy. Formwork for these roofs is very simple. Such hypar shell roofs are divided into two groups, namely,

1. Shallow shells
2. High rise shells

In shallow shells, the rise at the corner (as shown in Figure 26.1) should be less than about 1/5th the longer side of the *rectangle* (with sloping rides from two sides, this will give the slab a slope of 1 in 2.5). In high rise shells, the rise is more than 1/2 times the total side length. In the design of shallow shells, the dead load on the shell can be considered as a uniform loading over the horizontal projection. In high rise shell roofs, corrections have to be made for the gravity loading on the shell. However, due to the restrictions allowed in slopes of surfaces for laying of concrete in foundations, hypar shells used in foundations will fall under the category of shallow shells only. The type of loading through a filling of compact soil with blinding concrete on top also helps us to apply the *theory of shallow shells* subjected to uniform loading to hypar footings also. (High rise shells are commonly used for ornamental roofs as for churches, halls, etc. and are described in books on shells roofs.)

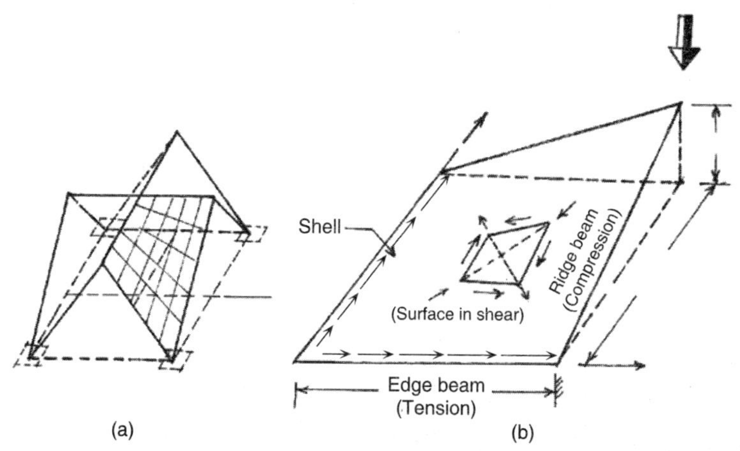

Figure 26.1 Analysis of hypar shells: (a) Deep hypar shell used as roofs, (b) Shallow hypar shell used as foundation. (*Note:* Tension in edge beams and compression in ridge beams. Four of the hypar shells shown in (b) are used for a column foundation)

Shell analysis and design can be approached in two ways: (i) by exact mathematical analysis and (ii) by conceptual design studying its physical behaviour using simple statics to calculate the force. Many of the shells such as the famous St. Peter's dome were constructed very long before the mathematical theory of shells were published by such methods. In this chapter, we also use the physical modelling for analysis and design. The design procedure described in this chapter follows those described in references [1] [2] [3] and [4].

26.2 NATURE OF FORCES IN HYPAR SHELLS

To compare the hypar foundation with roofs, let us look at the hypar footing upside down (in the direction the ground pressures are acting). It can be seen that the hypar footing is made up of four hypar shells with the centre at a higher level than the base. Each hypar consists of the following parts, as shown in Figure 26.2.

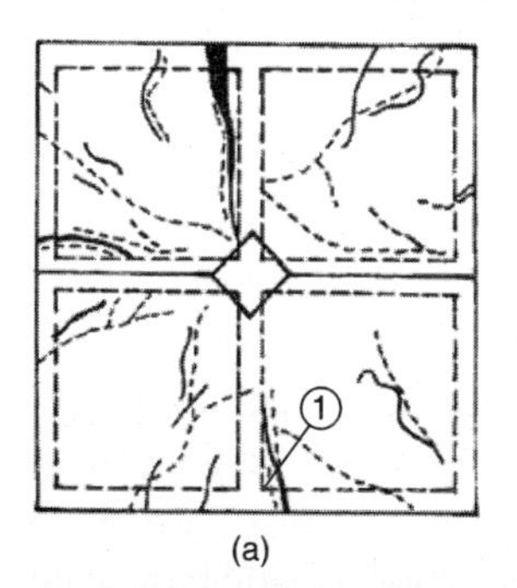

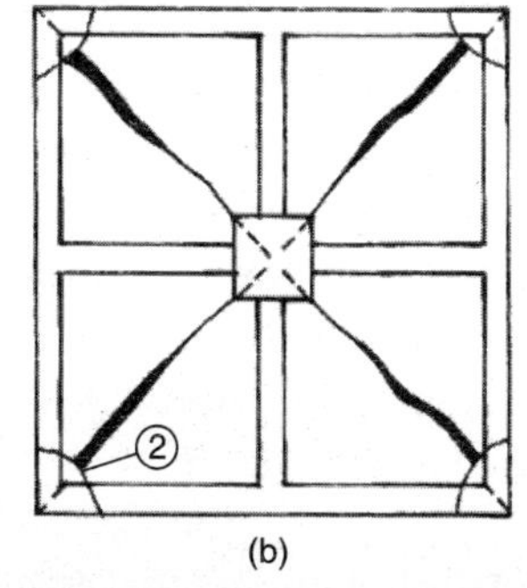

Figure 26.2 Patterns of failure of the four units of the hypar footings at ultimate loads showing places to be specially detailed in design: (a) Junctions of edge beams: (1) Cracking due to tension in edge beam, (b) Cracking of shell due to diagonal tension caused by shear (2) Cracking of corners.

- Shell
- Ridge beams (These are the sloping members that support the column)
- Edge beams (These are the beams on the ground along the edges)

26.3 DESIGN OF VARIOUS MEMBERS

We examine the nature of the stresses in each of them when the shell is loaded by a uniform load from the ground. It should be remembered that in a roof, the shell is loaded from top downwards and the low and high points of the roof shell are reversed as compared to a hypar footing.

26.4 MEMBRANE FORCES IN HYPAR FOUNDATION

The unique structural property of the hyperbolic paraboloid (hypar) shell is that under vertical loads, the middle or shell surface of a hypar (with reference to its X- and Y-axes as shown in Figure 26.1) will be subjected to *only uniform shear force* of the following magnitude. This is specially true when they are shallow hypar shells.

$$s = \frac{q}{2} \times \left(\frac{ab}{h}\right) = \frac{q}{2 \times (\text{warp})} \text{ in kN/m}$$

where,

q = ground pressure in kN/m^2

a, b = sides of hypar

h = rise

(h/ab) = warp

Direct forces $N_x = N_y = 0$ (For membranes $M = 0$)

In a *hypar shell roof* where the load acts down, this shear acts from the *lowest level to the higher level*. Hence, in a *foundation shell* where the ground pressure acts upwards and the column point is above the foundation level, the shear will be acting *in the shell* from the higher to the lower level as shown in Figure 26.1.

We have also the complementary shear as shown. We know that these shears produce equivalent tension and compression along the diagonals. These tensions and compressions can be compared to the forces in two sets of parabolas, each parallel to the diagonals, a concave parabola from the lower to the higher level acting in compression due to load from below and a convex parabola parallel to the other diagonals acting in tension again due to the load from below. The tension in the shell has to be resisted by steel placed in the shell. In fact, we provide a mesh of steel, as shown in Figure 26.3, to take care of this tension.

At the edge of the hypar, we must provide suitable structural members to take up the shearing forces acting from the shell. Thus, the shell is bounded at the edges by the edge beams and the ridge beams. These forces will be opposite to the shear forces, and the total force in the members will be the sum of the shears along these edges.

26.4.1 Forces in the Ridge Beams and the Edge Beams

In a *hypar foundation*, the forces in the ridge beams boundary members will be acting from the *lower to the higher points along the ridge beams* so that the ridge will be in compression. The total force in each ridge beam will be the sum of these forces in each shell on its sides. (That the forces in the ridges are in compression is obvious as we can see from statics that it is the vertical component of these forces that balances the vertical load from the columns.)

The forces in the edge beams will be the forces acting along the edge beam of each shell (and valley beams of multiple shells). This force in the edge beam will be equal to the sum of the shear forces along the edge of these members and it will obviously be in tension.

Thus, we have tension and compression in the shell proper, compression in the ridge beam and tension in the edge beams. These should be calculated and members properly designed.

26.5 MAGNITUDE OF FORCES

Based on the nature of forces described above we can write down the values of the forces in the various elements of the shell as follows:

1. *Stresses in the shell*

   The shell surface is in pure shear which produces tension and compression as shown in Figure 26.1.

   Shear = Tension = Compression in shell

$$s = t = \frac{q}{2}\left(\frac{ab}{h}\right) = \frac{q}{2 \times \text{Warp}} \tag{26.1}$$

2. *Tension in edge beam*

$$\text{Max tension} = \text{Sum of shear along length} = a \times s \tag{26.2}$$

   where,

   a = Length of edge member or side of shell
   = [1/2 the base length of foundation]

   Maximum tension occurs at the junction of the edge beam and ridge beam as shown in Figure 26.1.

3. *Compression in ridge beam* These compression members should be designed to be sufficient by rigid and should not have more than 5% compression steel in it.

   Compression for each steel = $L \times s$

   where

$$L = \sqrt{a^2 + h^2}$$

   As two shells from each side of the ridge beam meet along the ridge, the total compression is the sum of forces from the two shells.

$$C = 2Ls \tag{26.3}$$

4. *Check for column load* It is also advisable at this stage to check whether the vertical component of the compressions balances the column load.

$$C = (\text{Col. load}/4)\ (\text{Length of ridge beam/Rise}) = \frac{PL}{4h}$$

26.6 PROCEDURE IN DESIGN OF HYPAR SHELL FOUNDATION

The following steps are used to design a hypar shell under footing under a column:

Step 1: Find the area of base required from the safe bearing capacity and arrive at a layout as indicated in Sec. 26.7.

Step 2: Calculate membrane forces in the shell.

Step 3: Design the shell part for tension and compression.

Step 4: Find tension in edge beam and design edge beams.

Step 5: Find compression in ridge members and steel for compression.

Step 6: Detail the steel and check the special locations specified in Sec. 26.8 to avoid premature failure [Figure 26.2].

26.7 EMPIRICAL DIMENSIONING OF HYPAR FOOTING

Initially, the footing consists of four hypars, each of sides $a \times b$ placed together with ridge beams in between. The total base width will be $2a \times 2b$.

The minimum thickness of concrete used for shell roofs depends on practical considerations such as easiness of good concreting. Only a small thickness of 75–100 mm (3–4 inches) of concrete is usually used for large shell roofs. For spans up to say, 30 meters of reinforced concrete domes, cylindrical shells, etc., use only 75–100 mm shells with 0.2% steel. But we use larger thickness in foundation shells. Loads on foundation elements are much larger than in roofs and also from considerations of durability (as foundation structures are buried in soil), the elements of the shells will have larger sizes. We should also remember that foundation concrete is easier to lay in smaller slopes as otherwise very expensive formwork and method of placement will be required.

The following thumb rules can be used as a rough guide to choose the dimensioning of hypar footings for estimating as well as preliminary planning and design.

1. **Rise of shell.** The rise of the shell should not be more than the slope at which concrete can be placed and compacted, which is not more than 1 in 1.5 (say about 33.7 degrees). In addition, for a hypar to be considered shallow, the slope should not be more than 1 in 2.5 of each of the side of four hypar. Generally, a maximum slope of 1 in 2 with respect to the side of each hypar can be adopted. (This will be 1 in 4 of the side of foundation.)
2. **The thickness of the shell.** The thickness of the shell footing should be more than that used for roofs as we have to meet the needs of cover for foundations. Usually,

shells are cast on mud mat with a minimum cover of 50–75mm of 1 : 1½ : 3 concrete, and steel placed at the middle of the thickness will have to be 120–150 mm. "A thickness to length ratio" of 1/12–1/16 can be adopted. The shell surface is in pure shear and hence subjected to pure tension and compression. (Some recommend a minimum percentage steel of 0.5% to reduce cracking of the shell.)

3. **Edge beams.** The edge beams at the base are in tension. The thickness of the edge bems is made half the size of the column. Its depth should be about 1/6 the total length of the two hypar ($2a$) which form the base length. The percentage steel of not more than 5% is recommended. Nominal ties should be also provided. We should remember that this beam is in pure tension.
4. **Ridge beams.** The four inclined ridge beams are in compression and their vertical component of compression should carry the column. Their breadth is made equal to the size of the column and of enough depth to make it a rigid short column member and also to extend into the shell proper. The percentage of steel need not be more than 5%. Nominal ties as in columns should be also provided.

Extensive laboratory tests show that the ultimate failure pattern of hypar shell footings is as shown in Figure 26.3 [2]. The failures are known as *ridge failure* and *diagonal tension failure*.

26.8 DETAILING OF HYPAR FOOTINGS

Detailing of the steel should be with reference to the pattern of failure shown in Figure 26.2. The shell, edge beams and the ridges should be properly detailed for the structure to act together and prevent any premature failure. This is especially necessary as there are many parts of the shear tension. Ultimate load tests on models of hypar shells have shown that the following three points of the shell need special detailing to prevent premature failure.

1. **Junction of the column with shell and ridge beams.** The column should properly stand on top of the ridge beam junction and the column bars should be properly anchored equally into the ridge beams. Also, the shell should be properly joined to the column. Proper fillets should be used at the junction.
2. **Junction between edge beam and ridge beam.** This junction should be tied together as shown so that the section of maximum tension does not fail prematurely. Model studies show that this is the first point to fail if proper detailing is not carried out.
3. **Corners of the shell.** As the two edge beam members that meet at the corners are in tension, there is a resultant tension at the corner and hence a tendency to split along the diagonal. A suitable fillet with nominal steel will greatly assist in resisting the cracking and premature failure in tension.

The detailing of hypar shells is shown in Figure 26.3.

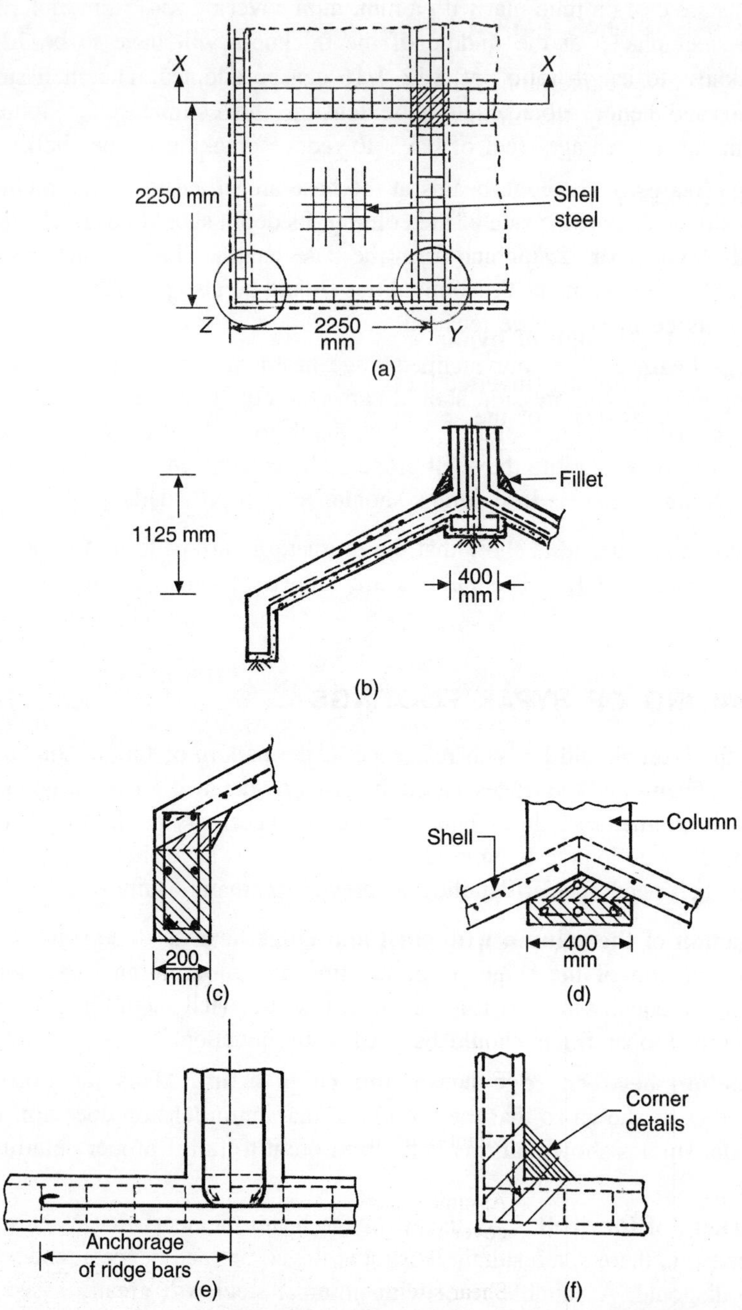

Figure 26.3 (and Example 26.1) Detailing of hypar shell footings: (a) General arrangement plan, (b) Section *X–X*, in Figure (a) above, (c) Section along edge beam, (d) Detail below column, (e) Detail at junction *YY*, (f) Detail at corner *Z*.

26.9 EXPRESSIONS FOR ULTIMATE BEARING CAPACITY

IS 9456 (1980) gives expressions for Ultimate Strength Capacity of hypar shell foundations and may be used for finding the ultimate strength of the shell foundation.

26.10 SUMMARY

This chapter explains the design and detailing of hypar foundations as recommended in IS 9456 (1980).

EXAMPLE 26.1 (Design of hypar shell footing)

Design a hypar shell footing (inverted umbrella type hypar shell) for a column carrying 100 tons if the safe bearing capacity of the soil is 50 kN/m^2.

Reference	*Step*	*Calculation*
	1	*Find shell dimensions*
		Load = 1000 kN, SBC = 50 kN/m^2
		Required base area $= \dfrac{1000}{50} = 20 \text{ m}^2$
		Adopt a 4.5 m square base = 20.25 m^2 (area)
Figure 26.1 and Figure 26.2		Adopt 4 hypar shell units, each 2.25 m in size to form a column foundation
		Adopt rise of 1 in 2, Rise $= \dfrac{2.25}{2} = 1.125 \text{ m}$
Eq. (26.1)		Warp of shell $= \dfrac{\text{Rise}}{\text{Area}} = \dfrac{1.125}{2.25 \times 2.25} = 0.222 \text{ m}^{-1}$
		Base pressure $= \dfrac{1000}{4.5 \times 4.5} = 49.38$. Assume 50 kN/m^2
	2	*Calculate membrane shear on factored load*
		Factored pressure = 1.5 × 50 = 75 kN/m^2
Sec. 26.3.1		Membrane shear $= \dfrac{q}{2 \times \text{Warp}} = \dfrac{75}{2 \times 0.222} = 170 \text{ kN/m} = s$
		Assume a shell thickness of 120 mm (with 10 mm steel and 50 mm cover on both sides)
		Shear stress $= \tau = \dfrac{170 \times 1000}{1000 \times 120} = 1.42 \text{ N/mm}^2$
IS 456		[Allowable shear for M_{20} = 2.8 N/mm^2. However, we have to provide for the tension and compression produced by the shear along the diagonals.]
	3	*Design the steel in shell (Find area of steel for tension due to shear)*
		Tension = Shear = 170 kN/m. Thickness t = 120 mm

Reference	Step	Calculation
		Steel required $= \dfrac{(170 \times 1000)}{0.87 \times 415} = 471\ \text{mm}^2/\text{m}$
		Percentage of steel $= \dfrac{471 \times 100}{1000 \times 120} = 0.4\%$
Sec. 26.6		(Some recommend 0.5% as minimum steel to reduce crack width in the slab.)
		(This steel is more than the minimum 0.12% for shrinkage)
		Provide 12 mm @ 225 mm giving 502 mm²/m
		(Maximum spacing is less than 2 × thickness)
		[*Provide this steel parallel to the sides of the shell. The diagonal component of effective steel* $= 2 \times \dfrac{A_s}{\sqrt{2}} = 1.4A_s$]
	4	*Check compression in concrete in the shell*
		Compression stress = Tension = Shear = 1.42 N/mm²
		This is very much less than $0.4f_{ck} = 0.4 \times 20 = 8$ N/mm²
	5	*Find tension in edge beam and area of steel as in beams*
Sec. 26.4		Max tension (each shell) = Shear × Length
		Area of steel required $= \dfrac{383 \times (10)^3}{0.87 \times 415} = 1060\ \text{mm}^2$
		Provide 4 Nos. 20 mm bars = 1257 mm²
		Assume width = 1/2 size of the column = 200 mm
		Assume depth = 300 mm
		Percentage of steel $= \dfrac{1257 \times 100}{300 \times 200} = 2.1\%$
Sec. 26.5		Good percentage for a beam. Not more than 5%. Also provide nominal ties of 6 or 8 mm @ 200 mm spacing.
	6	*Find compression in ridge beam and provide steel as in column*
		Inclined length of ridge beam $= \sqrt{(2.25)^2 + (1.125)^2}$ $= 2.516$ m
		Compression = [Shear × Length] (2 from two sides) $= 170 \times 2.516 \times 2 = 856$ kN
		Compare the above compression as calculated from the column load.
Sec. 26.5		Comp. $= \dfrac{PL}{4h} = \dfrac{1500 \times 2.576}{4 \times 1.125} = 858.7$ kN
Figure 26.2		Make width of beam = that of column = 400 mm

Reference	Step	Calculation
		Total beam area = (400 × 100 mm Rectangle) + (400 × 100 mm Triangle) = 60,000 mm^2
		(As the compression member is attached with the shell, we need not check *L/d* ratio.)
		$C = 0.4f_{ck}A_c + 0.67f_yA_s$
		$A_s \text{ required} = \dfrac{(858\times 10^3) - 0.4\times 20\times 60\times 10^3}{0.67\times 415}$
		$= 1359\text{ mm}^2 = \dfrac{1359\times 100}{60{,}000} = 2.26\%$
		Provide 4 rods (3 at the bottom of the rectangle and one at top of the triangle) of 25 mm giving 1923 mm area. Provide ties 6 mm @ 200 mm.
		(It is better to over-design this ridge member so that its vertical component can support the column load with ample safety margin.)
	7	*Detail special section to avoid premature failure*
		(a) *Corners at base*
Figure 26.3		At corners of the base, provide corner fillets equal to the width of edge members with nominal ties of 10 mm @ 100 mm spacing.
		(b) *Junction between column and ridge beams*
		Equal numbers of column steel are continued into ridge beams and lapped with ridge beam steel. The vertical component of the compression in the ridge beam should be more than balance of the column load.
		(c) *Junction of ridge and edge beams*
		Tie ridge and edge beams by extending steel for a length at least equal to the full development length. Provide also corner fillets.

REFERENCES

[1] IS 9456, 1980 (Amendment No. 1, 1982), *Code of Practice for Design and Construction of Conical and Hyperbolic Paraboloidal Types of Steel*, BIS, New Delhi.

[2] Kurien, Ninan P., *Shell Foundation*, Narosa Publishing House, New Delhi, 2006.

[3] Varghese, P.C. and S.S. Kaimal, Field test on a combined hyperbolic paraboloid shell foundation, Bulletin of the International Association for Shell Structures, No. 44, Dec. 1970.

[4] Kurien, N.P. and P.C. Varghese, "The Ultimate Strength of Reinforced Concrete Hyperbolic Paraboloid Footings", *The Indian Concrete Journal*, Dec. 1972.

Design of Conical Shell Foundation

27.1 INTRODUCTION

Conical shells are commonly used as *sub-structure* at the bottom of towers above the foundation to widen the base for resting on an annular raft. *Conical shell foundations* can be used as individual footings under single columns. They are used only in special cases. Such columns have a definite size and hence the column foundation will be on a frustum of a cone as shown in Figure 27.1. As foundation soil has to be shaped, covered with blinding concrete and the structural concrete cast without formwork, the rise of these shell foundations, unlike in roofs, is limited usually to about 2 horizontal to 1 vertical. These aspects limit its general use for foundations. In this chapter, we briefly deal with the conventional design of conical foundation under a single circular column base. The procedure follows those described in references [1] [2] and [3].

27.2 FORCES IN THE SHELL UNDER COLUMN LOADS

In general, stresses in a conical shell will depend on the way the shells are loaded and supported. For example, for a conical shell roof, we use a full cone supported at the centre forming an *umbrella roof.* Alternatively, it can be supported at the edges as a roof. For a footing foundation, the conical shell is loaded at the apex and buried in the ground. It is easy to visualize that the shell will be *subjected to compression* along the cone and *also hoop tension* along the circumferences *with no shear*. The stress distribution is very simple.

It is easy to calculate by simple statics the maximum compression which will occur at the place of contact (junction) between the column and the shell. The vertical component of the compression should balance the column load. If d is the diameter of the column in metres, its load P is in kN and N_c the compression per metre length. Taking d as the diameter of the shell at the junction, we have

$$(\pi d\ N_c) \cos\ \theta = \text{Column load } P$$

$$N_c = \frac{P}{\pi / d \cos \theta} \text{ kN/m} \qquad (27.1)$$

This compression reduces very rapidly as we go along the length of the shell as the area of the section increases. The horizontal component of the compression in the shell has to be resisted by the column base. If the conical shell supports a cylindrical shell instead of a solid column, as in Figure 27.1, a ring beam also has to be given to withstand the horizontal forces.

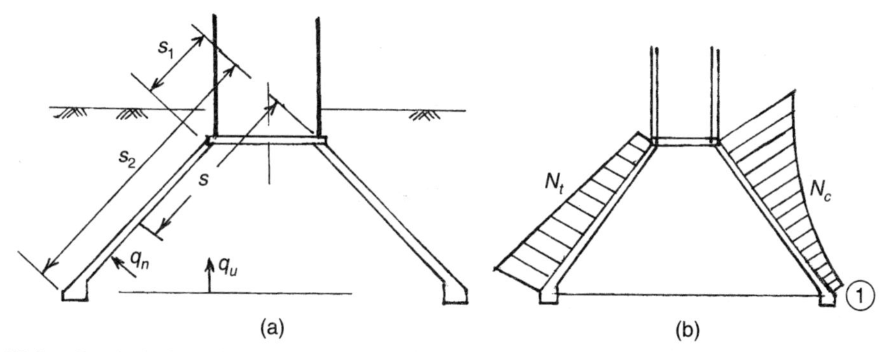

Figure 27.1 Conical shell foundation: (a) s values, (b) Variation N_c (meridional compression) and N_t (Hoop tension along the shell) [(1) provision of edge thickening (beam) at the ends.].

27.3 RESULT OF SHELL ANALYSIS

As already seen, the stresses in the shell will depend on the support condition and the loading on the shell. In the case of a conical footing, the inside of the cone should consist of non-shrinkable soil over which the blundering concrete and the shell concrete are cast. The load from the column is transmitted through this compact mass of soil inside the cone, to the base of the foundation. As concreting in steep slopes is difficult, the shells used in the foundation will be shallow.

As already stated, the shell is subjected only to meridional compression and hoop tension with no shear. The following expressions can be derived from *shell analysis* [2], [3] with the given boundary conditions with reference to Figure 27.1, with length s measured from apex along the length of the shell (For brevity, they are not derived here.)

Let

Half angle of cone = θ

Distance of column bases from apex = s_1

Distance of bottom of cone from apex = s_2

The *vertical base pressure* = q_v

The pressure *normal to surface of shell* = q_n

In shallow shells, we take $q_v = q_n$.

In terms of q_v, we have for point s:

$$N_c = \text{Meridional compression per unit length} = \frac{q_v}{2s} \tan\theta \,(s_2^2 - s^2) \qquad (27.2)$$

$$N_t = \text{Hoop tension per unit length} = q_v s \sin^2\theta \tan\theta$$

In terms of q_n, we have

$$N_c = \frac{q_n}{2s} \tan\theta \,(s_2^2 - s^2) \text{ (as above)} \qquad (27.3)$$

$$N_t = q_n s \tan\theta$$

(*Note:* As the value of N_t by Eq. (27.3) is larger than that obtained by Eq. (27.2), we will use Eq. (27.3) in our calculations. We will also check the maximum value of N_c (compression) obtained by the above equation with the value of N_c got by the above Eq. (27.1) derived from simple statics.

27.3.1 Nature of Forces

It can be clearly seen that:

1. Maximum compression is at the junction between the column and the shell and decreases to zero at the end non-linearly.
2. Minimum tension is at the column end and increases linearly to the base. The method of design of these shells is shown by Example 27.1.

We design for the maximum compression and tension and also find where only nominal steel will be required.

27.4 DETAILING OF STEEL

A truncated conical shell is the simplest form of the shell for placing of steel reinforcement. We place the necessary compression steel along the length of the shell and the tension along the circumference around the shell.

The junction between the column and the conical footing should be also carefully detailed to form a smooth transition from the circular column base to a cone. Necessary hoop tension steel should be provided at the junction. In all cases, where the conical shell forms the foundation for a hollow cylindrical column a *ring beam has to be also provided at the top* (at the transition).

A *ring beam at the bottom of the shell* also is usually provided as it increases the stiffness of the shell and delays cracking at the base where we have the maximum tension. It also increases the ultimate strength by preventing premature tension failure of the shell.

27.5 SUMMARY

This chapter explains briefly the use of conical shell as foundation for solid and hollow columns. Methods of designing and detailing the reinforcements are also explained.

EXAMPLE 27.1

Design a conical foundation for a 400 mm diameter column carrying a load of 1100 kN if the safe bearing capacity of the soil is 75 kN/m^2.

Reference	Step	Calculation
	1	*Find base diameter based on safe bearing capacity*
		$$\frac{\pi D^2}{4} \times 75 = 1100 \text{ kN gives } D = 4.3 \text{ m}$$
		Provide 4.5 m cone. Radius = 2.25 m
		Assume concreting slope of 1 to 2 horizontal
		Rise = 2.25/2 = 1.125 m
	2	*Find shell parameters* s_1, s_2 and θ
		(s_1 = distance from apex to column, s_2 = distance from apex to end of shell, and θ = half central angle)
		θ = 1/2 central angle = $\tan^{-1} 2 = 63.44°$
		s_1 = Length to base of column from apex (1 : 2 slope)
		$$= \sqrt{(0.2)^2 + (0.1)^2} = 0.224 \text{ m}$$
		Length of the end of the shell from the apex
		$$s_2 = \sqrt{(2.25)^2 + (1.125)^2} = 2.516 \text{ m}$$
		$\sin\theta = 0.894$, $\cos\theta = 0.447$, $\tan\theta = 2$
	3	*Find vertical pressure* q_v *for factored load*
		$$q_v = \frac{1.5 \times 1100 \times 4}{\pi \times (4.5)^2} = 103.8 \text{ kN/m}^2$$
Eq. (27.3)	4	*Maximum compression per metre is at base of column at the top of the cone.* $s = s_1$
		$$N_c = \frac{q_v}{2s_1} \tan\theta\,(s_2^2 - s_1^2)$$
Eq. (27.1)		$$= \frac{103.8 \times 2 \times (2.516^2 - 0.224^2)}{2 \times 0.224} = 2910 \text{ kN/m} = 2910 \text{ N/mm}$$
		Check by statics Eq. (27.1)
		$$N_c = \frac{1.5 \times P}{\pi d \cos\theta} = \frac{1.5 \times 1100}{3.14 \times 0.4 \times 0.447} = 2939 \text{ kN/m}$$
		(These two values tally)
		Compression varies from maximum to zero at the tip.
	5	*Design for compression*
		With minimum thickness for cover $\not<$ 50 mm for steel etc. least thickness = 120 mm

Reference	*Step*	*Calculation*
		Provide 140 mm thickness (minimum)
Step 4		Area of steel required for 2910 N/mm compression placed along the circumference of the column.
		Total compression in concrete = 2910 × π × 400 = 3655 kN
		Safe value of comp. in concrete = 0.4 × 20 × (π × 400) × 140 = 1407 kN
		Area of steel reqd. $= \dfrac{(3655 - 1407) \times 10^3}{0.67 \times 415} = 8084 \text{ mm}^2$
		Provide 26 Nos. of 20 mm rods equally spaced which gives 8170 mm^2.
	6	*Check where no compression steel is required (let it be s from both apex)*
		Comp. taken per mm $= \dfrac{103.8 \times 2(2.516^2 - s^2)}{2s} = \dfrac{657 - 103.8s^2}{s} \text{ N/mm}^2$
		Equating it to 1407 kN, we get s = 0.45 m beyond which no steel is theoretically required. However, we will extend the steel to the base and also provide a nominal ring beam at the bottom.
	7	*Check percentage of steel at bottom (where compression is least) assuming constant thickness of shell. (We must provide minimum steel.)*
		$p = \dfrac{8170 \times 100}{3.14 \times 4500 \times 140} = 0.41\%$
		This is more than the minimum required. However, spacing = π × 4500/26 = 544 $\ngtr 3t$. Add extra steel to control spacing as spacing for steel controls cracking.
	8	*Design for maximum hoop tension @ s_2*
Eq. (27.3)		$N_t = q_v s_2 \tan\theta$ = 103.8 × 2.516 × 2 = 522.3 kN/m
		A_{st} per metre length $= \dfrac{523.3 \times 10^3}{0.87 \times 415} = 1450 \text{ mm}^2/\text{m}$
		Provide 18 mm @ 175 mm (1453 mm^2/m)
		Check stress in concrete by elastic design
		Stress $= \dfrac{522300}{A_c + (m-1)A_s}$; $A_c = 140 \times 1000$ and $m = 14$
		$= \dfrac{522300}{(140000 + 13 \times 1453)} = 3.28 \text{ N/mm}^2$
		This must be less than $f_{ck}/10$ = 25/10 = 2.5 N/mm^2.

Reference	Step	Calculation
		This is slightly higher. We may increase the thickness of concrete or use stronger concrete M_{30}.
	9	*Design for hoop tension N_t at place where column and steel meet,* $s_1 = 0.224$ m
Eq. (27.3)		$N_t = q_v s_1 \tan \theta = 103.8 \times 0.224 \times 2 = 46.50$ kN/m (N/mm)
		Area of tension steel required $= \dfrac{46.5}{0.87 \times 415} = 0.128$ mm²/mm
		Area per metre length = 128 mm²/m (very low)
		Provide 12 mm @ 150 mm giving 754 mm²/m. This will restrict cracking.
	10	*Check elastic stress in tension [elastic design]*
		Tension = 46500/1.5 = 31000 N/m
		Area of concrete = 140 × 1000 = 140000 mm²/m
		Area of steel = 754 mm²/m. Assume $m = 13$.
		Eq. area = 140000 + (13 – 1)754 = 149048 mm²
		Stress $= \dfrac{31 \times 10^3}{149 \times 10^3} = 0.20$ N/mm²
		As tension = $f_{ck}/10$. This is allowable.
	11	*Detailing*
		(a) *Tapering of the shell* We can also reduce the thickness of concrete of the cone from 140 mm at the top to the bottom to 120 mm.
		(b) *Bottom / ring beam* A projection (at the bottom of the cone on the shell surface), of 80 mm × 80 mm can be provided as a ring beam at the end. The nominal steel provided at the bottom of the cone can be bent to suit this ring beam reinforcement.
		(c) *Placing of steel* Steel is placed at the middle surface to have maximum cover.
		(d) *Column cone junction* The reinforcements from the column should be distributed properly to the cone and properly tapped. A fillet, as shown in Figure 26.3, with at least nominal steel will greatly help the transition.

REFERENCES

[1] IS 9456, 1980 (with Amendment No. 1, 1982), *Code of Practice for Design and Construction of Conical and Hyperbolic Paraboloidal Types of Shell Foundations*, BIS, New Delhi.

[2] Kurien, Niran P., *Shell Foundations*, Narosa Publishing House, New Delhi, 2006.

[3] Fischer, L., *Theory and Practices of Shell Structures*, Wilhelm Ernst and Sohn, Berlin, 1968.

Effect of Earthquakes on Foundation Structures

28.1 INTRODUCTION

When we plan foundations of structures for earthquake forces, we have to consider the following:

1. What are the base shears for structural design of foundations. This is especially important when planning pile foundations on liquefiable soils, where all the base shears have to be taken by the piles as columns will not be laterally supported by soils.
2. What are the specific recommendations regarding design of foundations in IS 1893 (2002) "Criteria for earthquake design of structures"?
3. Is the foundation liable to liquefy due to the earthquake? What is the value of the local acceleration to be taken for examining liquifaction?
4. Is the foundation soil liable to settle?

In this chapter, we first deal with the general aspects of earthquakes with respect to structural design and then with the above problems.

28.2 GENERAL REMARKS ABOUT EARTHQUAKES

The following terms used in earthquake engineering are important.

28.2.1 Magnitude and Intensity of an Earthquake

Earthquakes are formed by the movement of the rock surfaces along the geological faults on the earth's surface. They release a large amount of in-built strain energy and can occur on land or sea. Powerful earthquakes deep under the sea can produce *Tsunami*, and on land earthquakes affect man-made structures, natural slopes, etc. It is very important to note that the magnitude of the earthquake will depend on the *length of the fault* that can move, its depth, etc. There are two types of measurement of earthquake, namely *magnitude* and *intensity* as described below.

There are many ways in which the magnitude of an earthquake is measured. We deal with three of them very briefly. (Details can be had from books dealing with earthquake engineering.) These three are M_W, M_L or M and MMI scales.

1. **Moment magnitude M_W.** Moment magnitude represented by M_W is a measure of the energy released by the earthquake, which will depend on the length and other characteristics of the fault. The following is one of the many expressions used to find M_W.

$$M_W = 4.86 + 1.32 \log L \tag{28.1a}$$

where L is the length of the fault in km (see Example 28.1). Table 28.1 gives some of the data regarding moment magnitudes and duration in relation to fault length.

TABLE 28.1 Relation between Length of Fault Magnitude M_W and Duration of Earthquake [1]

Length of fault (km)	M_W	*Duration* (sec)	
		Rock sites	*Soil sites*
2.0	5	4	8
7.5	6	8	16
42.0	7	16	32
233.0	8	31	62
680.0	8.5	43	86

2. **Richter magnitude scale (M_L or M).** This is based on the measurement of the effect of earthquake on the surface of the earth. Richter proposed that an earthquake be considered of zero magnitude if its effect produces a record of amplitude of 1 mm on a standard seismograph (Wood Anderson Seismograph) at a distance of 100 km from the epicentre. Another earthquake that will produce a record of amplitude of 10 mm will be called of magnitude 1 as log 10 = 1. It is written as $M_L = 1$ (M local) or $M = 1$. As it is a log scale measurement, an earthquake of M_5 will be 100 times of magnitude than M_3.

   It is obvious that M_W and M_L need not be the same. However, up to a moment magnitude of $M_W = 6.5$, its value is approximately equal to M_L; at higher values, they can be different. For example, it is stated that the San Francisco (1906) and Chile (1960) earthquakes were both of Richter scale 8.3, but the M_W values have been estimated as 7.9 in the former and 9.5 in the latter.

3. **Modified Mercalli intensity (MMI) of an earthquake** (I_0). In this case, the intensity is not measured by any instrument but it is a measure of how it affects people, structures, etc. It is measured as Modified Mercalli Intensity scale as follows:

I–IV (Very minor shocks): All people do not feel it and can be detected by instruments only.

V (Minor earthquake): Most people feel it.

VI (Light earthquake): People find it difficult to walk.

VII (Moderate earthquake): People cannot stand.

Up to XII (Severe to very severe earthquakes): Leading to total damage and destruction.

The relation between M_W and intensity I_0 is usually represented as,

$$M_W = 2/3\ I_0 + 1 \tag{28.1b}$$

For example, for I_0 = VI, $M_W = (2/3 \times 6) + 1 = 5$.

28.2.2 Peak Ground Acceleration (PGA)

Peak ground acceleration is the largest horizontal acceleration felt at a place. It is a very important quantity by which the earthquake effects are determined on structures and on the ground. In some cases, it is peak horizontal velocity (PGV) that is important. The relationship between velocity (v), acceleration (a) and displacement (y) can be taken as follows (see Example 28.1):

$$v = \frac{T}{2\pi} a = \frac{2\pi}{T} y \tag{28.2}$$

The relation between I_0 and acceleration a_{max} for that earthquake in cm/sec^2 is taken as,

$$\log a_{max} = I_0/3 - 1/2 \tag{28.3}$$

(*Notes:* 1. Assuming g is 98 cm/sec^2, we can find $a_{max}/g = \alpha_0(g)$.

2. As we will see later in Sec. 28.3.1, we usually design ordinary buildings for only 1/2 the above value and important buildings for more than 1/2 the value.)

28.2.3 Zone Factor (*Z*)

In 1970, the International Conference of Building Officials recommended the adoption of a Uniform Building Code (UBC). Many countries like USA and India accepted the 1991 UBC recommendations in their building codes. It recommended to divide a large country like India with different geology and different likelihood of earthquakes into different zones and assign a zone factor Z [3], [4]. Accordingly, India has been divided into four zones II, III, IV and V (zone I being omitted) corresponding UBI. This division of zone is shown in Figure 28.1.

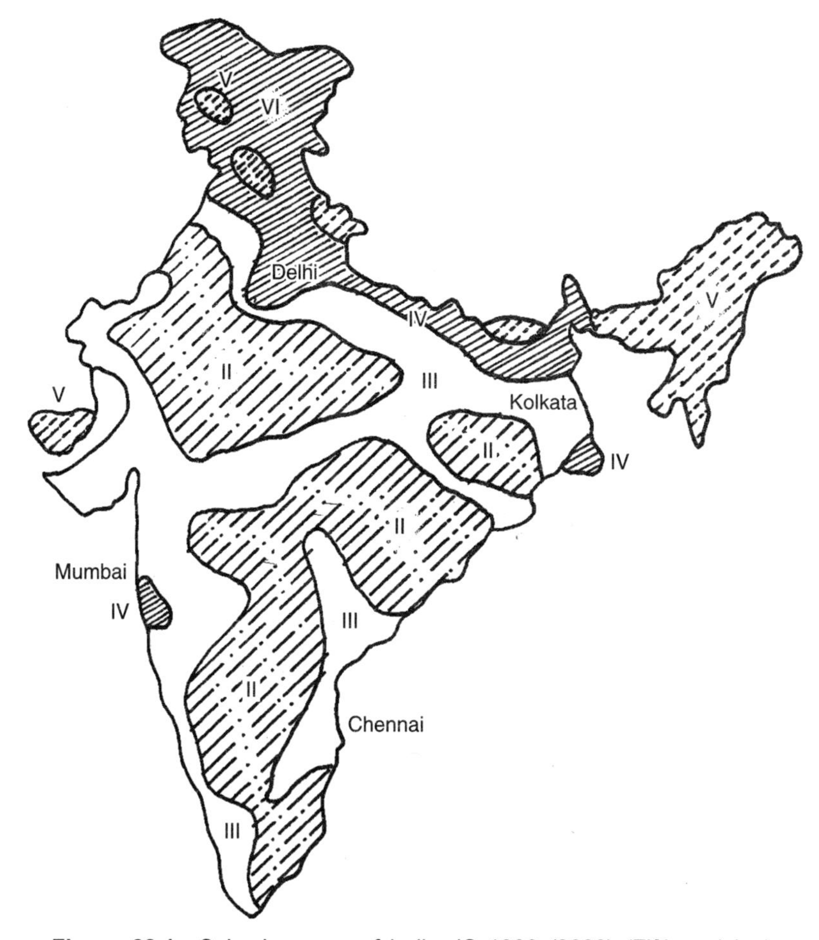

Figure 28.1 Seismic zones of India. IS 1893 (2002) (Fifth revision).

The value of Z gives us a reasonable estimate of the Effective Peak Ground Acceleration (EPGA) of a Maximum Credible Earthquake (MCE) that can occur at that place. When considering Design Basis Earthquake (DBE), we divide Z by 2 (see Sec. 28.3).

28.2.4 Relation between Various Factors [3]

The relation between the various items in earthquake engineering.

TABLE 28.2 Relation between Various Items in Earthquake Engineering

Description	*Zone*	M_L	$MMI = I_0$	$PGA\ (a_{max}/g)$	$\alpha_0(g)$	Z
Very low	I	< 5	< V	–	0.01	–
Low	II	5–6	VI	0.06–0.07	0.02	0.10
Moderate	III	6–6.5	VII	0.10–0.15	0.04	0.16
Severe	IV	6.5–7	VIII	0.25–0.30	0.05	0.24
Very severe	V	Over 7	IX	0.50–0.55	0.08	0.36

Notes: 1. Z values are 4–5 times the old α_0 values. [α_0 = Horizontal seismic coefficient we use in static method.]
2. Max values of $\alpha_0 = 0.08$ was adopted in the earlier Indian codes based on Assam Earthquake of 1893 (M.8.7).
3. Each successive zone factor Z is 1.5 times its lower one.
4. Figure 28.2 shows the average acceleration spectra recommended by IS for use till 2002, this has now been replaced by Figure 28.3 of the revised code.

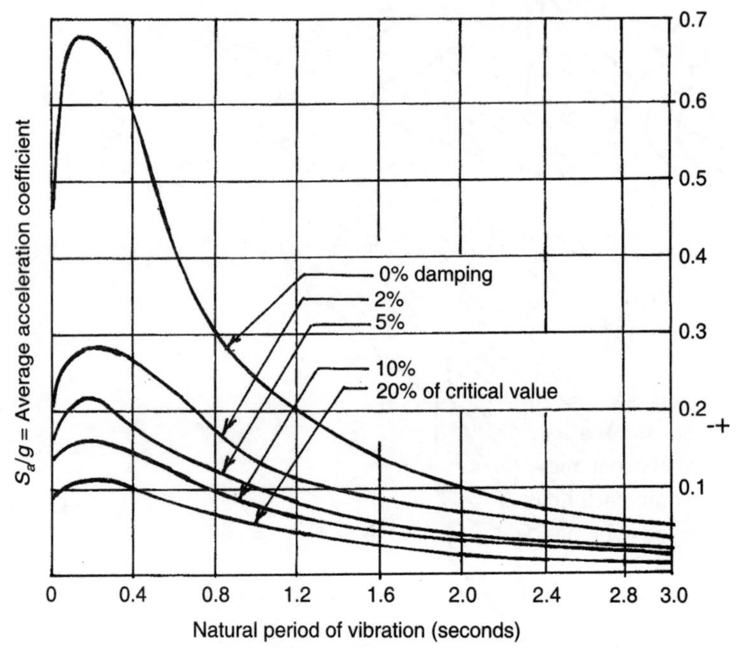

Figure 28.2 Average acceleration spectra prescribed by IS 1893, 1984 (Fourth revision) (diagram based on spectrum suggested by Housner) for static design of structure.

28.2.5 Response Spectrum [4]

UBC has recommended two approaches to find the effect of earthquake on buildings. One is the static approach which has now been replaced by the dynamic approach using *design response spectra.*

A *response spectrum represents the frequency dependant amplification of the ground motion by different local conditions for a given (usually 5%) damping factor.* These normalized spectra are presented for three surface profiles, with greater acceleration for softer deep soil profiles and less for rock profiles. The response spectra, suggested by UBC and also accepted in IS 1896 (2002), is given in Figure 28.3.

The natural frequency of a framed building depends, to a large extent, on its height or number of storeys. A rough value is $0.1n$ seconds, where n is the number of storeys of the building. The natural period of the system and its damping factor are the important items that influence the effect of the earthquake.

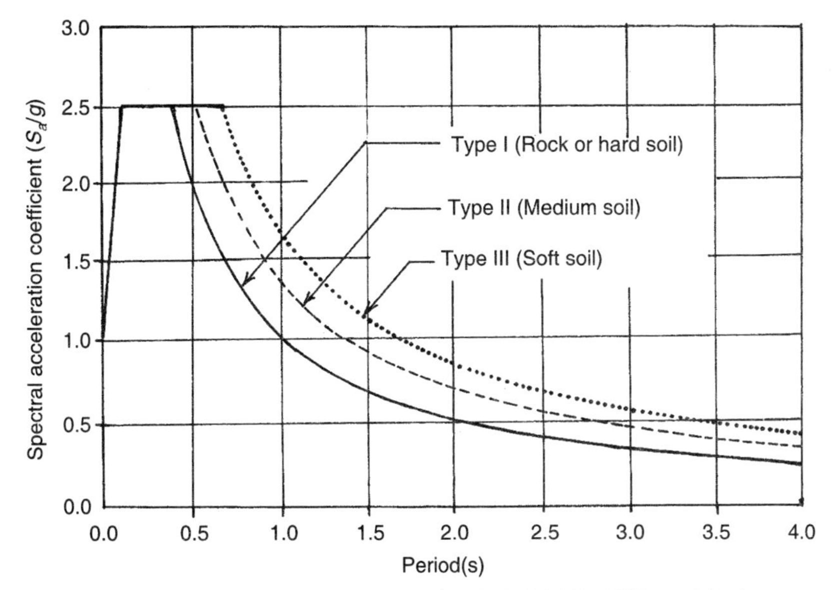

Figure 28.3 Seismic zones in IS 1893 (2002) (Fifth revision).

It has been found that for structural systems, a system which has a natural period up to 0.5 second, can be considered as rigid, and the earthquake effects are amplified in these rigid structures. But at higher frequency, the amplification is reduced. The effect of this phenomenon was first noticed in some earthquakes where buildings of 3–4 storeys, which can be considered as rigid, were destroyed whereas very tall buildings survived the earthquakes.

28.2.6 Damping Factor

The damping factor (like base isolation of buildings) is another factor which has a great effect on the response spectrum. In structural design of conventional foundations, we take the damping factor as 0.5 per cent. The design response spectrum in IS 1896 is based on 5% damping.

28.2.7 Design Horizontal Seismic Coefficient

In earlier days, the base shear on a building was calculated by the static method by the simple formula

$$V_B = \text{(Design horizontal seismic coeff.)} \times \text{(Seismic wt.)} = \alpha_0 W$$

As can be seen in Table 28.1, IS 1893 recommended its use till as recently as 2002.

28.3 HISTORICAL DEVELOPMENT OF IS 1893

IS 1893 was first published in 1962 and then revised in 1966, 1970, 1975, 1984 and 2002. The base shear calculated by the various versions is shown in Table 28.3. Reference can be made to books on R.C. Design for a fuller treatment of the subject [5], [6].

TABLE 28.3 Table Showing Development of IS 1893 for Calculation of Base Shear for a Hospital Building in Zone IV (5)

No.	*IS code*	*Calculation of base shear in buildings*	*Base shear* in example*
1	1893 (1962)	$V_B = \alpha_h W$ $\alpha_h = K\alpha_0$ (α_0 = 0.04 for zone IV) k for flexibility depending n storey height	200 kN
2	1893 (1975)	$V_B = C\alpha_h W$ C = Factor for flexibility $\not> 1$ a_h = Horizontal seismic coefficient = 0.05 for zone IV. See Table 28.2.	500 kN
3	1893 (1966)	$V_B = C\alpha_h \beta W$ C = Factor for flexibility α_h = Table 28.2 β = Factor for foundation (rock = 1) [India adopts UBC recommendations of 1985 and 1991]	338 kN
4	1893 (1975)	$V_B = C\alpha_h W$ C = Factor for flexibility *Method 1*: α_h = 0.075 for zone IV, static method *Method 2*: $\alpha_h = \beta I F_0 (S_a/g)$, response spectrum method β = Factor for foundation I = Importance factor F_0 = Seismic zone factor S_a/g = Average response acceleration coefficient	675 kN
5	1893 (1984)	$V_B = KC\alpha_h W$ (Figure 28.2) K = Factor for structural framing system C = Factor for flexibility *Method 1*: $\alpha_h = \beta I a_0$ *Method 2*: $\alpha_h = \beta I F_0 (S_a/g)$	675 kN
6	1893 (2002)	$V_B = A_h W$ Using response spectrum $A_h = \left(\frac{Z}{2}\right)\left(\frac{I}{R}\right)\left(\frac{S_a}{g}\right)$ (Figure 28.3)	740 kN

**Note:* This indicates the steady increase of forces that is required to be resisted during the past years and also due to the adoption of UBC by India.

28.3.1 Philosophy of Design of Buildings according to IS 1893 (2002)

The present philosophy of design of buildings, as recommended by UBC, is as follows:

The Maximum Credible Earthquake (MCE) is the absolute maximum which can ever be expected to occur in the given site. The value of zone factoris a reasonable estimate of the *effective peak ground acceleration* (PGA). Complete protection against MCE is not economically advisable for ordinary buildings. Hence, assuming that we can depend on both *strength and ductility and also provision of special features such as shear walls in design*, we accept a lower value of Z in the design. Thus, we design for a Design Basis Earthquake so that:

1. The structure should have ample strength to resist all earthquakes less than DBE without any damage.
2. The structure should resist the DBE without structural damage though some non-structural damage that we can repair later can take place under DBE.
3. Ordinary structures should be able to resist MCE without total collapse so that there is no loss of life because of the inbuilt ductility and special features. Some structural damage is expected under an MCE.
4. Depending on the importance of the structure (I), we may increase the DBE to withstand higher earthquakes.
5. Vertical projections (such as chimneys, tanks, etc. in buildings) should be checked for five times the horizontal seismic coefficient. Similarly, horizontal projections should be checked for five times the 2/3 horizontal acceleration ($5 \times 2/3 = 10/3$ the horizontal acceleration, Cl. 7.12.2.1 and Cl. 2.2).

Thus, the computation of base shear for each zone is laid down specifically by IS 1893, Part I (2002).

28.3.2 Calculation of Base Shear by IS 1893 (2002)

Now let us take the first topic of an study mentioned in the Introduction (Sec. 28.1) namely the value of the base shear to be taken in design of superstructures and substructures. We must know this value will be different from what values we take for studying liquifaction (Sec. 28.5.4). The base shear for structural design of foundations will be obtained from the lateral forces acting on the superstructure.

IS 1893 recommends the estimation of this base shear by any of the following three methods [3]:

1. Equivalent static lateral force method
2. Response spectrum method
3. Time history method

The *equivalent static lateral force method* is the simplest and can be used for most small to medium sized buildings. It uses the following formulae:

$$V_s = A_h W$$

$$A_h = \left(\frac{Z}{2}\right)\left(\frac{I}{R}\right)\left(\frac{S_a}{g}\right) \tag{28.4}$$

where

A_h = Design horizontal seismic coefficient

Z = Zone factor

I = Importance factor (varies from 1–2, depending on the importance of the structure)

R = Response reduction factor depending on the type of detailing of structure (varies from 3–5)

S_a/g = *Spectral* acceleration coefficient from response spectrum, shown in Figure 28.3.

Reference can be made to specialized literature for the details of calculation of the base shear by other methods [5], [6].

28.4 IS 1893 (2002) RECOMMENDATIONS REGARDING LAYOUT OF FOUNDATIONS

Let us now examine the second aspect, namely the general recommendations in IS 1893 (criteria for Earthquake Design of structures) about planning for foundations for earthquakes. The following are the broad recommendations regarding types of foundations that can be used in different zones in India.

28.4.1 Classification of Foundation Strata

Foundation materials are divided into three types:

- Rock and hard soil (well graded sand with gravel, etc.) with SPT value $N > 30$
- Medium soil with $N > 15$
- Soft soil with $N < 10$

28.4.2 Types of Foundations Allowed in Sandy Soils

Zone II No special considerations are needed for ordinary buildings. However, for *important buildings*, the foundation should be as follows:

At depth 5 m, N should be ≥ 15

At depth 10 m, N should be ≥ 20

Zones III, IV and V for all types of structures

At depth 5 m, N should be ≥ 15

At depth 10 m, N should be ≥ 25

If the N values of the soil are less than those specified above, artificial densification should be adopted to achieve the above values (see Sec. 28.8). Otherwise, pile foundations going to stronger strata should be adopted.

28.4.3 Types of Foundations that can be Adopted and Increase in Safe Bearing Capacity Allowed

1. **Isolated footings without connecting beams.** These can be used in Zone II in all soils with $N > 10$. It should not be used if $N < 10$. No increase in bearing capacity is allowed in type 3 soils, but the bearing capacity can be increased by 50% in rocks and type 1 soils and by 25% in type 2 soils for earthquake loads.
2. **R.C. footings with connecting tie beams.** All footings in soils with $N < 10$ should be tied together with grade beams and/or plinth beams. When these ties are used, we can increase the allowable bearing capacity by 50% in rock and type 1 soil and by 25% in type 2 soils for earthquake loads.
3. **Raft foundations.** These foundations are suitable in all cases. An increase of 50% in the bearing capacity is allowed when earthquake forces are also considered.
4. **Pile foundation.** The piles should rest in good soils $N > 15$. They are very suitable for earthquake resistance. An increase of 50% in the bearing capacity is allowed for piled foundations when considering forces, including earthquake forces.

[*Notes:* 1. It can be seen from the above that in regions liable to liquefaction, shallow foundations should be tied together. R.C. perimeter footings and wall footings should be tied together into a grid. Rigid raft foundations will also behave well. It is also necessary that buried pipes, such as sewage and water mains should have ductile connections.

2. When designing pile foundations for important buildings, we should check whether the foundation soil can liquefy and if so, to what depth it will do so. The lateral resistance and bearing capacity of this region should be neglected. The pile will act as a column fixed below this lever of liquefaction and subjected to the lateral earthquake forces.]

28.4.4 Summary of IS 1893 Recommendations for Foundation Design for Earthquakes

The base shear should be calculated as specified. The safe bearing capacity for loads including earthquake forces, can be increased as specified. Special attention should be given to the possibility of liquefaction of soils, especially in sand deposits with N values less than 10 (see below).

28.5 LIQUEFACTION OF SOILS

The third aspect we consider in detail is whether the foundation soil is liable to liquefy and what is the PGA value to be adopted for checking liquifaction of the foundation soils. This occurs generally in sands. Liquefaction is the phenomenon in which the pore pressure in sand under the influence of shear stress produced by repeated cyclic stresses builds up to higher levels than *its effective pressure (acting on the strata)* so that the soil mass behaves like a liquid. If the soils below structures liquefy, the structures will undergo large ground settlement. We can divide this state into two classes, viz. *flow liquefaction* and *cyclic mobility*. A detailed study of liquefaction can be made from specialized literature on the subject. In this chapter, we only examine the method of identification of soils that are liable to undergo liquefaction during earthquakes and also methods to treat these soils against liquefaction.

28.5.1 Soils Susceptible to Liquefaction

In soil mechanics, we classify soil into clay, silt, sand, etc. as shown in Table 28.4. (Fines are commonly defined as that portion of soil below 0.074 mm.)

TABLE 28.4 Soil Fractions

Clay	*Silt*			*Sand*			*Gravel*
	Fine	*Medium*	*Coarse*	*Fine*	*Medium*	*Coarse*	

mm 0.002 0.006 0.02 0.06 0.2 0.6 2.0

Clays remain non-susceptible to liquefaction even though sensitive soft clays can lose strength during earthquakes. The most susceptible soils are fine grained soils with fractions up to fine sands. Geologically speaking, uniform fluvial deposits in alluvial fans, alluvial plains, beaches and estuarine deposits are the most likely to be affected, especially if they are near geological fault zones and with high ground water level. From geotechnical considerations, *fine grained soils that satisfy the following so-called Chinese criteria* are considered liable to liquefaction:

1. Fraction finer than 0.05 mm equal or less than 15%
2. Liquid limit less than 35%
3. Natural water content not be less than 90% of liquid limit
4. Liquidity index $= \dfrac{w - \text{PL}}{\text{LL} - \text{PL}}$ less than 0.75

The modified Chinese code for liquefaction proposes clay content to be equal or less than 5 per cent.

We also add the requirement that $PT(N) value should be less than 10. In short, liquefaction occurs mostly in fine to medium sands, with the void ratio more than the critical void ratio. The soil should be also saturated, i.e. the ground water level should be high. With increasing overburden the effective pressure increases and the chances of liquefaction decrease, so generally, liquefaction occurs in the top 10 m only. The grain size distribution of soils usually expected to liquefy is shown in Figure 28.4.

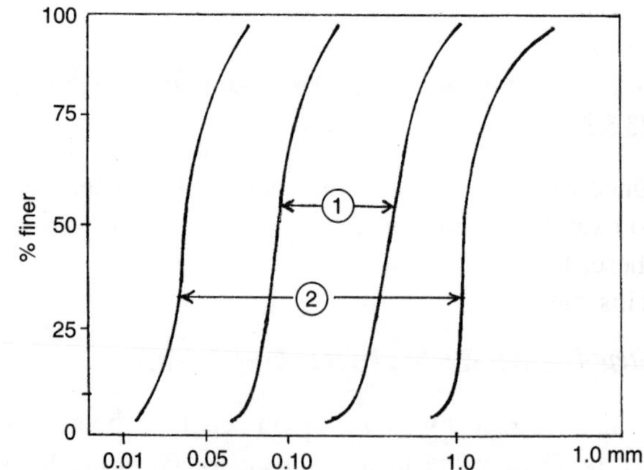

Figure 28.4 Range of grain size distribution of soils considered liable for liquefaction. (1) Boundaries for most liable soils, (2) Boundaries of potentially liquefiable soils.

28.5.2 Field Data on Liquefaction

The best way to study liquifaction is to examine past records. The field investigation of the past liquefaction has shown that the limiting epicentral distance beyond which liquefaction has not been observed in earthquakes of different magnitudes can be as shown in Figure 28.5.

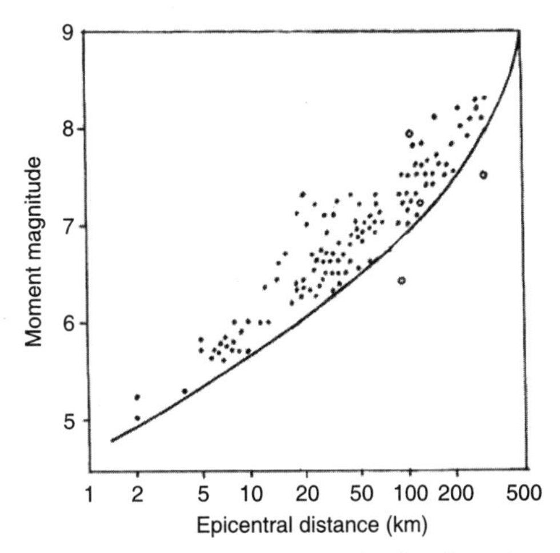

Figure 28.5 Relationship between moment magnitude of earthquake (see Sec. 28.2.1) and limiting epicentral distance of sites at which liquefaction has been observed. Deep earthquakes with focal depth > 50 km have produced liquefaction at greater distances as reported by N. Ambrasays.

It can be seen that with low magnitude of the earthquake, the distance within which liquefaction can take place is small, but the distance to which liquefaction can be expected increases very rapidly with the increase in magnitude of the earthquake [2]. The distances of the site from the fault and possible magnitude of earthquake are very important factors to be considered when we examine liquefaction potential of soils at a given site.

28.5.3 Cyclic Stress Ratio (CSR) Method of Prediction

Once we have estimated the peak ground acceleration (PGA) at the site, there are many methods to examine the possibility of liquefaction that have been proposed by various investigators. Of these, the following CSR (cyclic stress ratio) method is the one most commercially used [2]. This method is based on the *peak horizontal acceleration* expected at the given site.

Step 1: Find cyclic shear stress (CSS). The average shear stress τ_{max} produced by an earthquake of peak acceleration (α_{max}/g) at the ground level can be assumed to be equal to 65% of the maximum shear stress produced. This is called the cyclic shear stress (CSS) = τ_c produced by the earthquake loading. Its value is calculated as follows:

$$\text{CSS} = \tau_c = 0.65\left(\frac{\alpha_{max}}{g}\right)(\sigma \times r_d) \tag{28.5}$$

where

σ = Total vertical overburden stress in the sand layer under consideration
r_d = Depth correction factor = $(1 - 0.015d)$

Step 2: Find cyclic stress ratio CSR. The ratio of cyclic shear stress τ_c at the site to the effective stress σ_0 at the site is called cyclic stress ratio (CSR). It is given by the following expression.

$$\text{CSR} = \frac{\tau_c}{\sigma_0} = 0.65\left(\frac{\alpha_{max}}{g}\right)\left(\frac{\sigma}{\sigma_0}\right) r_d$$

Thus, $\text{CSR}_{M=7.5}$ means cyclic stress ratio for M 7.5 earthquake.

From the actual field observations, the relation between CSR and corrected SPT values for 60% efficiency $(N_1)_{60}$ for an earthquake of magnitude 7.5 at which liquefaction will take place is shown in Figure 28.6 [2]. [Go to step 3]

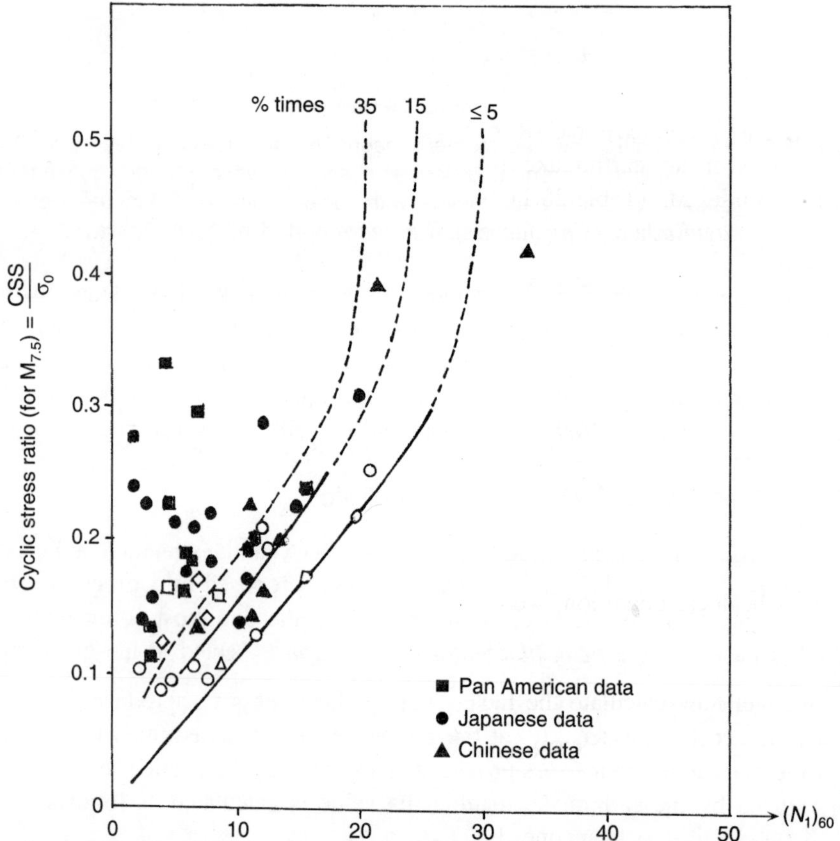

Figure 28.6 Relationship between cyclic stress ratios (cyclic shear stress/σ_0) causing liquefaction and S.P.T. $(N_1)_{60}$ values for sandy soils for $M_{7.5}$ earthquakes (Reported by H.B. Seed).

[*Note:* $(N_1)_{60}$ is obtained as follows. $(N_1)_{60}$ is the corrected SPT value, corrected for 60% efficiency of free fall energy of the standard equipment, the depth of overburden pressure, fine content, etc. The efficiency of a good automatic trip hammer can be taken as 80%. The formula used for the calculation of N_{60} is as follows:

$$N_{60} = N_m C_N \frac{E_m}{0.6E_{ff}}$$

where

N_m = SPT for the actual energy of hammer E_m

C_N = Depth correction factor

E_{ff} = Theoretical free fall hammer energy.

For example, if the SPT value with a field equipment delivering 70% of the drop energy is N_1 equal to and the *depth correction factor (from books on Soil Mechanics) is* 1.08, then

$$N_{60} = 9 \times 1.08 \times \left[\frac{0.70E_{ff}}{0.60E_{ff}}\right] = 11.3\,]$$

Step 3: The data presented for CSR against N in Figure 28.6 are values for which liquifaction occurs for an earthquake of magnitude 7.5. For earthquakes of other moment magnitudes M_w (Table 28.1), we have to scale up or scale down the M_{75} values by the factor *magnitude scaling factor* (MSF) as given in Table 28.5 [2].

TABLE 28.5 Magnitude Scaling Factor (MSF)

Value of M_W	*MSF*
5.0	1.50
6.0	1.32
6.75	1.13
7.5	1.00
8.5	0.89

For the local condition, we calculate

$$CSR_L = CSR_{7.5} \times MSF$$

Step 4: We can now calculate the factor of safety against liquefaction as,

$$FS = \frac{\text{CSR required for liquefaction at the site}}{\text{CSR produced at the site}} \tag{28.6}$$

If the FS is less than one, liquefaction can take place. This procedure is shown in Example 28.2.

28.5.4 Value of (α_{max}/g) to be used for a Given Site—Site Effects

We are able, to some extent, estimate the PGA (maximum value) for a given earthquake for structural design. But when examining the liquefaction effects at a given site, we have to estimate the local value of (α_{max}/g) to be used for the given site *which may be very distant from the epicentre*. The local value of a_{max} will depend on many factors.

1. Magnitude of the earthquake
2. Distance of the site from the epicentre
3. Effect of the local site condition, whether the soil deposit will *amplify* or attenuate (deamplify) the (a_{max}/g) values.

Amplification is defined as an increase in seismic ground motion intensity greater than that is expected from type I (rock or firm soils) soils. Amplification occurs often at sites overlain by thick, soft soil, especially when the period of earthquake motion matches the predominant period of the site [7], [8], [9] and [10].

The opposite of amplification is *attenuation or deamplification*. Figure 28.7 gives an idea of the decrease in PGA with distance in km from the epicentre. The intensity dies down quickly.

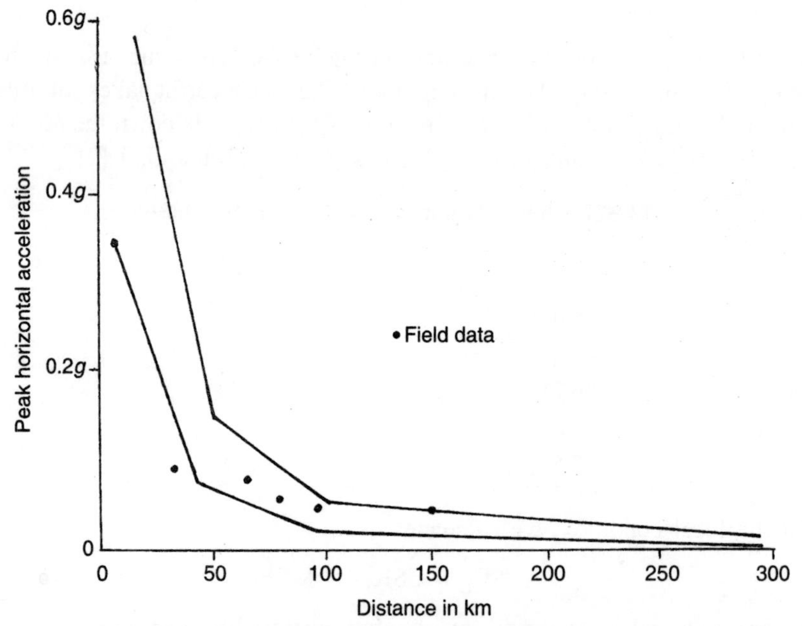

Figure 28.7 Attenuation (reduction) of peak horizontal (or ground) acceleration with distance from fault rupture.

Various formulae and software (e.g. FINSIM) are available to estimate the reduced magnitude of the earthquake at a given place at a specified distance from the active fault zone [2]. Similarly, the amplification of this reduced PGA due to the local soil site can be estimated by software such as SHAKE 2000. Reference can be made to specialized literature for this subject [2], [7], [8], [9], and [10].

28.6 AMPLIFICATION OF PEAK GROUND PRESSURE OF ROCK MOTION BY SOIL DEPOSITS

Table 28.6 gives roughly the soil amplification rating based on amplification susceptibility. When we meet soils which are prone to amplify the ground motion, we should be careful.

TABLE 28.6 Soil Category and Amplification

Soil category	*Description*	*Wave velocity* (m/sec)	*Susceptibility rating for amplification*
A	Compact rock	> 760	Nil
B	Stiff cohesionless soil	760–360	Low
C	Sands and clays	360–180	Moderate
D	Soft to medium clay	< 180	High
E	Peats, organic clay and soft clay	3–7	Very high

28.7 GROUND SETTLEMENT

The fourth problem, referred in our introduction that is important, is the settlement of foundation soils. Clays do not settle with vibration. Settlement of *sands which are loose* is a problem in earthquakes. The settlement is due to horizontal motion rather than vertical vibration. These can cause large differential settlement and cracking of buildings. The possibility depends on the magnitude of the vibration and the looseness of the sand deposits. *Sand deposits whose void ratio is less than its critical void ratio are liable to settle.* The denseness of the sand can be easily estimated by the N values of the deposit. The best precaution against settlement is to prevent the probability of its occurence.

28.8 METHODS TO PREVENT LIQUEFACTION AND SETTLEMENT

As we have already seen, liquefaction usually occurs in loose sands with $N < 10$, with void ratio much larger than the critical void ratio along with high ground water level. The obvious measures to prevent liquefaction are densification by any of the methods that can be easily used at the given site. Some of the methods used are as follows:

1. Compaction of the soil by procedures like
   - (a) Compaction piles (sand piles can be installed by driving hollow tubes with expendable driving tips, filling the tubes with sand, using stage compaction, and with drawing the tubes can be used)
   - (b) Vibro floatation
   - (c) Compaction with blasting
   - (d) Stone columns
   - (e) Dynamic compaction by falling weights

2. Stabilization of soil by grouting with cement slurry micropiles, or chemicals suitable for the purpose. A case study for improving the bearing capacity is given in Ref. [11].
3. As liquefaction needs high water level, drainage of water to greater depths will prevent liquefaction.
4. Increasing the value of the ground pressure on the liquefiable deposit by surcharge can also assist in prevention of liquefaction.

The effect of soil improvement undertaken should be always verified by in site tests such as SPT or cone penetration tests.

28.9 SUMMARY

The four basic considerations, namely the base shear in buildings for earthquake design, type of foundations to be used at a given site, liquefaction potential of foundation soil and the ground settlement possibility of the foundation, have been covered briefly in this chapter. Specialized literature given in the reference should be referred for a detailed study. Base shears have to be calculated according to IS 1893 (2002) and structural designs of superstructure and foundation have to be made on these calculations. However, no Indian Code has yet been evolved on estimation of liquefaction potential of a site. It is very important that a detailed soil report indicating the distance of the fault from the site, attenuation due to the distance of the site from fault, amplification (or attenuation) due to the site local soil condition have to be arrived at for a correct estimation of *the peak ground acceleration at the site.* Then only can we have an accurate assessment of the liquefaction potential of the soil at the given site. Otherwise, we will be overestimating disaster possibility for a given site.

Broad recommendations can be easily made from the grain size, index properties, void ratio—N values, ground water level, etc. of the site for cases *where liquefaction will not occur.* However, for the cases where these properties show that liquefaction is possible, further investigation to determine the a_{max} values for the site is necessary to predict that liquefaction will take place with a probable earthquake due to an active nearby fault. There are many cases where decisions made simply with respect to soil conditions without the estimation of correct a_{max} with respect to the nearest active fault have led to very expensive foundations even for ordinary type of buildings such as residential buildings.

EXAMPLE 28.1 (Determination of magnitude, intensity and peak horizontal acceleration due to a fault of known length)

The length of a fault at a place is 45 km. Estimate the moment magnitude (M_W), Richter magnitude (M_L) and peak horizontal acceleration (a_{max}).

Reference	*Step*	*Calculation*
	1	*Calculate moment magnitude* M_W
Eq. (28.1a)		$M_W = 5.08 + 1.16 \log L$
& Table 28.1		$= 5.08 + 1.16 \log 45 = 7$

Reference	Step	Calculation
	2	*Estimate magnitude by Richter scale M_L*
		$M_W = 7$ may show up on surface only as $M_L = 6.5$ or 7 depending on the soil profiles.
	3	*Find maximum intensity I_0 (based on feeling)*
Eq. (28.1b)		$M_W = \frac{2}{3}I_0 + 1; 7 = \frac{2}{3}I_0 + 1$
		$I_0 = 9$ (This is the maximum at the fault. At distances away from the epicentre, it can be less.)
	4	*Find peak ground acceleration a_{max}*
		$\log a_{max} = I_0/3 - 1/2$ (cm/sec^2)
Eq. (28.3)		$\log a_{max} = 9/3 - 1/2 = 2.5$
		$a_{max} = \log 2.5 = 316/981 = 0.32g$
		(*Note:* This corresponds to *Z* value obtained from Table 28.2. It is the maximum probable value.)

EXAMPLE 28.2 (Calculation for liquefaction susceptibility)

A deposit of medium to fine deposit has a corrected SPT value of 11.7. The water table at the site is 1.5 m below the ground level. If it is subjected to peak horizontal acceleration of 0.16*g* of M_5 earthquake, estimate its factor of safety against liquefaction at 6.2 m depth in the soil. Assume fines are less than 5%.

Reference	Step	Calculation
	1	*Find the total pressure at 6.2 m depth (σ)*
		Let dry unit weight be 18 kN/m^3 and saturated unit weight, 21 kN/m^2
		$\sigma = (1.5 \times 18) + (4.7 \times 21) = 125.7$ kN/m^3
	2	*Find cyclic shear stress produced by earthquake with* $(\alpha_{max}/g) = 0.16$
Eq. (28.5)		$\tau_c = 0.65\left(\frac{\alpha_{max}}{g}\right)\sigma r_d$
		$r_d = (1 - 0.015d) = (1 - 0.015 \times 6.2) = 0.91$
		$\tau_c = 0.65\ (0.16)\ (125.7)\ (0.91) = 11.90$
		Eff. pressure, $\sigma_0 = (1.5 \times 1.8) + (4.7 \times 11) = 78.5$ kN/m^2
		Cyclic stress ratio (CSR) $= \frac{11.90}{78.7} = 0.151$
	3	*Find cyclic resistance ratio to produce liquefaction (with respect to effective pressure) for M_5 earthquake for $N = 11.7$*
Eq. (28.6) Figure 28.6		Cyclic stress ratio = CSR($M_{7.5}$)/σ_0 for $N = 11.7$ with percentage of fines less than 5% from Figure 28.6 = 0.13.

Reference	Step	Calculation
Table 28.5		$(CSR)_{M_5} = [CSR]_{M_{7.5}} \times 1.5 = 0.13 \times 1.5 = 0.195$
	4	*Find FS against liquefaction for cyclic shear stress of Step* 2
Eq. (28.6)		$FS = \dfrac{\text{CSR required for liquefaction}}{\text{CSR produced at the site}} = \dfrac{0.195}{0.151} = 1.29$
		This means that the cyclic stress induced by the earthquake is lower than the cyclic stress required to initiate liquefaction. We already know that liquefaction does not happen for M_5 with $N > 10$ and the site value is 11.7.

REFERENCES

[1] Monograph for Planning and Design of Tall Buildings, Vol. C.L., Chapter CL2, Council of Tall Buildings, New York, 1969.

[2] Kramer, Steven L., *Geotechnical Earthquake Engineering*, Prentice Hall International Series, 1996. Published in India by Dorling Kindersley, Delhi.

[3] Srivathsava, L.S., A Note on Seismic Zoning of India, *Bulletin of the Indian Society of Earthquake Technology*, Roorkee, 1969.

[4] IS 1893 (Part I) 2002, *Criteria for Earthquake Design of Structures*, BIS, New Delhi.

[5] Varghese, P.C., *Advanced Reinforced Concrete Design*, 2nd Edition, Prentice-Hall of India, New Delhi, 2006.

[6] Agrawal, P. and M. Shrikhanda, *Earthquake Resistance Design of Structures*, Prentice-Hall of India, New Delhi, 2007.

[7] Neelima, Satyam D. and K.S. Rao, Estimation of Peak Ground Acceleration for Delhi Region using FINISH, Proceedings Indian Geotechnical Conference, Chennai, Vol. II, 2006, pp. 819–822.

[8] Anbazhagan, A., T.G. Sitharam and C. Divya, Site Amplification and Liquefaction Study for Bangalore City, Proceedings Indian Geotechnical Conference, Chennai, Vol. II, 2006, pp. 823–826.

[9] Suganthi, A. and A. Boominathan, Seismic Response Study of Chennai City, Proceedings Indian Geotechnical Conference, Chennai, Vol. II, 2006, pp. 831–832.

[10] Premalatha, K. and P. Anbazhagan, Microzonation of Liquefaction and Amplification Potential of Chennai City, Proceedings of the International Workshop Risk Assessment in Site Characterization and Geotechnical Design, Bangalore, India, Nov. 2004.

[11] Mirjalidi, Mojtaba and Amin, Askarinej, Preformed Micropile and Integral Packer Grouting for improving Bearing Capacity—A case study, Proceedings Indian Geotechnical Conference, Vol. II, Chennai, 2006, pp. 623–626.

A

Geotechnical Data

A.1 INTRODUCTION

In this section, we examine how field observations are to be recorded during soil investigation and also study the empirical methods used to estimate the safe bearing capacity of the soil from the field data. The background of these methods and additional data can be obtained from standard books on Soil Mechanics and Foundation Engineering.

A.2 CONVERSION UNITS

1 ton = 1000 kg

1 kg/cm^2 = $\frac{1}{10}$ N/mm^2 = 100 kN/m^2 = 10 t/m^2

1 N/mm^2 = 10 kg/cm^2 and 1 MN/m^2 = 100 t/m^2 = 10 t/sq ft (approx)

1 kg/cm^2 $\approx$ 1 t/sq ft = 10 t/m^2 (approx)

1 pascal (Pa) = 1 N/m^2 and 1 MPa = 1 N/mm^2

1 kN/m^2 = 1/10 t/m^2

A.3 STRENGTH OF MATERIALS

(Unit as N/mm^2 used by structural engineers is used here.)

S.No.	*Material*	*Strength*
1	High tensile steel wires	1570 N/mm^2 (ultimate strength)
2	High strength ribbed bars	415 N/mm^2 (yield strength)
3	Mild steel bars	250 N/mm^2 (yield strength)
4	High strength concrete	> 50 N/mm^2 (cube strength)
5	Structural concrete	≥ 20 N/mm^2 (cube strength)
6	Timber (for timbering of excavations)	≈ 10.5 N/mm^2 (sal)
	Allowable bending	≈ 10.0 N/mm^2 (Indian teak)
7	Bricks—wire cut	10–15 N/mm^2 (crushing strength)
8	Bricks—country-made	3–5 N/mm^2 (crushing strength)
9	Hard clay (N = 20) (SPT value)	0.2 N/mm^2 (q_u) [$c = q_u/2$]
10	Medium clay (N = 10)	0.1 N/mm^2 (q_u) [$c = q_u/2$]
11	Rock (granite)	3.2 N/mm^2 (allowable)
12	Limestone, sand stone	1.6 N/mm^2 (allowable)
13	Shale	0.9 N/mm^2 (allowable)
14	Concrete	$0.45f_{ck}$ (ultimate in bearing)

A.4 USUAL UNIT WEIGHT AND VOID RATIO OF SOILS

Soils	*Unit weight* (kN/m^3)		*Void ratio*	*Saturated w/c* – %
	Saturated	*Dry*		
Sands				
Uniform dense	21	17.5	0.51	19
Uniform loose	19	14.5	0.85	32
Clays				
Organic clays	15.8	–	1.90	70
Dense soft clays	17.7	–	1.20	45
Stiff clay	20.7	–	0.60	22

A.5 CLASSIFICATION OF SOILS

Classification according to grain size (M.I.T. system)

				Sand		
Grain size (mm)	< 0.002	to 0.06	to 0.2	to 0.6	to 2.0	Gravel
Classification	Clay	Silt	Fine	Medium	Coarse	

A.6 METHOD OF DESCRIBING SOIL AT SITE (FIELD CLASSIFICATION OF SOILS)

Field Observation

1. **Visual examination.** If more than 50% of the soil grain is visible to the naked eye we call it coarse grained soil. Describe soil as indicated in Sec. A.5.
2. **Dilatency test or shake test** to distinguish between silt and clay or fine sand. Saturated sand is taken on left hand, which is shaken by blows by right hand (or vice versa). Appearance of water on top distinguishes silt, fine sand from clay.
3. **Toughness test.** Roll sample between the palms. Silty soils break up at plastic limit. Hard sample indicates high plastic clays.
4. **Dry strength test.** This distinguishes organic clays and clays of low plasticity.
5. **Organic content.** Smell the soil for organic matter.

Field Description of Soils in Field Report

1. **State colour.** Black, brown, light grey, etc.
2. **Describe constituents.** Sand, clay, silt, etc.
3. **Describe structure.** Loose, dense, hard, soft, etc.
4. **Describe inclusions.** Calcareous, micacious with shells, etc.
5. **For sands specify size and grading.** Coarse, fine, cemented, well graded, poorly graded (uniform).
6. **Specify natural state.** For sands, dense, loose, etc. For clays, soft, medium, hard, etc.
7. **Indicate probable laboratory classification.** ML, CH, etc.

Examples of Visual Classification

Short description of soil

Sand Well compacted, brown, poorly graded, medium alluvial sand with about 10% gravel and 15% non-plastic fines (SM)

Clay Greyish brown, hard clay of low plasticity with kankar modules and coarse particles of in-situ rock (CL)

Field identification of rocks It is important that we should always specify *the type of rock we meet with such as granite, limestone, laterite, etc. Describe the state of weathering as follows*:

1. In-situ rock (solid rock)
2. Slightly weathered rock > 90% rock

3. Moderately weathered rock > 50% rock as large pieces
4. Highly weathered rock, can be excavated by tools
5. Decomposed rock, completely weathered rock
6. In-situ soil

Always soak rock in water for 48 hours and test whether it disintegrates into soil.

Cole and Stroud Classification (see Table 18.4) The plasticity chart used for classification of clays is given in Figure A.1.

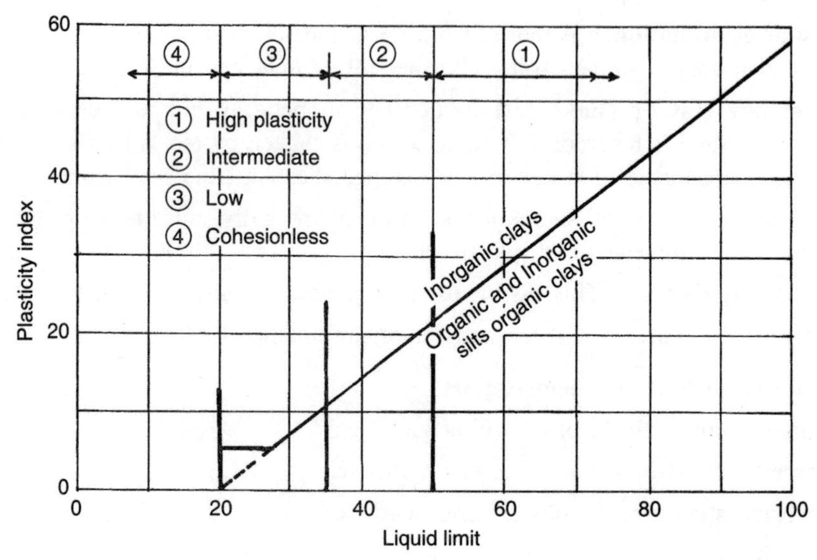

Figure A.1 Casagrandis plasticity chart for identification of fine-grained soils.

A.7 STANDARD PENETRATION TEST

In standard penetration test, the ideal procedure is to count the number of blows for every 75 mm (3 inch) penetration for six steps (75 × 6 = 450 mm) or at least record blows for three successive 150 mm penetrations (150 × 3 = 450 mm) and take the number of blows for the last 300 mm penetration only as N value.

Corrections

(a) Saturated fine sand — $N > 15$; $N = 15 + 0.5(N - 15)$

(b) Other correction $N_{cr} = N \times \eta_1 \times \eta_2 \times \eta_3 \times \eta_4 \times \eta_5$

η_1 = for overburden

η_2 = for hammer efficiency for N_{60} or 60% efficiency of blows

η_3 = for rod length

η_4 = for sampler (with or without liner)

η_5 = for bore hole diameter

Generally, we apply η_1 for overburden effects (or depth of test). We may also calculate N_{60} as explained in Sec. 28.5.3 (step 2) for 60% efficiency of hammer. Many data in earthquake engineering is based on N_{60} values.

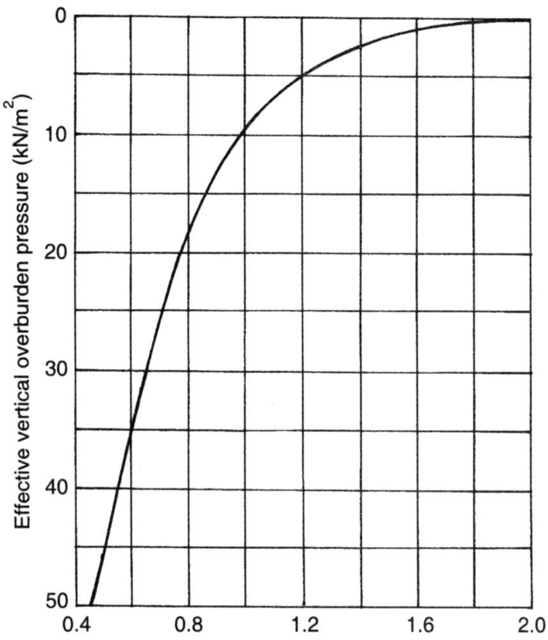

Figure A.2 Depth correction factor for SPT values.

A.8 STATIC CONE PENETRATION TEST (SCPT)

The result of SCPT is expressed as q_c in kg/cm^2. In soft clays, SCPT gives better results than SPT. The European practice is to use SCPT and the American practice is to use SPT. The test consists of finding the resistance in kg/cm^2 of a 60° cone of 35.7 mm diameter (10 cm^2 in area) being pushed into the soil at 5 mm/second.

Relation between SCPT and SPT (N values) [IS 2911]

Soil type	q_c/N (kg/cm^2)	
	IS	*Others*
Clays	1.5–2.0	–
Silts, slightly cohesive soils	2.0–2.5	2.0
Clean fine to medium sand	3.0–4.0	3.5
Coarse sand with gravel	5.0–6.0	5.0
Sandy gravels and gravel	8.0–10.0	6.0

A.9 IN-SITU VANE SHEAR TEST

The vane shear test is a very reliable test for clays whose SPT value is less than 8. A 50 mm diameter, 100 mm length (length/diameter = 2) is pushed into the clay with top of vane also inside the soil. Bjerrum proposed a correction as shown in Figure A.3. This correction is suggested on the assumption that the vane remoulds the soil and the value we get by this test is only the remoulded strength, which will be less than the real strength.

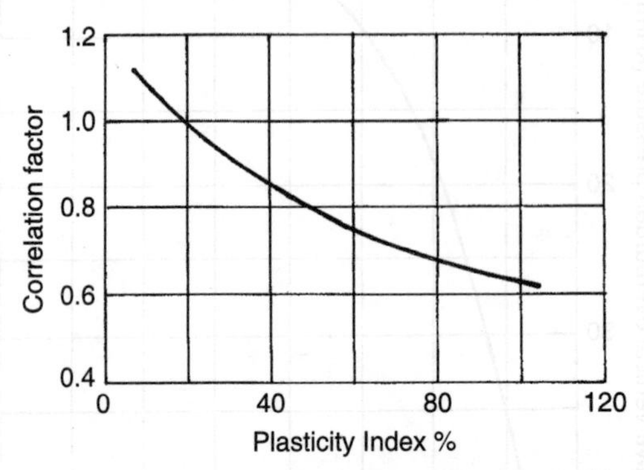

Figure A.3 Correction factor for remoulding effect in field vane shear test. (In-situ Shear strength = Field value/Correction factor.)

A.10 SPT VALUES AND NATURE OF SOILS

1. Clays

SPT value	*Nature of soil when moulded in the hand*	*Consistency index*
0–2	Very soft, extrudes between fingers in the fist	0.5
2–4	Soft, easily moulded between fingers	0.5–0.75
4–8	Medium to firm, needs some pressure to mould between fingers	0.75–0.85
8–16	Stiff, cannot be moulded between fingers	0.85–1.0
16–32	Very stiff above plastic limit	1.0–1.5
> 32	Hard (solid)	> 1.5

Notes: 1. Consistency index = (LL – w)/PI. When w/c is at LL, then consistency index = 0. When w/c is at PL, consistency index = 1.

2. Unconfined strength of clays, $q_u = \frac{N}{10}$ kg/cm^2 is a conservative value for clays.

3. Cohesion, $c = q_u/2$.

2. Sand

SPT	*Denseness*	*Relative density*	*Friction value* (degrees)	*SCPT value* (kg/cm²)	*Unit weight* (kN/m³)
<4	Very loose	<15	–	<20	<15
4–10	Loose	15–35	<30	20–40	15–17
10–30	Medium	35–65	30–36	40–120	17–19
30–50	Dense	65–85	36–40	120–200	18–21
>50	Very dense	85–100	>40	>200	>21

Note: We can take $\phi = 0.3N + 27$ or $\sqrt{20N} + 15$ degrees

A.11 MODULUS OF ELASTICITY OF SOILS

Modulus of elasticity of soils E_s is sometimes referred also as *modulus of compressibility*. This quantity is used to calculate immediate settlement of foundation. E_s should not be confused with *coefficient of volume compressibility* m_v we use to calculate final consolidation settlement in clays (see Sec. A.15).

The values recommended for E_s by IS 290 are as follows.

Material	E_s = *Value* × 10^6 N/m² (Range)
(a) Cohesionless soil	
Loose sand (round)	20–50
Loose sand (angular)	40–80
Medium dense sand (round)	50–100
Medium dense sand (angular)	80–100
Gravel	150–300
Silt	5–20
(b) Cohesive soils	
Very soft clay	2–15
Soft clay	5–25
Medium soft	15–50
Hard	50–100
Sandy	25–250

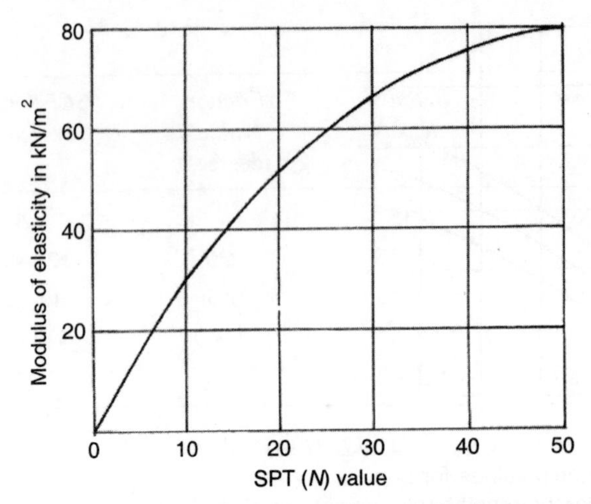

Figure A.4 Relation between SPT(N) values and elastic modulus of sand deposits (according to Schultze and Muhs).

A.12 POISSON'S RATIO OF MATERIALS

Poisson's ratio is as follows:

Sand:	0.3–0.4
Clay:	0.1–0.3
Concrete:	0.15–0.25
Sled:	0.28–0.31

A.13 SAFE BEARING CAPACITY (SBC) OF SOILS [1]

We define safe bearing capacity as ultimate bearing capacity divided by factor of safety. Allowable bearing capacity (also called allowable soil pressure) is bearing capacity we allow with respect to both safe bearing capacity and allowable settlement of the structure.

1. **For clays.** SBC from ultimate failures (approx. value)

$$\text{SBC} = \frac{5.7c}{(\text{FS} = 2.85)} = 2c = q_u = N/10 \text{ kg/cm}^2 = 10N \text{ in kN/m}^2$$

Effect of depth is shown in Figure A.5. (Settlement analysis is also needed)

2. **For sands.** SBC from maximum settlement 25 mm

$$= N/10 \text{ kg/cm}^2 = 10N \text{ in kN/m}^2$$

Thus, the **important thumb rule** is that from ultimate failure conditions in clays and maximum settlement of 25 mm consideration in sands, the *safe bearing capacity of soils* can be taken as 1/10 (SPT value) in kg/cm^2 or $10N$ in kN/m^2. (Figure A.7)

[More exact calculations will give slightly higher safe bearing capacity values.]

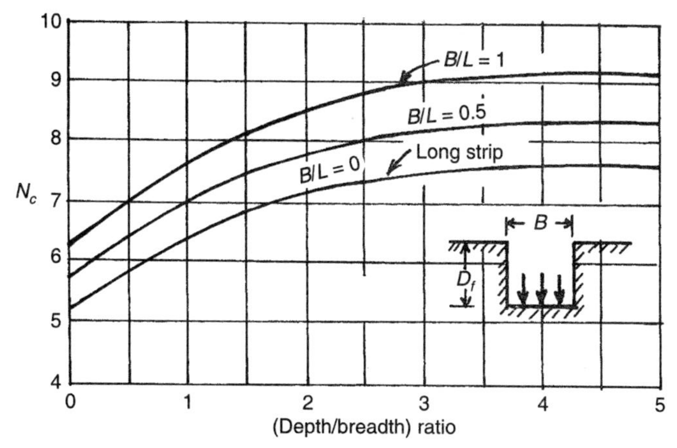

Figure A.5 Skempton's values for bearing capacity factor N_c for cohesive soils, varying with (depth/breadth) ratios (IS 6403–1981).

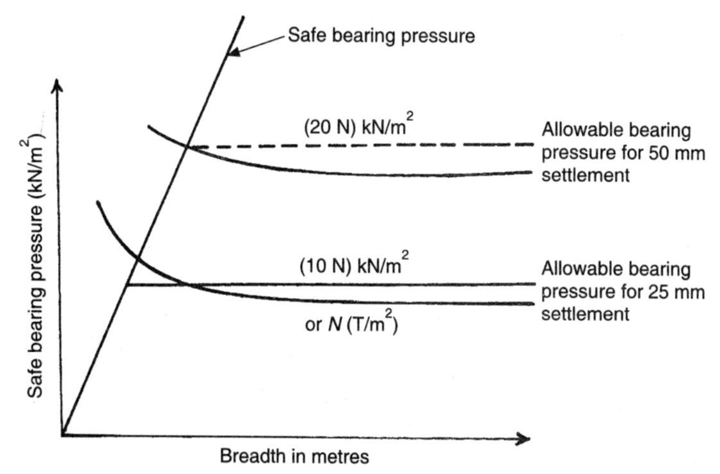

Figure A.6 Relation between safe bearing pressure and allowable bearing pressure as a function of allowable settlement and breadth of footing.

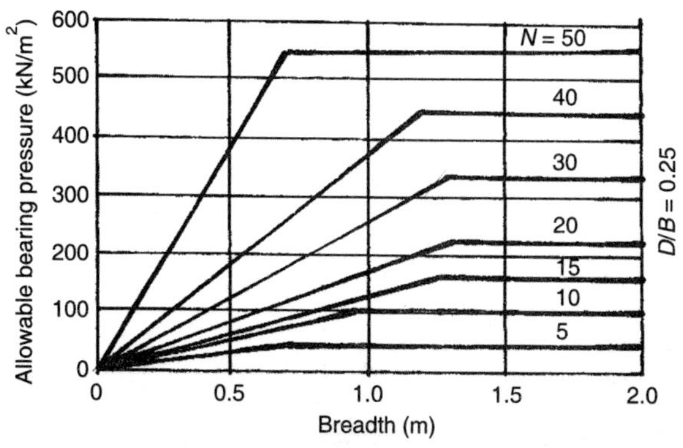

Figure A.7 Allowable bearing pressure for footings in granular soils based on SPT(N) values for 25 mm allowable settlement. (After Peek, Hansen and Thornburn: For other D/B ratios see Ref. 1)

A.14 IDENTIFICATION OF EXPANSIVE SOILS

We estimate the expansive character of soil from the data of the % of clay fraction and plasticity index as shown in Figure A.8.

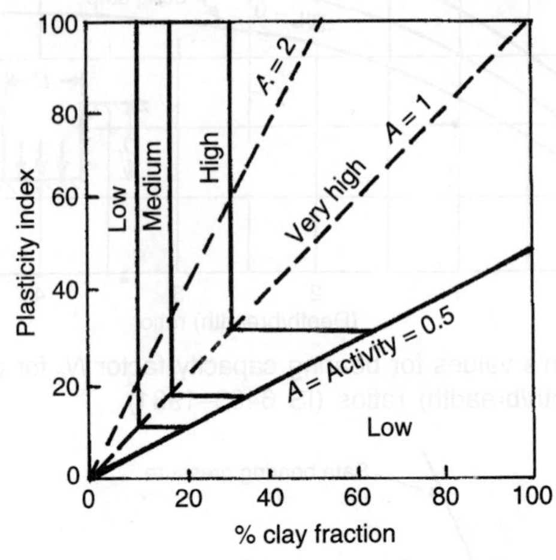

Figure A.8 Identification of expansive soils by acitivity chart.

A.15 CONSOLIDATION OF CLAYS

The following are the terms most commonly used in the theory of consolidation:

1. Coefficient of compressibility $a_v = \dfrac{\Delta e}{\Delta \sigma}$

2. Coefficient of volume compressibility $m_v = \dfrac{a_v}{1 + e_0}$ (Also called coefficient of volume change)

3. Compression index $C_c = \dfrac{\Delta e}{\log (\sigma_1 / \sigma_0)}$

4. Coefficient of consolidation $C_v = k/r_w m_v$ (This coefficient is used to find time rate of consolidation)

For the calculation of settlement, we use two methods.

Method 1: By using $m_v = \dfrac{\text{Strain}}{\text{Stress}} = \left(\dfrac{\Delta e}{1 + e_0}\right)\left(\dfrac{1}{\Delta \sigma}\right)$ is inverse of E in cm²/kg

Settlement, $\Delta = m_v H_0 \Delta \sigma$

The value of m_v for a given soil depends on the stress range. BS 1377 specifies the use of the coefficient m_v calculated for stress increment of 1 kg/cm² (100 kN/m²) over the existing stress on the clay layer. The value of m_v may be estimated from Figure A.1. It is more accurate to determine it from tests.

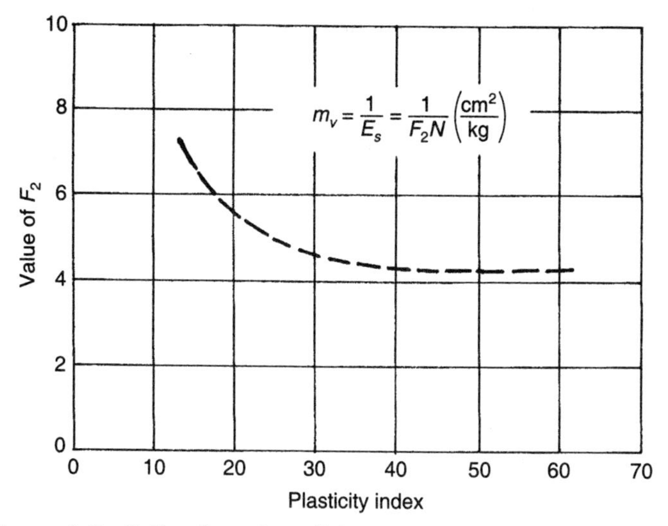

Figure A.9 Estimation of coefficient of volume compressibility of clays from plasticity index and SPT(N) values.

Method 2: By using compression index C_c obtained from log-log plot.

$$\text{Settlement } \Delta = C_c \frac{H_0}{(1+e_0)} \log \frac{\sigma_0 + \Delta\sigma}{\sigma_0}$$

For undistributed clay, $C_c = 0.009$ (LL – 10)

For remoulded clay, $C_c = 0.007$ (LL – 7)

A.16 CLASSIFICATION OF DAMAGES IN MASONRY BUILDINGS

Estimation of settlements and the allowable maximum and differential settlement in buildings are dealt with in Secs. 14.5 and 14.6 respectively. The following table gives the commonly accepted classification of damages in masonry buildings. We should note that wall cracking is only one of the many items to be considered in reporting damage of buildings.

Degree of damage	*Crack width*	*Description*
zero (Negligible)	< 0.1 mm	Vanishes on painting
1 (Very slight)	$\ngtr 0.1 < 0.5$ mm	Treaded during repair
3 (Slight)	$\ngtr 0.5$ mm	Needs special repair
4 (Moderate)	5 to 15 mm $\ngtr$ 3 Nos.	Requires opening up and patching
5 (Severe)	15 to 25 mm not many	Can be repaired with special care near opening
6 (Very severe)	> 25 mm large number	Requires major repair or reconstruction

EXAMPLE A.1 (Rough estimation of total settlement in clays)

A raft 6 m square loaded at 160 kN/m^2 is located in a clay deposit 15 m in depth. If E_s = 50 MN/m^2 and m_v from consolidation test is 0.13 m^2/MN, *estimate* the total settlement (Ref: IS 8009, Part 1).

Reference	*Step*	*Calculation*
	1	*Estimate immediate settlement*
		$\Delta = \dfrac{qB(1-\mu^2)}{E_s} \times I_f$
		B = 6 m, I_f = 0.87, μ = 0.25
Sec. A.11		$\Delta_{\text{elastic}} = \dfrac{160 \times 10^3 \times 6 \times 0.87 \times 0.94}{50 \times 10^6} = 15 \text{ mm}$
	2	*Consolidation settlement (m_v)*
		$\Delta_{\text{consol}} = m_v H \Delta\sigma$
		Assume 2 to 1 distribution. Find average $\Delta\sigma$.
		$\Delta\sigma = \left(160 + \dfrac{160 \times 36}{21 \times 21}\right) \times 0.5 = 8.74 \text{ kN/m}^2$
		$\Delta_c = \dfrac{0.13 \times 15 \times 86 \times 10^3 \times 10^3}{10^6} = 167 \text{ mm}$
IS 8009 Part I		(This has to be corrected for clay layer resting on sand or rock, A = 0.35)
	3	*Find total settlement*
		Δ_c = 0.35 × 167 = 58 mm
		Total settlement = 15 + 58 = 74 mm (approx.)

REFERENCES

[1] Varghese, P.C., *Foundation Engineering*, Prentice-Hall of India, New Delhi, 2005.

[2] IS 8009-code of practice for calculation of settlement of foundations Part 1. Shallow foundations Part (2) Deep foundations B.I.S., New Delhi.

Extracts from SP 16 for Design of Reinforced Concrete Members

TABLE B.1 Flexure-reinforcement Percentage, p_1 for Singly Reinforced Sections using Fe 415 Steel and M20 Concrete*

M_U/bd^2 (N/mm^2)	*Steel (%)*	M_U/bd^2	p	M_U/bd^2	p	M_U/bd^2	p
0.35	0.099	1.25	0.376	2.06	0.662	2.42	0.806
0.40	0.114	1.30	0.392	2.08	0.670	2.44	0.814
0.45	0.128	1.35	0.409	2.10	0.678	2.46	0.823
0.50	0.143	1.40	0.426	2.12	0.685	2.48	0.831
0.55	0.158	1.45	0.443	2.14	0.693	2.50	0.840
0.60	0.172	1.50	0.460	2.16	0.701	2.52	0.848
0.65	0.187	1.55	0.477	2.18	0.709	2.54	0.857
0.70	0.203	1.60	0.494	2.20	0.717	2.56	0.866
0.75	0.218	1.65	0.512	2.22	0.725	2.58	0.874
0.80	0.233	1.70	0.530	2.24	0.733	2.60	0.883
0.85	0.248	1.75	0.547	2.26	0.741	2.62	0.892
0.90	0.264	1.80	0.565	2.28	0.749	2.64	0.901
0.95	0.280	1.85	0.584	2.30	0.757	2.66	0.910

(Contd.)

*For other values of strength of steel and concrete, see SP 16.

TABLE B.1 Flexure-reinforcement Percentage, p_1 for Singly Reinforced Sections using Fe 415 Steel and M20 Concrete* (contd.)

M_U/bd^2 (N/mm^2)	*Steel* (%)	M_U/bd^2	p	M_U/bd^2	p	M_U/bd^2	p
1.00	0.295	1.90	0.602	2.32	0.765	2.68	0.919
1.05	0.311	1.95	0.621	2.34	0.773	2.70	0.928
1.10	0.327	2.00	0.640	2.36	0.781	2.72	0.937
1.15	0.343	2.02	0.647	2.38	0.790	2.74	0.946
1.20	0.359	2.04	0.655	2.40	0.798	2.76	0.955

*For other values of strength of steel and concrete, see SP 16.

TABLE B.2 IS Rules for Minimum Steel in R.C. Members

No.	*Member*	*Recommendation*
1	Beams	1. Tension steel $\not<$ 0.25% 2. Shear steel minimum stirrups to take 0.4 N/mm^2 for wet area. Max spacing 0.75d or 300 mm
2	Ordinary slabs	1. Tension steel $\not<$ 0.12% of whole area for shrinkage (It is better to have $\not<$ 0.25% of effective area for foundations). Spacing for main steel $\not>$ 3d or 300 mm Spacing for secondary steel $\not>$ 5d or 450 mm Spacing in flat slabs $\not>$ 2d 2. For slabs > 1 m thick minimum area of steel per metre length in each direction $\not<$ 300 mm^2
3	Flat slabs	1. Tension steel $\not<$ 0.25% spacing $\not>$ 2d 2. Distribution steel as in ordinary slabs 3. Integrity steel over columns should be given
4	Foundation slab	Thickness not less than 150 mm in soils and 300 mm on top of piles (IS, Cl. 34.1.2)

TABLE B.3 L_d/ϕ Development Length for Bars in Tension*

Grade of steel	*Grade of concrete*		
	M20	M25	M30
Fe 250	45	39	36
Fe 415	47	40	36
Fe 500	57	49	45

(*We require 25% less for development lengths of compression bars.)
End anchorage = 12ϕ or depth of member (larger value) [We must have a clear idea of end anchorage and development length. End anchorage is part of development length]

TABLE B.4 Design Shear Strength of Concrete τ_c N/mm²
(IS 456 (2000), Table 19)

p_t	f_{ck}, N/mm²					
	15	20	25	30	35	40
≤ 0.15	0.28	0.28	0.29	0.29	0.29	0.30
0.20	0.32	0.33	0.33	0.33	0.34	0.34
0.30	0.38	0.39	0.39	0.40	0.40	0.41
0.40	0.43	0.44	0.45	0.45	0.46	0.46
0.50	0.46	0.48	0.49	0.50	0.50	0.51
0.60	0.50	0.51	0.53	0.54	0.54	0.55
0.70	0.53	0.55	0.56	0.57	0.58	0.59
0.80	0.55	0.57	0.59	0.60	0.61	0.62
0.90	0.57	0.60	0.62	0.63	0.64	0.65
1.00	0.60	0.62	0.64	0.66	0.67	0.68
1.10	0.62	0.64	0.66	0.68	0.69	0.70
1.20	0.63	0.66	0.69	0.70	0.72	0.73
1.30	0.65	0.68	0.71	0.72	0.74	0.75
1.40	0.67	0.70	0.72	0.74	0.76	0.77
1.50	0.68	0.72	0.74	0.76	0.78	0.79
1.60	0.69	0.73	0.76	0.78	0.80	0.81
1.70	0.71	0.75	0.77	0.80	0.81	0.83
1.80	0.71	0.76	0.79	0.81	0.83	0.85
1.90	0.71	0.77	0.80	0.83	0.85	0.96
2.00	0.71	0.79	0.82	0.84	0.86	0.88
2.10	0.71	0.80	0.83	0.86	0.88	0.90
2.20	0.71	0.81	0.84	0.87	0.89	0.91
2.30	0.71	0.82	0.86	0.88	0.91	0.93
2.40	0.71	0.82	0.87	0.90	0.92	0.94
2.50	0.71	0.82	0.88	0.91	0.93	0.95
2.60	0.71	0.82	0.89	0.92	0.94	0.97
2.70	0.71	0.82	0.90	0.93	0.96	0.98
2.80	0.71	0.82	0.91	0.94	0.97	0.99
2.90	0.71	0.82	0.92	0.95	0.98	1.00
≥ 3.00	0.71	0.82	0.92	0.96	0.99	1.01

[*Note:* If the shear is only one half the value of the above table and in members of minor structural importance such as lintels, no shear reinforcement is deemed to be necessary (IS 456, Cl. 26.5.1.6).]

TABLE B.5 Increased Shear in Slabs Less than 300 mm — IS 456, Cl. 40.2.1 ($\tau_c^1 = k\tau_c$)

Depth of slab (in mm)	300 *or more*	275	250	225	200	175	150 *or less*
k	1	1.05	1.10	1.15	1.20	1.25	1.30

TABLE B.6 Max Allowable Values of Shear Allowed in Reinforced Concrete (τ_{max}) with Shear Reinforcements

Grade	20	25	30	35	40 or more
τ_{max}	2.8	3.1	3.5	3.7	4.0

TABLE B.7 Shear-vertical Stirrups (Table 62 of SP 16) (Values of V_s/d for Two-legged Stirrups, kN/cm, Eq. (2.4)

Stirrup spacing (cm)	f_y = 415 N/mm² *Diameter* (mm)				
	6	8	10	12	16 .
5	4.083	7.259	11.342	16.334	29.037
6	3.403	6.049	9.452	13.611	24.197
7	2.917	5.185	8.102	11.667	20.741
8	2.552	4.537	7.089	10.208	18.148
9	2.269	4.033	6.302	9.074	16.132
10	2.042	3.630	5.671	8.167	14.518
11	1.856	3.299	5.156	7.424	13.199
12	1.701	3.025	4.726	6.806	12.099
13	1.571	2.792	4.363	6.286	11.168
14	1.458	2.593	4.051	5.833	10.370
15	1.361	2.420	3.781	.445	9.679
16	1.276	2.269	3.545	5.104	9.074
17	1.201	2.135	3.336	4.804	8.540
18	1.134	2.016	3.151	4.537	8.066
19	1.075	1.910	2.985	4.298	7.641
20	1.020	1.815	2.836	4.083	7.259
25	0.817	1.452	2.269	3.267	5.807
30	0.681	1.210	1.890	2.722	4.839
35	0.583	1.037	1.620	2.333	4.148
40	0.510	0.907	1.418	2.042	3.629
45	0.454	0.807	1.260	1.815	3.226

Minimum shear steel is necessary in beams for the following reasons:

1. To guard against any sudden failure of a beam if concrete cover bursts and the bond to the tension steel is lost.
2. To prevent brittle shear failure which can occur without shear steel.
3. To prevent failure that can be caused by tension due to shrinkage and thermal stresses and internal cracking in the beams.
4. To hold the reinforcements in place while pouring concrete, and act as the necessary ties for the compression steel and make them effective.

Steel Reinforcement Data

TABLE C.1 Areas of Bars (mm²) **(Refer SP 16, Table 93)**

No. of bars \ *Diameter of bars* (mm)	6	8	10	12	16	20	25	32
1	28.3	5.3	78.5	113.1	201.1	314.2	490.0	804.2
2	56.5	100.5	157.1	226.2	402.1	628.3	981.7	1608
3	84.8	150.8	235.6	339.3	603.2	942.5	1473	2413
4	113.1	201.1	314.2	452.4	804.2	1257	1963	3217
5	141.4	251.3	397.2	565.5	1005	1571	2454	4021
6	169.6	301.6	471.2	678.6	1206	1885	2945	4825
7	197.9	351.9	549.8	791.7	1407	2199	3436	5360
8	226.2	402.1	628.3	904.8	1608	2513	3927	6434
9	254.5	452.4	706.9	1018	1810	2827	4418	7328
10	282.7	502.7	785.4	1131	2011	3142	4909	8042
Perimeter of one bar (mm)	18.8	25.1	31.4	37.6	50.2	62.8	78.5	100.5

TABLE C.2 Areas of Bars at Given Spacings (mm²) (Refer SP 16, Table 96)

Spacing (mm) \ Diameter of bars (mm)	6	8	10	12	16	20	25	32
50	565	1005	1571	2262	4021			
75	377	670	1047	1508	2680	4188	6545	
100	282	503	785	1131	2010	3141	4908	8042
125	226	402	628	904	1608	2513	3929	6434
150	188	335	523	754	1340	2094	3272	5361
175	151	287	448	646	1149	1795	2805	4595
200	141	251	392	565	1005	1570	2454	4021
225	125	223	349	502	893	1396	2181	3574
250	113	201	314	452	804	1256	1963	3217
275	102	182	285	411	731	1142	1785	2924
300	94	167	261	377	670	1047	1636	2680

TABLE C.3 Unit Weights and Weights at Specified Spacing of Bars

Unit weight			*Weights of bars* (kg/m²)								
Size (mm)	*Weight/* m (kg)	*Length/* *tonne* (m)	*Spacings of bars* (mm) 75	100	125	150	175	200	225	250	275
6	0.22	4505	2.960	2.220	1.776	1.480	1.269	1.110	0.987	0.888	0.807
8	0.395	2532	5.267	3.950	3.160	2.633	2.257	1.975	1.756	1.580	1.436
10	0.616	1623	8.213	6.160	4.928	4.107	3.520	3.080	2.738	2.464	2.240
12	0.888	1126	11.84	8.880	7.104	5.920	5.074	4.440	3.947	3.552	3.229
16	1.579	633	21.05	15.79	12.63	10.53	9.023	7.895	7.018	6.316	5.742
20	2.466	406	32.88	24.66	19.73	16.44	14.09	12.33	10.96	9.864	8.967
25	3.854	259	51.39	38.54	30.83	25.69	22.02	19.27	17.13	15.42	14.01
32	6.313	159		63.13	50.50	42.09	36.07	31.57	28.06	25.25	22.96
40	9.864	101			78.91	65.76	56.37	49.32	43.84	39.46	35.87

TABLE C.4 Development Length (L_d/ϕ) in Tension

Steel grade	M20	M25	M30	M35	≥ M40
Fe 250	45	39	36	32	29
Fe 415	47	40	38	33	30
Fe 500	57	49	45	40	36

Note: L_d/ϕ in compression = 0.8 × value in tension

End anchorage = 12ϕ or depth of member whichever is larger.

TABLE C.5 Approximate Consumption of Steel in Foundations (Fe 415 and M20 concrete)

Item	*Description*	*Total consumption of steel in kg/m³*	*Designed steel as per cent of area*
1	Footings	30–78	0.27–1.0
2	Cast in-situ piles	30–40	0.38–0.51
3	Rafts	60–120	0.80–1.54
4	Roof slabs	20–50	0.27–0.64

Note: 1. Wt. of steel = (% of steel) (78.5) kg/m^3.

2. Generally in foundation slabs we provide a minimum of 0.2% steel in each direction (total of 0.4%). Hence minimum weight of steel required works out to $0.4 \times 78.5 = 32$ kg/m^3. Maximum can be 1% in each direction equal to $2 \times 78.5 = 160$ kg/m^3.

EXAMPLE C.1

In the design of a square column footing 2.25 m in size and 0.5 m in depth, the percentage of steel required works out 0.4% in each (X and Y) directions. Estimate the amount of steel required for the footing.

Reference	*Step*	*Calculation*
	1	*Calculate steel required in kg/m³* Total % of steel = $2 \times 0.4 = 0.8$% Steel in kg/m^3 = $0.8 \times 78.5 = 62.8$ kg/m^3
	2	*Total steel required for footing* Steel required = (kg/m^3) × [Volume of concrete] Volume = $2.25 \times 2.25 \times 0.5 = 2.53$ m^3 Wt. of steel required = 62.8×2.53 = 160 kg (approx.)

D

Design Charts of Centrally Loaded Columns and Footings

Chart D.1 Design of Columns and Footings

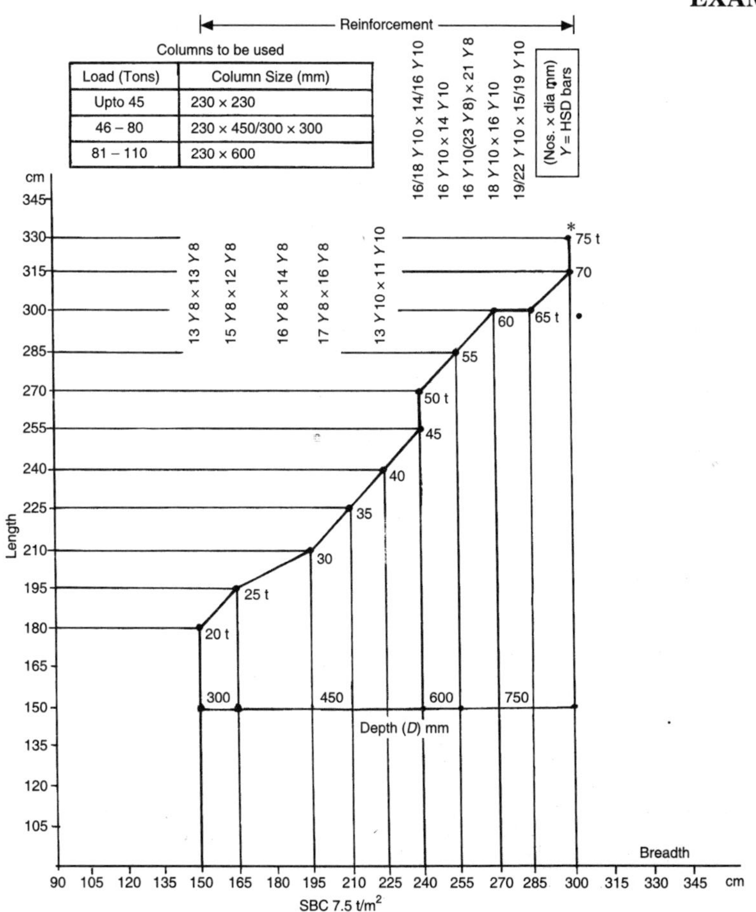

Load (Tons)	Column Size (mm)
Upto 45	230 × 230
46 – 80	230 × 450/300 × 300
81 – 110	230 × 600

EXAMPLES (S.B.C. 7.5 t/m^2)

(1) Col. load 45T

From table

Col. size 230 × 230 mm

Footing 255 × 240 cm

Depth 600 mm

Steel 16 × 14.Y.10

(2) Col. load 50 T

Col. size 300 × 300 mm

Footing 270 × 240 cm

Depth 600 mm

Steel 18 × 16.Y.10

* Column characteristics loads

Chart D.2 Design of Columns and Footings

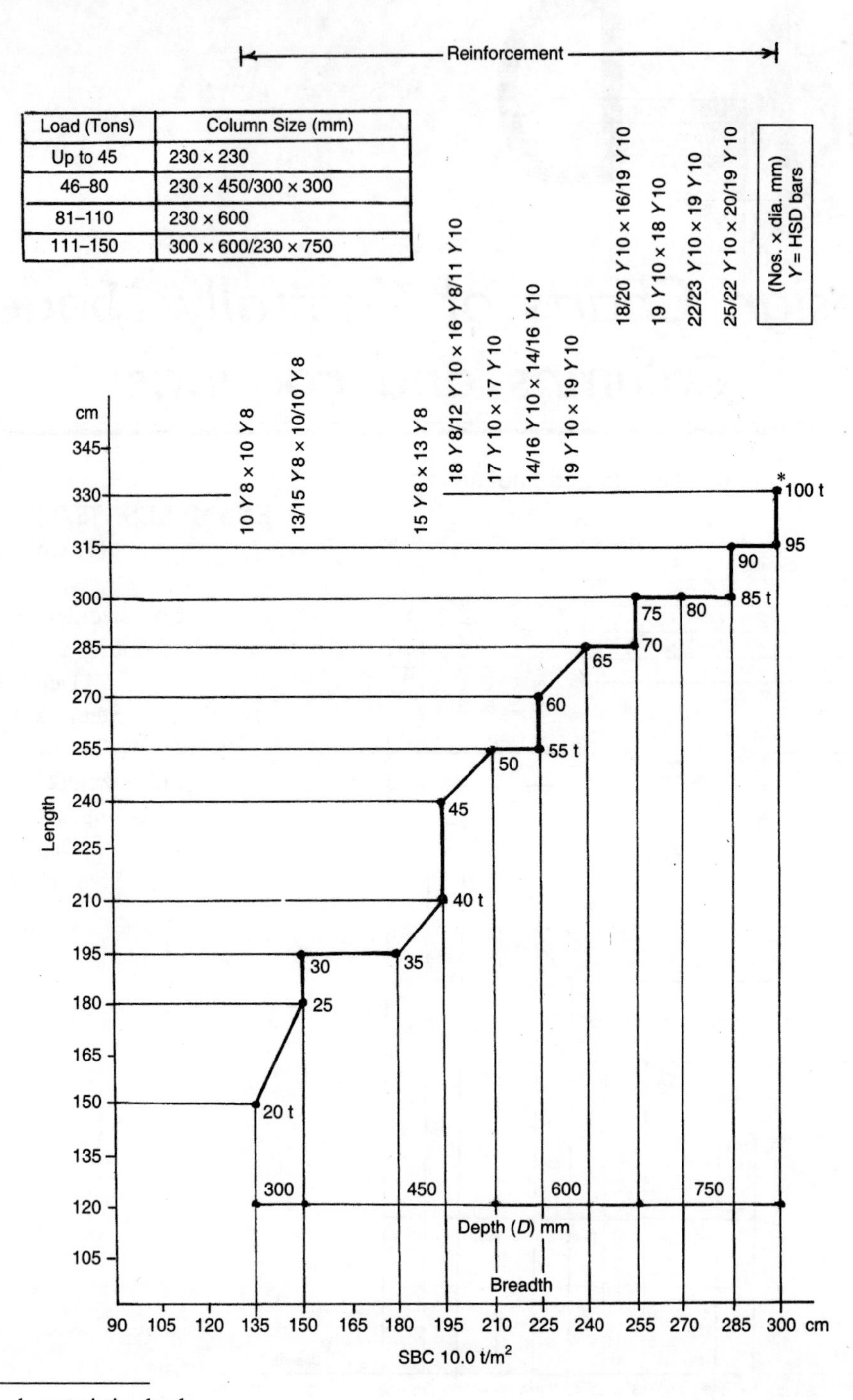

Load (Tons)	Column Size (mm)
Up to 45	230 × 230
46–80	230 × 450/300 × 300
81–110	230 × 600
111–150	300 × 600/230 × 750

* Column characteristics loads

Chart D.3 Design of Columns and Footings

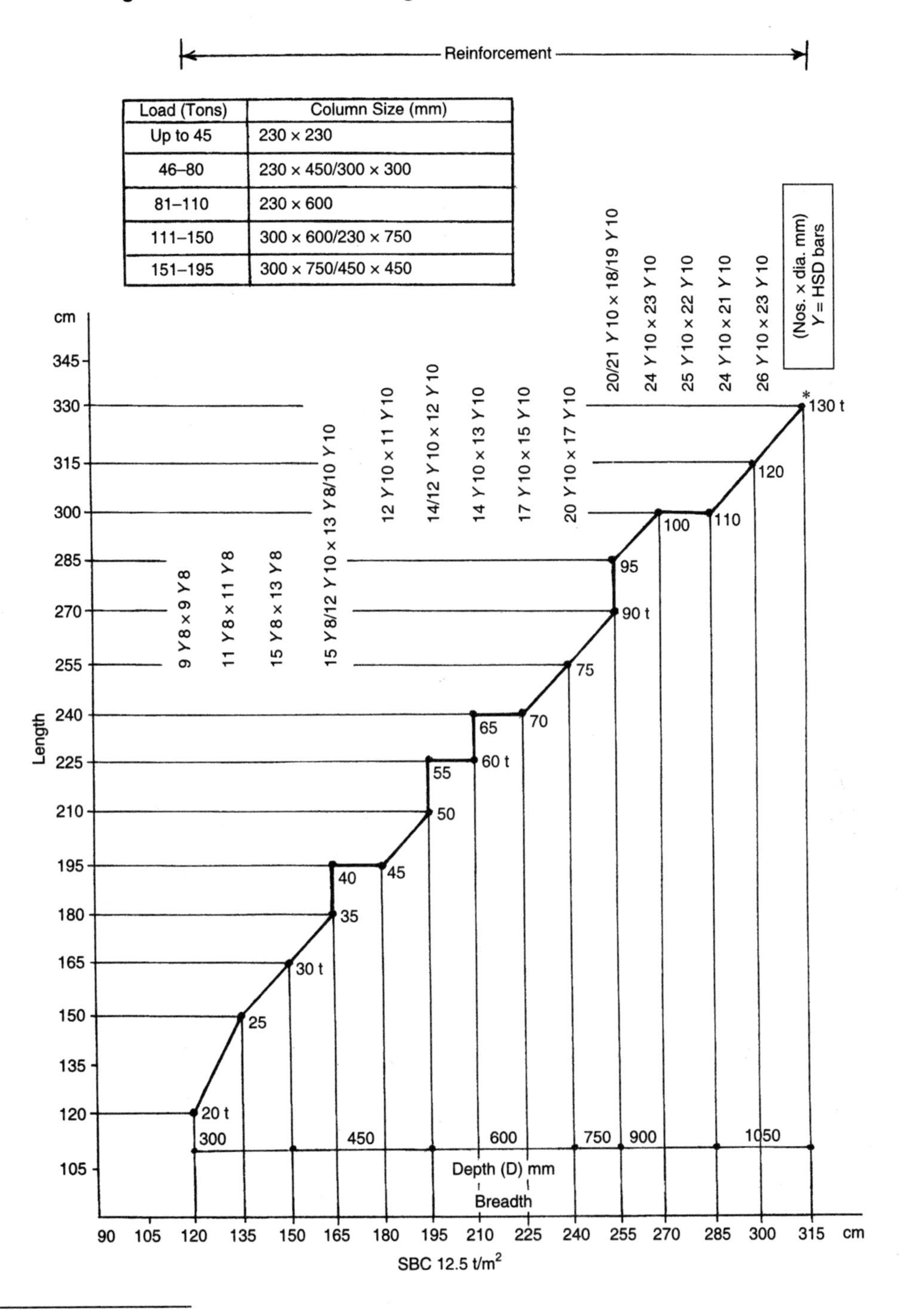

Load (Tons)	Column Size (mm)
Up to 45	230 × 230
46–80	230 × 450/300 × 300
81–110	230 × 600
111–150	300 × 600/230 × 750
151–195	300 × 750/450 × 450

* Column characteristics loads

Chart D.4 Design of Columns and Footings

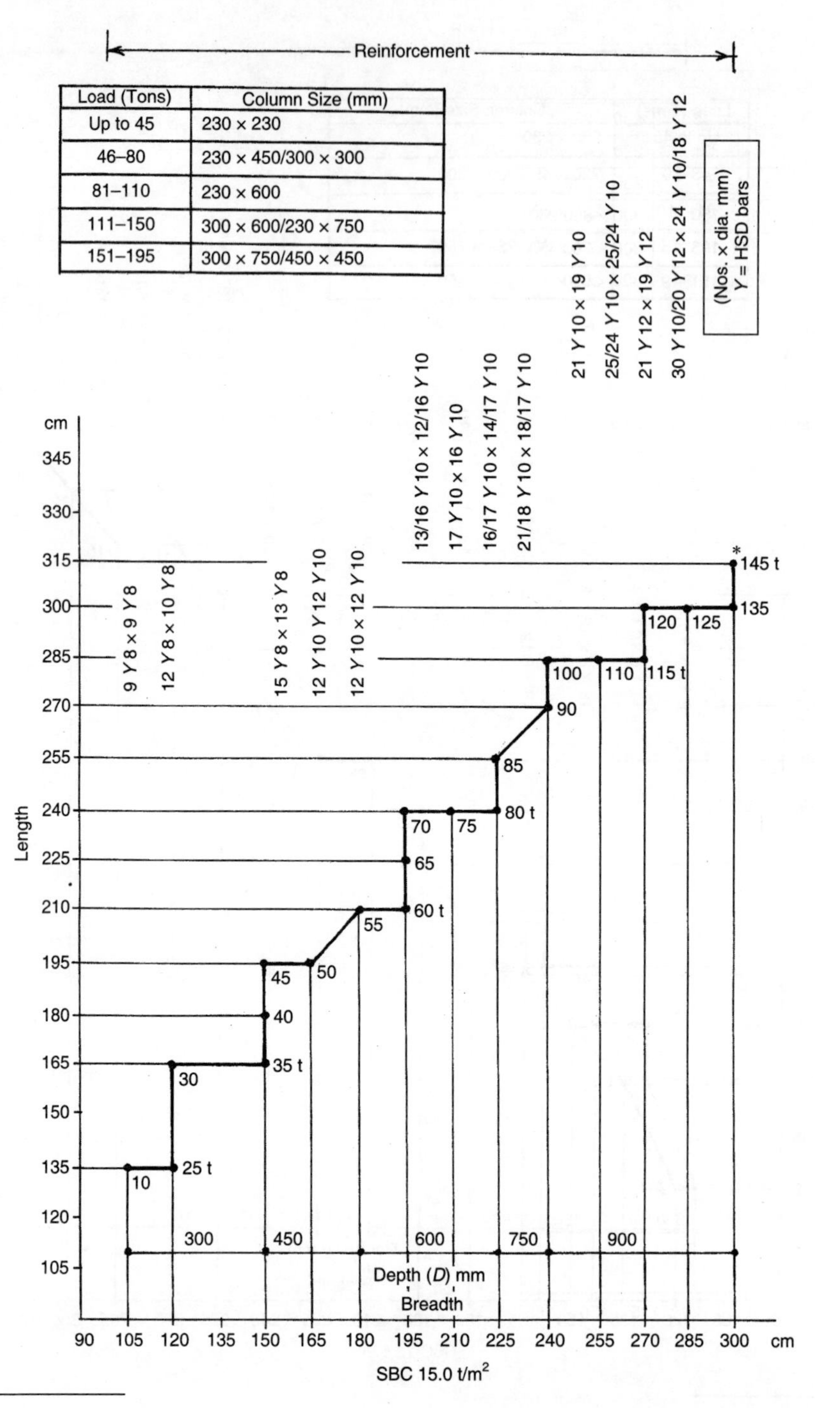

Load (Tons)	Column Size (mm)
Up to 45	230 × 230
46–80	230 × 450/300 × 300
81–110	230 × 600
111–150	300 × 600/230 × 750
151–195	300 × 750/450 × 450

* Column characteristics loads

Chart D.5 Design of Columns and Footings

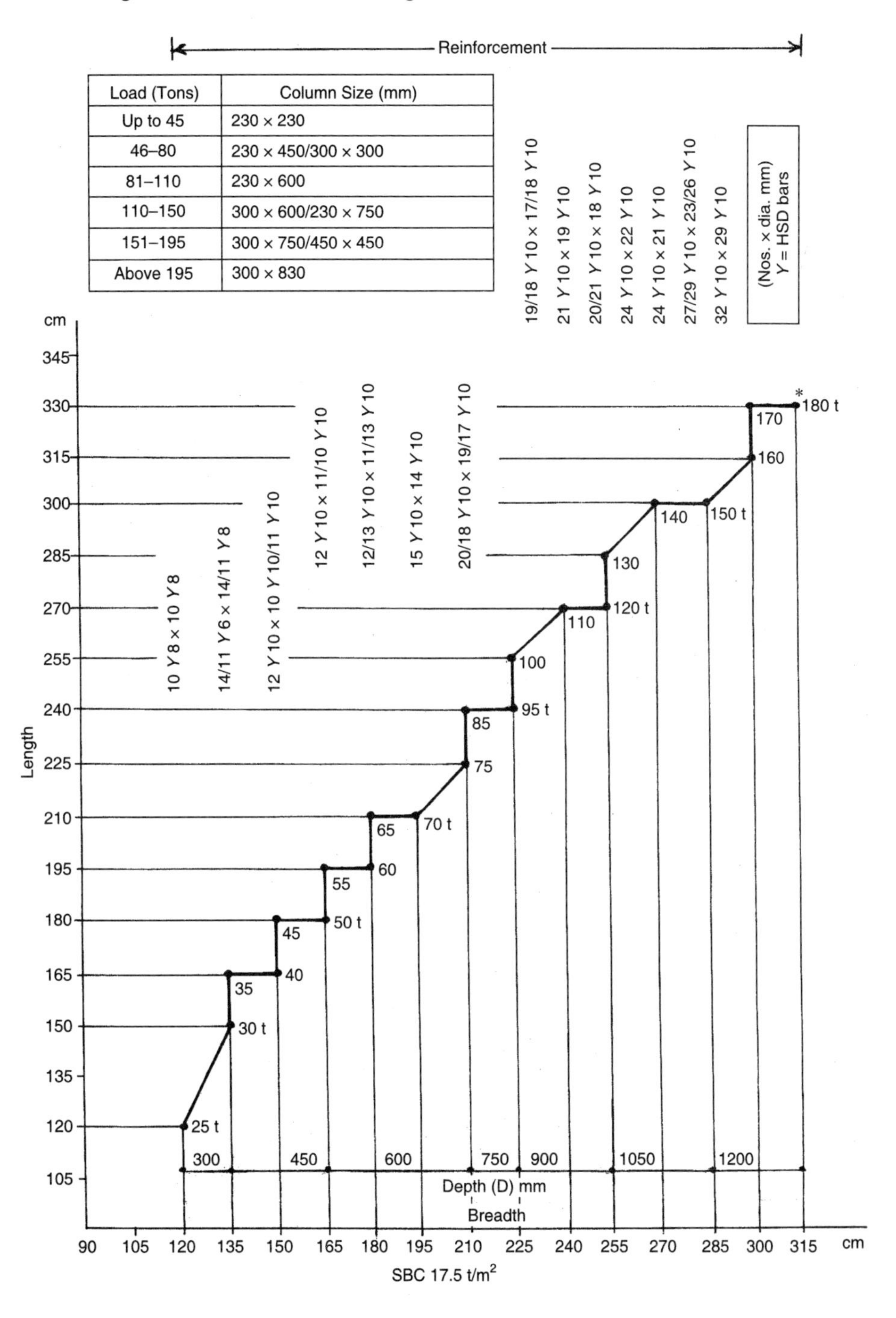

Load (Tons)	Column Size (mm)
Up to 45	230 × 230
46–80	230 × 450/300 × 300
81–110	230 × 600
110–150	300 × 600/230 × 750
151–195	300 × 750/450 × 450
Above 195	300 × 830

* Column characteristics loads

Chart D.6 Design of Columns and Footings

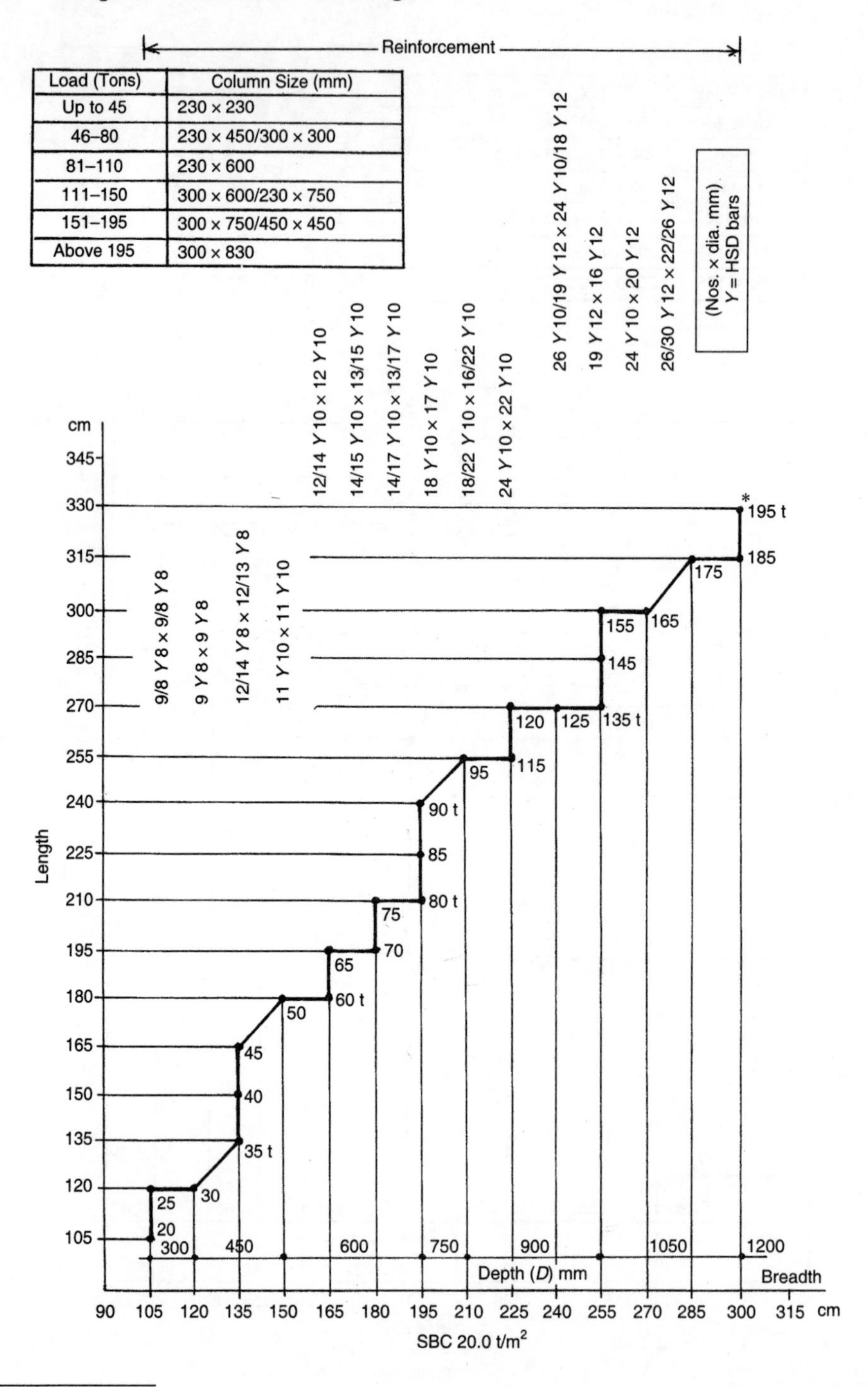

Load (Tons)	Column Size (mm)
Up to 45	230 × 230
46–80	230 × 450/300 × 300
81–110	230 × 600
111–150	300 × 600/230 × 750
151–195	300 × 750/450 × 450
Above 195	300 × 830

* Column characteristics loads

Chart D.7 Design of Columns and Footings

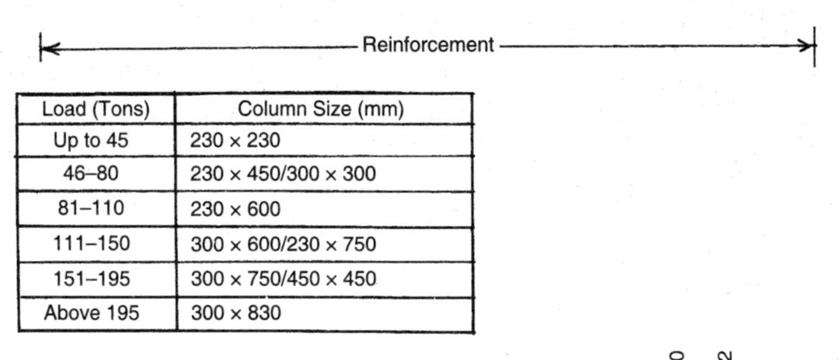

Load (Tons)	Column Size (mm)
Up to 45	230 × 230
46–80	230 × 450/300 × 300
81–110	230 × 600
111–150	300 × 600/230 × 750
151–195	300 × 750/450 × 450
Above 195	300 × 830

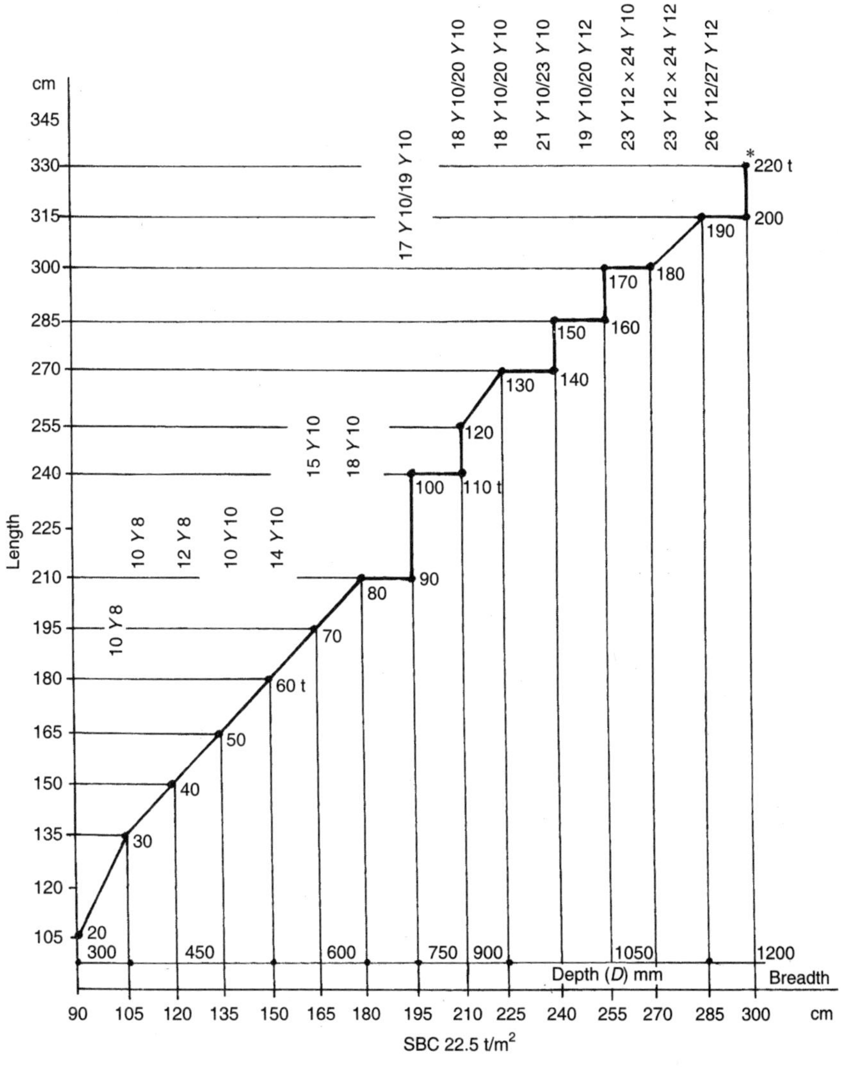

* Column characteristics loads

Chart D.8 Design of Columns and Footings

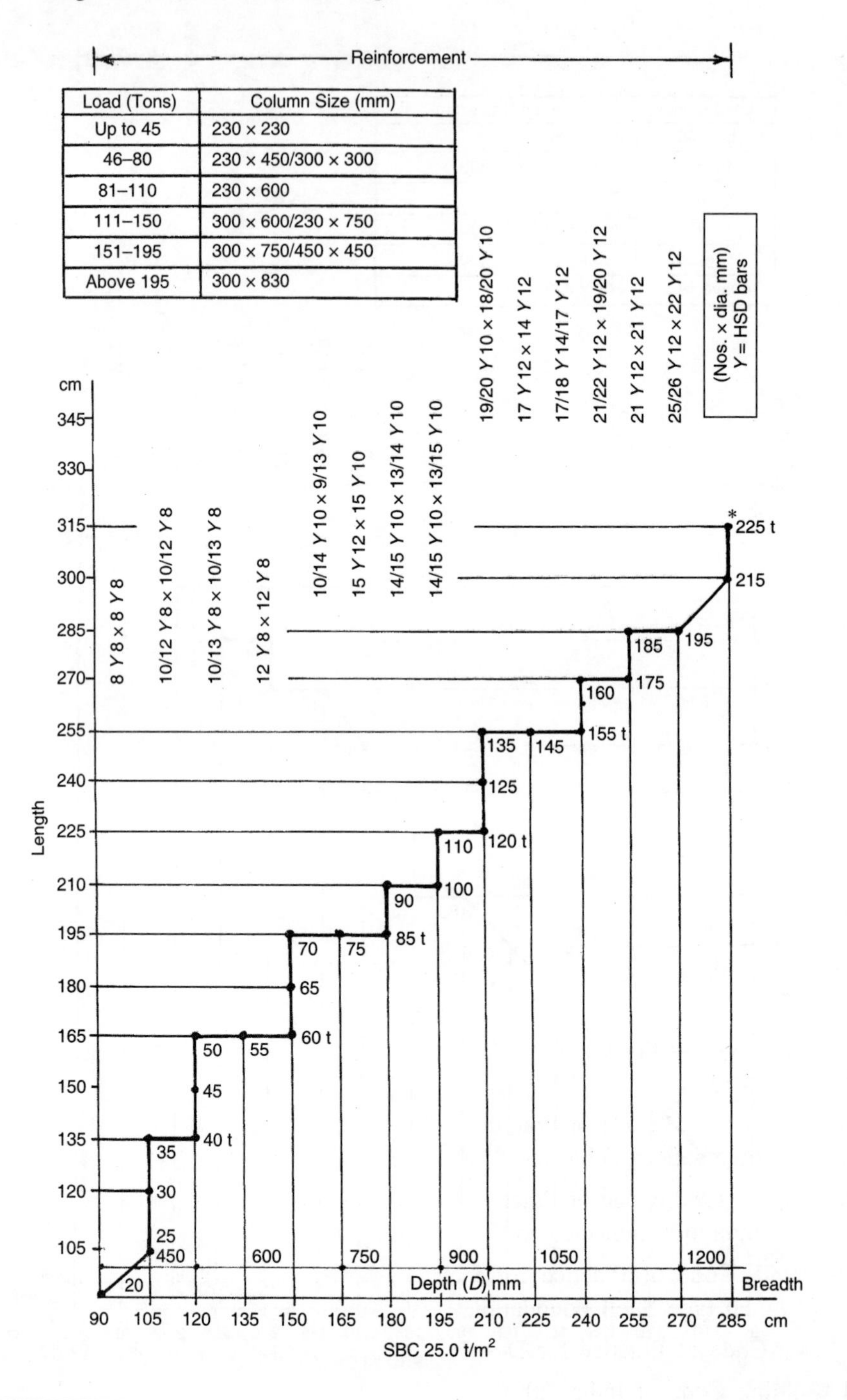

Load (Tons)	Column Size (mm)
Up to 45	230 × 230
46–80	230 × 450/300 × 300
81–110	230 × 600
111–150	300 × 600/230 × 750
151–195	300 × 750/450 × 450
Above 195	300 × 830

* Column characteristics loads

Bibliography

I. Codes of Practice

IS 1080, 1980, Code of Practice for Design and Construction of Simple Spread Foundation.

IS 1904, 1978, Code of Practice for Structural Safety of Buildings, Shallow Foundations.

IS 2950 (Part I), 1981, Code of Practice for Design and Construction of Raft Foundation (2nd Revision).

IS 6403, 1981, Code of Practice for Determination of Bearing Pressure on Shallow Foundations (First Revision).

IS 8009 (Part I), 1976, Code of Practice for Calculation of Settlement of Foundations, Part 1, Shallow Foundations subjected to Symmetrical Static Loads.

IS 8009 (Part I), 1980, Code of Practice for Calculation of Settlement of Foundations, Part 2, Deep Foundations subjected to Symmetrical Static Vertical Loads

IS 9456, 1980, Code of Practice for Design and Construction of Conical and Hyperbolic Paraboloidal type Shell Foundations with Amendment No. 1 (1982).

11089, 1984, Code of Practice for Design and Construction of Ring Foundation.

Natural Building Code of India, 2006 (Section on Foundations).

IRS, 78, 1983, Standard Specifications and Code of Practice for Road Bridges, Sec. VII, Foundations and Substructure.

Indian Railways, Standard of Practice for the Design of Substructures and Foundations of Bridges (Bridge Substructure and Foundation Code).

IS 456, 2000, Code of Practice for Plan and Reinforced Concrete.

II. Books

Bell, B.J., *Reinforced Concrete Foundations* (2nd ed.), George Godwin Limited, London, 1981.

Das, B.M., *Principles of Foundation Engineering* (5th ed.), Thompson Asia Pvt. Ltd., Singapore, 2004.

Gambhir, M.L., *Design of Reinforced Concrete Structures*, Prentice-Hall of India, New Delhi, 2008.

Kramer S.L., *Geotechnical Earthquake Engineering* (Published in India by Dorling Kindersley, New Delhi), 1996.+

Kurien, N.P., *Design of Foundation Systems: Principles and Practices* (3rd ed.), Narosa Publishing House, New Delhi, 2005.

Kurien N.P., *Shell Foundations*, Narosa Publishing House, New Delhi, 2006.

Manohar, S.N., *Tall Chimneys*, Tata McGraw Hill, New Delhi, 1985.

National Building Code, 2006, B.I.S., New Delhi.

Reynolds, C.E. and J.C. Steedman, *Examples of Design of Buildings—A View Point Publication*, Cement and Concrete Association, London, 1978.

Saran, Swami, *Analysis and Design of Substructrues*, Oxford and IBH Publishing Company, New Delhi.

Thomlinsion, M.J., *Foundation Design and Construction*, Longman Group Ltd., Singapore, 1986.

Varghese, P.C., *Foundation Engineering*, Prentice-Hall of India, New Delhi, 2005.

Varghese, P.C., *Limit State Design of Reinforced Concrete* (2nd ed.), Prentice-Hall of India, New Delhi, 2002.

Veerappan and Pragadeswaran, *Design of Foundations and Detailing*, Association of Engineers, P.W.D., Tamil Nadu, 1991.

Winterborn, H.F. and Hsai-Yang Fang (Eds.), *Foundation Engineering Handbook*, Van Nostrand Reinhold, New York, 1975.

Index